T0190386

'This lavishly illustrated volume makes a major contribution to our understanding of the Space Age. It expands historiography beyond the superpowers and radically reconceptualizes outer space. *Militarizing Outer Space* obliges us to think of "outer space" as a zone beyond the confines of the earth, produced by cultural, political and technological interventions that embed it in earthly projects and respond to a multitude of hopes and anxieties. Space is not a remote, inaccessible realm, but a nearby "non-space" that can be populated by technological infrastructures advocated by the military and appropriated by the market, colonized by earthlings fleeing Armageddon or the disasters of climate change, and filled with utopian aspirations or dystopian fears, but always appropriated by multiple stakeholders who imagine new worlds and ways of being in response to critical contingencies in everyday life. Readers will discover new and unexpected features of their life worlds presented in outstanding essays framed by a superb introduction and conclusion.'

—John Krige, Georgia Institute of Technology, USA

'In this very fine last part of a trilogy that meritoriously orbits around the concept of "astroculture", one is reminded of the centrality of military technologies to modernization. The fourteen fascinating chapters offer a rich and welcome contribution to the history of outer space and globality. Popular imaginaries are tied to promises of supremacy, while the fuzzy boundaries between civilian and military use are interrogated. In a global age we would be wise to re-visit these manifold projections and dreams of space technology and its cultural repercussions, as they have much to teach us about the present. A very important book.'

—Nina Wormbs, KTH Royal Institute of Technology, Sweden

'*Militarizing Outer Space* is a compellingly original collection of essays that breaks out of the conventional mold of interpreting space races and arms races narrowly as products of the Cold War. Long before we could reach it, humans imagined space as a realm of war populated with laser-wielding heroes, orbital fortresses and extraterrestrials ripe for conquest. From the moral thought of C. S. Lewis to Ronald Reagan's Strategic Defense Initiative, the authors offer a deeply researched analysis of the connections between security, fantasy and technopolitics. Although no war has ever occurred outside the earth's atmosphere, this volume convincingly shows how military anxieties more than a desire to reach the stars drove the development of spaceflight. For anyone interested in the rise of militant astroculture and actual warfare, *Militarizing Outer Space* is a must-read.'

—Joe Maiolo, King's College London, UK

Palgrave Studies in the History of Science and Technology

Series Editors
James Rodger Fleming
Colby College
Waterville, ME, USA

Roger D. Launius
Auburn, AL, USA

Designed to bridge the gap between the history of science and the history of technology, this series publishes the best new work by promising and accomplished authors in both areas. In particular, it offers historical perspectives on issues of current and ongoing concern, provides international and global perspectives on scientific issues, and encourages productive communication between historians and practicing scientists.

More information about this series at
http://www.palgrave.com/gp/series/14581

Other Publications by the Emmy Noether Research Group 'The Future in the Stars: European Astroculture and Extraterrestrial Life in the Twentieth Century'

IMAGINING OUTER SPACE
European Astroculture in the Twentieth Century
(European Astroculture, vol. 1)

LIMITING OUTER SPACE
Astroculture after Apollo
(European Astroculture, vol. 2)

ASTROCULTURE AND TECHNOSCIENCE

SOUNDS OF SPACE

BERLINER WELTRÄUME IM FRÜHEN 20. JAHRHUNDERT

ROCKET STARS
Astrocultural Genealogies in the Global Space Age (forthcoming)

Alexander C. T. Geppert
Daniel Brandau
Tilmann Siebeneichner
Editors

Militarizing Outer Space

Astroculture, Dystopia and the Cold War

European Astroculture
Volume 3

Editors
Alexander C. T. Geppert
New York University
New York, USA

NYU Shanghai
Shanghai, China

Daniel Brandau
Freie Universität Berlin
Berlin, Germany

Tilmann Siebeneichner
Humboldt-Universität zu Berlin
Berlin, Germany

ISSN 2730-972X ISSN 2730-9738 (electronic)
Palgrave Studies in the History of Science and Technology
European Astroculture, Volume 3
ISBN 978-1-349-96063-7 ISBN 978-1-349-95851-1 (eBook)
https://doi.org/10.1057/978-1-349-95851-1

Cover image: © Gösta Röver
Cover design by Thomas Howey

This Palgrave Macmillan imprint is published by the registered company Springer Nature Limited
The registered company address is: The Campus, 4 Crinan Street, London, N1 9XW, United Kingdom

Contents

Acknowledgments

'Space is a war-fighting domain, just like the land, air, and sea,' US President Donald Trump declared in March 2018. 'Space Force all the way!,' he tweeted a few months later, reaffirming his intentions to establish a new branch of the military designated to secure American hegemony beyond earth. Instantly noting how closely Trump's self-proclaimed 'great idea' resembled Ronald Reagan's 1983 Strategic Defense Initiative (SDI), commonly remembered as 'Star Wars,' critics across the political spectrum were far from convinced. What had started as a belligerent fantasy both then and now, they feared, effectively fueled the ongoing militarization of outer space. Massively underestimating the technological challenges of missile defense, both presidents seemed to favor Hollywood's striking imaginings of space wars instead, in wide circulation long before the beginning of the Space Age.[1]

Trump's overblown rhetoric evoked a long-established arsenal of images and artifacts, media and practices aiming to assign and extract meaning from outer space. Fantasies of space war both between nations on earth and against alien worlds have captured the imagination of artists, engineers, intellectuals and politicians, and spurred their extraterrestrial agendas throughout the twentieth century. Because images and notions of violence and conflict figure prominently in all variants of astroculture around the globe, the need for a volume on the military underpinnings of outer space was apparent long before the 45th US president gave the topic its most recent twist. Early versions of most contributions were presented at an international symposium *Embattled Heavens: The Militarization of Space in Science, Fiction, and Politics*, convened by the Emmy Noether research group 'The Future in the Stars' at Freie Universität Berlin in April 2014. At the time of Trump's space war 2018 twitter barrage, publication of this volume was well under way, leaving both editors and contributors wondering what to make of the topic's regained currency and unexpected relevance in day-to-day politics.

Militarizing Outer Space constitutes the third and final volume of the *European Astroculture* trilogy. While *Imagining Outer Space*, the first volume, set out to establish and contour the historical field of 'astroculture' largely in the 1950s and 1960s, *Limiting Outer Space*, the second volume, zeroed in on a single decade, the post-Apollo crisis-ridden 1970s.[2] Given the interplay between military and civilian rationales in the history of spaceflight, notions of crisis and confrontation also serve as a starting point for this third volume. Unlike its two predecessors, the book extends the collective inquiry into the early 1980s, up to Ronald Reagan's 'Star Wars' scenario and beyond. Constituting a preliminary climax of space militarization, SDI heavily influenced Cold War dynamics of deterrence and détente, with apocalyptic scenarios of imminent doom looming large in the popular imagination. The underlying imaginaries, however, were much older. A closer look reveals the extent to which they were grounded in early Space Age utopias. Popular notions of space exploration and conquest were more than mere rhetoric, being deeply intertwined with military strategies, technoscientific ambitions and social fears throughout the twentieth century. Scrutinizing belligerent imaginaries, popular narratives and widespread space war scenarios, from early European astroculture to Star Wars, *Militarizing Outer Space* links the cultural history of outer space more explicitly to conventional Cold War politics than the two preceding books in this trilogy. At the same time it challenges the conventional assumption that the Cold War context is a both necessary *and* sufficient framework to explain the making and ever-intensifying militarization of outer space.

Coming to terms with a subject as vast as outer space could easily have been overwhelming for a research group as small as ours, and we are enormously grateful to everyone who helped us not get lost in space. This includes, first and foremost, the Deutsche Forschungsgemeinschaft (DFG) which generously funded the Emmy Noether research group 'The Future in the Stars: European Astroculture and Extraterrestrial Life in the Twentieth Century' during the six years of its existence from 2010 through 2016.[3] Group members Jana Bruggmann, Ralf Bülow, Ruth Haake, Gilda Langkau, Friederike Mehl, Tom Reichard, Katja Rippert and Magdalena Stotter were there to make it happen. Conference speakers, commentators and participants who shaped the outcome even if their contributions could, alas, not be integrated in this volume include Colleen Anderson, Norman Aselmeyer, Jordan Bimm, Thore Bjørnvig, Katherine Boyce-Jacino, David Edgerton, Greg Eghigian, Danilo Flores, Paweł Frelik, Bernd Greiner, Jörg Hartmann, Matthias Hurst, Joe Maiolo, Markus Pöhlmann, Robert Poole, Alex Roland, Diethard Sawicki, Isabell Schrickel, Kai-Uwe Schrogl, Eva-Maria Silies, Simon Spiegel, Dierk Spreen and Patryk Wasiak. Anonymous reviewers offered invaluable criticism and pointed advice.

As with the two companion volumes, Gösta Röver's brilliant designs form the basis of the book cover. Once again, photographer Hubert Graml helped prepare the more than 50 illustrations for publication, many never before shown and arguably never in such a carefully curated context. As numerous times before, Katja Rippert assisted with her excellent Russian language skills. At Palgrave Macmillan, we are indebted to Molly Beck for overseeing the long and complex publication process of the entire trilogy with calm and vigor. Meanwhile, cooperating with project manager Kayalvizhi Saravanakumar and her team of professionals was as delightful as prior. Audrey McClellan produced yet another index imbued with her impressive mixture of perceptiveness and attention to detail. We would also like to extend our gratitude to those who came along later, including Michel Dubois, Grégoire and Janine Durrens, Michael Najjar and NYU Shanghai's Xinyi Xiong. Last but not least, we are once again profoundly indebted to Ruth Haake. Without her infectious optimism and indefatigable assistance in securing obscure copyright permissions, tireless and astute fact-, manuscript- and footnote-checking, both this volume and its editors would be in very different shape. Although constituting the last volume in the *European Astroculture* trilogy, *Militarizing Outer Space* does not purport to be the final say on past space futures, either in Europe or among the stars. We are too well-adjusted a crew to abort our mission midstream. Hence stay tuned and keep watching the skies. Klaatu barada nikto.

New York and Berlin
July 2020

Alexander C. T. Geppert
Daniel Brandau
Tilmann Siebeneichner

Notes

1. See Christina Wilkie, 'Trump Floats the Idea of Creating a "Space Force" to Fight Wars in Space,' *CNBC* (13 March 2018), https://cnb.cx/2Xo2pre; Donald J. Trump on Twitter, 9 August 2018, https://twitter.com/realdonaldtrump/status/1027586174448218113?); 'Trump in Space,' *New York Times* (27 July 2018), A18. All Internet sources were last accessed on 15 July 2020.
2. Alexander C. T. Geppert, ed., *Imagining Outer Space: European Astroculture in the Twentieth Century*, Basingstoke: Palgrave Macmillan, 2012 (2nd edn, London: Palgrave Macmillan, 2018) (= *European Astroculture*, vol. 1); idem, ed. *Limiting Outer Space: Astroculture after Apollo*, London: Palgrave Macmillan, 2018 (= *European Astroculture*, vol. 2).
3. A detailed conference program can be found at http://heavens.geschkult.fu-berlin.de. For comprehensive reports, see Norman Aselmeyer, 'Stellare Kriege,' *Technikgeschichte* 81.4 (2014), 371–8; Katherine Boyce-Jacino, 'Embattled Heavens: The Militarization of Space in Science, Fiction, and Politics,' *Foundation: The International Review of Science Fiction* 118 (2014), 96–100; Paweł Frelik,

'"Embattled Heavens" Conference,' *Science Fiction Studies* 41.2 (July 2014), 446–7; Ulf von Rauchhaupt, 'Als der größte Großraum zum Schlachtfeld wurde: Die Raumfahrt zwischen Politik, Technik und Science-Fiction,' *Frankfurter Allgemeine Zeitung* (16 April 2014), N3; Tom Reichard, 'Battlefield Cosmos: The Militarization of Space, 1942–1990,' *NASA History News & Notes* 31.3 (2014), 20–1; idem, 'Embattled Heavens: The Militarization of Space in Science, Fiction, and Politics,' *H-Soz-u-Kult* (8 August 2014), online available at https://www.hsozkult.de/conferencereport/id/tagungsberichte-5496; and Stephan Töpper, 'Krieg in den Sternen: Wie Konflikte auf der Erde unsere Vorstellungen vom Weltraum prägten,' *Der Tagesspiegel* (13 April 2014), B6. For further information on the Emmy Noether research group 'The Future in the Stars: European Astroculture in the Twentieth Century,' consult http://www.geschkult.fu-berlin.de/astrofuturism.

List of Figures

Within the past 60 years over 70 nations have launched more than 8,000 spacecraft, more than half for military purposes. The cover image portrays one of them, an idealized reconnaissance satellite orbiting planet Earth.

Abbreviations

ABM	Anti-Ballistic Missile
AFES	AG Friedensforschung und Europäische Sicherheit
AFRA	Arbeitsgemeinschaft für Raketentechnik
AGARD	Advisory Group for Aerospace Research and Development
ASAT	Anti-Satellite Weapon
ASTP	Apollo-Soyuz Test Project
ATV	Automated Transfer Vehicle
BBC	British Broadcasting Corporation
BDS	BeiDou Navigation Satellite System
BIS	British Interplanetary Society
BMD	Ballistic Missile Defense
CA	Canada
CAT	Computer of Average Transients
CCD	Charge-Coupled Device
CERN	Conseil Européen pour la Recherche Nucléaire
CH	Switzerland
CIA	Central Intelligence Agency
CNES	Centre National d'Etudes Spatiales
COSPAR	Committee on Space Research
CRV	Coordinate Remote Viewing
CSG	Centre Spatial Guyanais
DAG	Deutsche Astronautische Gesellschaft
DDR	Deutsche Demokratische Republik
DE	Deutschland
DEFA	Deutsche Film-Aktiengesellschaft
DFG	Deutsche Forschungsgemeinschaft
DGRR	Deutsche Gesellschaft für Raketentechnik und Raumfahrt
DIA	Defense Intelligence Agency
DK	Denmark
DLR	Deutsches Zentrum für Luft- und Raumfahrt
DoD	Department of Defense
DOS	Long-Term Orbital Station

DRG	Deutsche Raketengesellschaft
DVL	Deutsche Versuchsanstalt für Luftfahrt
EEC	European Economic Community
ELDO	European Launcher Development Organization
ELF	Extremely Low Frequency
ESA	European Space Agency
ESDAC	European Space Data Acquisition Centre
ESOC	European Space Operations Centre
ESP	Extrasensory Perception
ESRO	European Space Research Organisation
ESTEC	European Space Technology Centre
ESTRACK	ESA Tracking Stations
EU	European Union
EURATOM	European Atomic Energy Community
FAA	Federal Administration Agency
FDJ	Freie Deutsche Jugend
FGB	Functional Cargo Block
FOIA	Freedom of Information Act
FR	France
FTD	Foreign Technology Division
GfW	Gesellschaft für Weltraumfahrt
GNSS	Global Satellite Navigation System
GPS	Global Positioning System
GSOC	German Space Operations Center
IAC	International Astronautical Congress
IAF	International Astronautical Federation
ICBM	Intercontinental Ballistic Missile
IGY	International Geophysical Year
INSCOM	Intelligence and Security Command
IRBM	Intermediate-Range Ballistic Missile
ISS	International Space Station
IT	Italy
ITAR	International Traffic in Arms Regulations
JP	Japan
JPL	Jet Propulsion Laboratory (NASA)
KH-1	Keyhole-1
KSI	Information Return Capsule
LEO	Low-Earth Orbit
LEOP	Launch and Early Orbit Phase
LORAN	Long-Range Navigation
MAD	Mutual Assured Destruction
MBB	Messerschmitt-Bölkow-Blohm
MIDAS	Missile Defense Alarm System
MIT	Massachusetts Institute of Technology
MOL	Manned Orbiting Laboratory
MOU	Memorandum of Understanding
MRBM	Medium-Range Ballistic Missile
MSSS	Multi-Satellite Support System
MTR	Military-Technical Revolution
n.p.	No publisher/pagination

NASA	National Aeronautics and Space Administration
NASM	National Air and Space Museum
NATO	North Atlantic Treaty Organization
NICE	National Institute for Co-Ordinated Experiments
NMD	National Missile Defense
NORAD	North American Air Defense Command
NRL	Naval Research Laboratory
NRO	National Reconnaissance Office
NSA	National Security Agency
NTS-1	Navigation Technology Satellite 1
NVA	Nationale Volksarmee
OPS	Orbiting Piloted Station
ORD	Office of Research and Development
OSI	Office of Strategic Intelligence
OST	Outer Space Treaty
OTRAG	Orbitale Transport- und Raketen Aktiengesellschaft
OTS	Office of Technical Service
PL	Poland
RAND	Research and Development
SAGE	Semi-Automatic Ground Environment
SAM	Surface-to-Air-Missile
SAMOS	Satellite and Missile Observation System
SCOS	Spacecraft Control and Operations System
SDI	Strategic Defense Initiative
SDIO	Strategic Defense Initiative Organization
SDS	Strategic Defense System
SED	Sozialistische Einheitspartei Deutschlands
SIPRI	Stockholm International Peace Research Institute
SPADATS	Space Detection and Tracking System
SPASUR	Space Surveillance System
SRI	Stanford Research Institute
STS	Space Transportation System
TALOS	Tactical Assault Light Operator Suit
TCBM	Transparency and Confidence-Building Measure
TCP	Technological Capabilities Panel
THAAD	Terminal High Altitude Area Defense
TKS	Transport Supply Spacecraft
UCL	University College London
UFO	Unidentified Flying Object
UK	United Kingdom
UN	United Nations
UNESCO	United Nations Educational, Scientific and Cultural Organization
UNISPACE	United Nations Conference on the Exploration and Peaceful Uses of Outer Space
URDF	Unidentified Research and Development Facility
USA	United States of America
USAF	United States Air Force
USSR	Union of Soviet Socialist Republics
VfR	Verein für Raumschiffahrt
WEU	Western European Union

Notes on Contributors

Daniel Brandau teaches at Freie Universität Berlin. After studying history and literature at Universität Bielefeld (BA, MEdu) and the University of Cambridge (MPhil), he joined the Emmy Noether research group 'The Future in the Stars: European Astroculture and Extraterrestrial Life in the Twentieth Century' at Freie Universität. Brandau completed his PhD in 2017 with a dissertation on the cultural history of rocketry, published as *Raketenträume: Raumfahrt- und Technikenthusiasmus in Deutschland, 1923–1963* (2019). His research interests include the didactics of history and public history. From 2016 to 2019 he was postdoctoral researcher in the 'Meta-Peenemünde' project at Technische Universität Braunschweig, focusing on the remembrance of technologies and former military sites in East Germany after the end of the Cold War.

Paul E. Ceruzzi is Curator Emeritus of Aerospace Electronics and Computing at the Smithsonian National Air and Space Museum in Washington, DC. He has written several books on the history of computing and aerospace including *Beyond the Limits: Flight Enters the Computer Age* (1989); *A History of Modern Computing* (1998); *Internet Alley: High Technology in Tysons Corner, 1945–2005* (2008); *Computing: A Concise History* (2008); and, together with Andrew K. Johnston, Roger D. Connor and Carlene E. Stephens, *Time and Navigation* (2014). His most recent book publication is *GPS: A Concise History* (2018).

Oliver Dunnett is a Lecturer in Human Geography at Queen's University Belfast. His research interests focus on the ways in which cultures of science, technology and outer space are connected to questions of place, landscape and identity. Oliver Dunnett has published in journals such as *Cultural Geographies*, *Geopolitics* and *Social and Cultural Geography*, on topics including the moral geographies of light pollution and understandings of tropicality in twentieth-century space science. He is also the author of the forthcoming book *Cultures of British Outer Space, 1900–2020*.

Anthony Enns is Associate Professor of Contemporary Culture in the Department of English at Dalhousie University in Canada. His work in media studies has appeared in such journals as *Senses and Society*, *Screen*, *Culture, Theory & Critique*, *Journal of Sonic Studies*, *Journal of Popular Film and Television*, *Quarterly Review of Film and Video*, *Popular Culture Review* and *Studies in Popular Culture*.

Christopher Gainor has written extensively on the history of space exploration and aeronautics, and studied the history of intercontinental ballistic missiles for his PhD dissertation in the history of technology from the University of Alberta. He is the editor of *Quest: The History of Spaceflight Quarterly*, has taught history at the University of Victoria and for the Royal Military College of Canada, and is writing a history of the Hubble Space Telescope. He is the author of five books, including *Arrows to the Moon: Avro's Engineers and the Space Race* (2001); *To A Distant Day: The Rocket Pioneers* (2008); and *The Bomb and America's Missile Age* (2018).

Alexander C. T. Geppert is Associate Professor of History and European Studies at New York University, jointly appointed by NYU Shanghai and NYU's Center for European and Mediterranean Studies with the Department of History. From 2010 to 2016 he directed the Emmy Noether research group 'The Future in the Stars: European Astroculture and Extraterrestrial Life in the Twentieth Century' at Freie Universität Berlin. In 2019–20 he held the Charles A. Lindbergh Chair in Aerospace History at the Smithsonian National Air and Space Museum in Washington, DC. In 2021–22 he will serve as the Eleanor Searle Visiting Professor in History at the California Institute of Technology and the Huntington Library in Los Angeles. His book publications include *Fleeting Cities: Imperial Expositions in* Fin-de-Siècle *Europe* (2010, 2013); *Wunder: Poetik und Politik des Staunens im 20. Jahrhundert* (2011, co-ed.); *Imagining Outer Space: European Astroculture in the Twentieth Century* (2012, 2018, ed.); *Obsession der Gegenwart: Zeit im 20. Jahrhundert* (2015, co-ed.); *Berliner Welträume im frühen 20. Jahrhundert* (2017, co-ed.); and *Limiting Outer Space: Astroculture after Apollo* (2018, 2020, ed.). At present Alexander Geppert is completing a cultural history of outer space in the European imagination, entitled *The Future in the Stars: Europe, Astroculture and the Age of Space*.

Patrick Kilian is a PhD candidate at the Research Institute for Social and Economic History at Universität Zürich and a graduate fellow at the Zentrum Geschichte des Wissens (ZGW) in Zurich. His dissertation project on 'Astronautic Bodies in the Cold War' is funded by the Swiss National Science Foundation. His research interests include the history of the Cold War, space history, science and technology studies, as well as the history of knowledge and French theory. Since 2015 Patrick Kilian has been a co-founder and co-editor of the interdisciplinary open-access journal *Le Foucaldien*. He is the author of *Georges Bataille, André Breton und die Gruppe Contre-Attaque* (2013). Recent publications include 'The Well-Tempered Astronaut' (co-authored with Jordan Bimm, 2017) and 'John C. Lilly auf Tauchstation' (2018).

Cathleen Lewis is Curator of International Space Programs and Spacesuits at the Smithsonian National Air and Space Museum in Washington, DC. She has completed both a bachelor's and master's degree in Russian and East European studies at Yale University and a PhD in history at George Washington University. Her current research is on the history of the public culture of Russian fascination with the idea of human spaceflight. Cathleen Lewis has written about artifacts in the Smithsonian's collection, and articles comparing Soviet and American approaches to exhibiting and portraying spaceflight. At present she is working on a history of the development of space suit gloves.

Natalija Majsova is a postdoctoral researcher at Université Catholique de Louvain (UCL), Belgium, where she investigates nineteenth- and early-twentieth-century visual mass media, and an assistant professor at the University of Ljubljana, Slovenia, where she contributes to courses in cultural studies and sociology. She earned her PhD (2015) and MA (2011) in cultural studies, and a BA in international relations (2010) from the University of Ljubljana. In 2014 she was a visiting scholar in the Emmy Noether research group 'The Future in the Stars: European Astroculture and Extraterrestrial Life in the Twentieth Century' at Freie Universität Berlin. In 2017–18 she carried out a postdoctoral project on the aesthetics of Soviet science-fiction cinema at the Research Centre for Visual Poetics at the University of Antwerp. Her recent book publications include *Konstruktor, estetika in kozmonavt: Vesolje v sodobnem ruskem filmu 2001–2017* (2017).

Michael J. Neufeld is a Senior Curator in the Space History Division of the Smithsonian National Air and Space Museum, where he is responsible for the early rocket collection and for Mercury and Gemini spacecraft. He is also the lead curator of the Destination Moon exhibit project. Born and raised in Canada, he has four history degrees, including a PhD from Johns Hopkins University. Michael Neufeld has written four books, *The Skilled Metalworkers of Nuremberg: Craft and Class in the Industrial Revolution* (1989); *The Rocket and the Reich: Peenemünde and the Coming of the Ballistic Missile Era* (1995); *Von Braun: Dreamer of Space, Engineer of War* (2007); and *Spaceflight: A Concise History* (2018). He has edited five others: *Planet Dora* (1997); *The Bombing of Auschwitz: Should the Allies Have Attempted It?* (2000); *Smithsonian National Air and Space Museum: An Autobiography* (2010); *Spacefarers: Images of Astronauts and Cosmonauts in the Heroic Era of Spaceflight* (2013); and *Milestones of Space: Eleven Iconic Objects from the Smithsonian National Air and Space Museum* (2014). In 2017 Secretary David Skorton gave him the Smithsonian Distinguished Scholar Award, the highest research award of the Institution.

Regina Peldszus is a policy officer with DLR Space Administration, Department of Space Situational Awareness, focusing on the intersection of European governance, operations, security and infrastructure for Space Surveillance and Tracking. From 2013 to 2015 she was a Research Fellow and then consultant at the European Space Agency (ESA), based at the European Space Operations Centre, Studies and Special Projects Division. Prior to this, she contributed to various future systems and safety projects in Europe, Russia and the United States. Regina Peldszus holds a PhD in crewed exploration mission scenarios and simulation from Kingston University, London. Her interests focus on resilience and foresight of complex systems in high-reliability domains including space, polar and nuclear.

Michael Sheehan is Professor of International Relations at Swansea University. His research focuses on international security, particularly theories and conceptualizations of security. He teaches space policy on both the BA and MA programs. His current research interests include Chinese popular understandings of, and attitudes towards, the Chinese space program, and the relationship between cyber-security and social exclusion in the European Arctic. He is the author of 13 books, including *International Security: An Analytical Survey* (2005); *The International Politics of Space* (2007); and *Securing Outer Space* (2009, co-ed.).

Tilmann Siebeneichner is a postdoctoral researcher at Humboldt-Universität zu Berlin. He holds a degree in philosophy and history, and graduated from Georg-August-Universität Göttingen in 2011 with a PhD thesis on the workers' militia in the DDR. From 2013 to 2016 he was a member of the Emmy Noether research group 'The Future in the Stars: European Astroculture and Extraterrestrial Life in the Twentieth Century' at Freie Universität Berlin. His recent publications include *Proletarischer Mythos und realer Sozialismus: Die Kampfgruppen der Arbeiterklasse in der DDR* (2014); 'Die "Narren von Tegel": Technische Innovation und ihre Inszenierung auf dem Berliner Raketenflugplatz, 1930–1934' (2017); *Berliner Welträume im frühen 20. Jahrhundert* (2017, co-ed.); and 'Spacelab: Peace, Progress and European Politics in Outer Space, 1973-85' (2018). His current research focuses on the militarization of outer space in the 1970s.

Philipp Theisohn is Professor of Comparative Literature and Modern German Literature at the German Department of Universität Zürich. He also directs the research project 'Conditio extraterrestris: The Inhabited Galaxy as the Space of Literary Imagination and Communication (1600–2000).' His current research interests include the history of the alien reader, the use of literature as a futurological medium as well as the history of literary property. Theisohn's book publications include *Die Urbarkeit der Zeichen: Zionismus und Literatur – eine andere Poetik der Moderne* (2005); *Die kommende Dichtung: Geschichte des literarischen Orakels 1450–2050* (2012); *Literarisches Eigentum: Zur Ethik geistiger Arbeit im digitalen Zeitalter* (2012); and *Des Sirius goldne Küsten: Astronomie und Weltraumfiktion* (2019, co-ed.).

Introduction

CHAPTER 1

Spacewar! The Dark Side of Astroculture

Alexander C. T. Geppert and Tilmann Siebeneichner

> The time has come to ask what the people of the Earth are going to do about Space. Are they to use it to make themselves masters of the Universe, or to destroy themselves?
>
> *Daily Mail*, 1959[1]

History may not repeat itself, a truism goes, yet it often rhymes. When the 45th US president first floated the idea of creating a new 'Space Force' in March 2018, many observers were bemused. Those who knew their space history could not help but recall a remarkably similar announcement another American president had made 35 years earlier, in March 1983.[2] In his 'Address to the Nation on Defense and National Security,' broadcast live on television and radio, then commander-in-chief Ronald Reagan proclaimed a 'long-term plan to make America strong again.' After a 'decade of neglecting our military forces,' the US president called for a comprehensive technological modernization program. Soon nicknamed the 'Star Wars' speech, after George Lucas's eponymous 1977 space opera, Reagan's announcement underscored the strategic significance of outer space as the battlefield of the future and (re)militarized international cosmopolitics (Figure 1.1).[3]

Alexander C. T. Geppert (✉)
New York University, New York, USA
NYU Shanghai, Shanghai, China
e-mail: alexander.geppert@nyu.edu

Tilmann Siebeneichner
Humboldt-Universität zu Berlin, Berlin, Germany
e-mail: tilmann.siebeneichner@hu-berlin.de

Alexander C. T. Geppert et al. (eds), *Militarizing Outer Space*
European Astroculture, vol. 3
https://doi.org/10.1057/978-1-349-95851-1_1

Figure 1.1 Even before SDI, the armament of outer space was perceived as imminent. A 1978 study conducted by the renowned Stockholm International Peace Research Institute (SIPRI) pictured the future of global warfare as dependent on satellite technology.
Source: Bhupendra Jasani, *Outer Space: Battlefield of the Future?*, London: Taylor & Francis, 1978, cover image. Courtesy of SIPRI.

Often described as a climax of Cold War confrontation, Reagan's bid for control of earth orbit constituted the preliminary endpoint of a longer historical development. For much of the twentieth century, outer space was a site of utopian thinking that drew upon prospects of peaceful expansion. Yet the development of modern spaceflight technology was equally grounded in violent and often outright dystopian scenarios of warfare. More than anything else, SDI illustrated that the so-called conquest of space, first envisioned (and termed as such) during the interwar period, was as much driven by futuristic fantasies of interplanetary expansion as by mundane aspirations of securing military control from out of space. 'So far as sovereign power is concerned [...], control of the moon in the interplanetary world of the atomic future could mean military control of our whole portion of the solar system. Its dominance could include not only the earth but also Mars and Venus, the two other possibly habitable planets,' spaceflight propagandist Edward Pendray (1901–87) speculated in 1946, bringing – and thinking – together peaceful outreach into space and its hegemonic benefits.[4]

Concentrating on weapons, warfare and violence beyond planet Earth, *Militarizing Outer Space* explores this military dimension of astroculture and zeroes in on the oscillations between peaceful and aggressive, imaginary and material, national and international dimensions of human and robotic spaceflight. Rather than invoking oft-repeated narratives of bipolar Cold War rivalry and an escalating Space Race between the two superpowers, *Militarizing Outer Space* examines the ways in which fantastic anticipation and political rationales, technological failures and apocalyptic threats were part and parcel of the legitimization and popularization of space exploration from the 1920s through the 1980s, from early space war imaginaries to Reagan's 'Star Wars' scenario.

I 'Dual use' and other technopolitical fictions

In the beginning there was war, both on earth and in the skies. Rockets were imagined as weapons of the future long before their first combat deployment in the early 1940s. When the popular German magazine *Die Gartenlaube* published an article on 'The Three Faces of the Rocket' in 1930, doubts about the technology's feasibility were widespread. Naming the so-called father of spaceflight Hermann Oberth (1894–1989) as author, the article went to great lengths to distinguish itself from science fiction, however it might have mimicked the genre's aesthetics. Lavishly illustrated by A. B. Henninger, it praised the benefits of liquid-fuel rocketry while elaborating three classes of objectives: scientific, belligerent and futuristic. While long-range missiles would serve a number of earth-bound purposes, for the author only the third type – the *Weltraumschiff* (spaceship) – constituted the ultimate goal of technological progress, as it would aim to leave the confines of earth behind.[5] In their colorful brutality the accompanying illustrations spoke a more candid language of threat and destruction. Here, a bleak picture of a nocturnal raid on New York City, featuring the Brooklyn Bridge in the foreground and the Manhattan Bridge behind, introduced a belligerent technology that would soon capture the minds of military theorists all over the world: the ballistic missile (Figure 1.2).

Figure 1.2 Depicting a nocturnal missile raid on New York City, this illustration by A. B. Henninger featured in a 1930 article published under Hermann Oberth's name anticipated the strategic value of rocketry for future warfare.
Source: A. B. Henninger, 'Nächtlicher Raketenüberfall auf New York,' *Die Gartenlaube* 43 (23 Oktober 1930), 887. Courtesy of Martin Kelter Verlag.

Oberth, his American counterpart Robert H. Goddard (1882–1945) and others had proposed using rockets as weapons as early as 1917.[6] But it was only after the First World War when public debate began to latch onto the idea of revolutionizing warfare by means of rocketry, 'death-rays' and other fantastic weapons. Historian Peter Bowler has recently shown that the logic of deterrence, commonly considered a Cold War product, was effectively conceived at that time.[7] A year before the *Gartenlaube* article was published, Oberth had already made headlines with *Wege zur Raumschiffahrt*, a revised edition of his seminal 1923 treatise *Die Rakete zu den Planetenräumen* with an added section on the conduct of future warfare. The use of rockets would enable their proprietors to strike against concentrated military facilities and civilian infrastructures such as ammunition dumps or railway junctions rather than squander them on individual combatants widely dispersed through the trenches. 'One would not go to war as easily if one knew: "The first one to be hit will be me",' Oberth cautioned his readers.[8] As a consequence, he concluded during a lecture held in Vienna in 1931, these 'deadly death-rockets [...] would force the world, in self-protection, to outlaw all war.'[9] Recognizing the simultaneously utopian and dystopian implications of this powerful new technology, Oberth contributed to the making of deterrence as a strategic concept, one which would only later rise to fame and serve to legitimize the militarization of outer space throughout the twentieth century.

Germany's defeat during the First World War prepared the ground for its fascination with rocketry. Most notably, members of the Verein für Raumschiffahrt (VfR), an amateur lobby group founded in 1927, rallied for the development of rocket technology by promising 'benefits of a kind that would immediately restore Germany's erstwhile international standing.'[10] The exact extent to which these early rocketeers were driven by futurist space-mindedness or rather advocated the ballistic missile as key to national rebirth is still a matter of debate. But there can be little doubt that ideas and images such as those featured in *Die Gartenlaube*, *B.Z. am Mittag* and other popular outlets made the new technology's appeal anything but utopian, innocent and immaculate.

Similar enthusiasm for all matters space and widespread interest in utilizing the third dimension for military purposes existed also in the United Kingdom, France, the United States and the Soviet Union after 1918. 'Rockets are in everybody's thoughts just now,' a British journalist observed. Popular science magazines such as *Science and Invention*, *Popular Science Monthly*, *La Science et la Vie* and *Everyday Science and Mechanics* simultaneously discussed both the likelihood of reaching the moon and the military potential of rocketry.[11] The deployment of aircraft during the First World War convinced military theorists Giulio Douhet (1869–1930), Hugh Trenchard (1873–1956), Billy Mitchell (1879–1936) and others that airpower would be the deciding factor in any future warfare, enabling the aggressor to achieve a 'knock-out blow' against which – very much in line with Oberth's missile raid scenario – defense was nigh on impossible.[12]

Early space imaginaries were anything but innocent projections, and indeed anticipated military applicability. From the interwar space craze to today, they continually oscillated between belligerent ambitions of space war and pacificatory hopes for a unification of humankind. The close relationship between science, industry and the military accrued throughout the Second World War and its two central technical innovations, nuclear energy and liquid-propellant rocketry, serve as a reminder how slippery the distinction between civilian and military technology is.[13]

Commonly evoked analytical categories such as 'dual use,' 'civilian' versus 'military spaceflight' or the realm of 'aerospace' were effectively the product of semantic interventions to mask the increasing military relevance of the third dimension. The Janus-faced character of rocketry has never been inherent in the technology per se; rather it is the outcome of a strategic move to erect an artificial distinction between the purportedly peaceful, civilian face versus the militaristic, potentially violent side of post-Second World War space technology. As political scientist Roger Handberg has argued, the concept of 'dual use' was invented in the 1950s to add a second layer of incontestable, morally "good" space exploration to the ensuing military-political competition between East and West. Similarly, the US Air Force created 'aerospace' in the 1950s to purport the technical and spatial unity of aeronautics and astronautics, effectively two entirely different geospatial spheres.[14] And US President Dwight D. Eisenhower's decision to establish the National Aeronautics and Space Administration (NASA) in 1958 as a civilian, non-military institution was likewise intended to signal to a wary American public that the nation as a whole would benefit from civilizational progress achieved in outer space, in addition to gaining technoscientific supremacy and ideological hegemony.[15] The European Space Agency (ESA), NASA's West European counterpart, only founded a quarter of a century later, went even further in propagating the technopolitical fiction that rocket technology could be used exclusively to advance civilian and utopian purposes. The Soviet government, on the other hand, never set up a central civilian space agency in the first place and kept its space activities entirely under military control, as does present-day China.[16]

In short, on closer inspection all three concepts conventionally employed to historicize the "dark side" of real and imagined space exploration – 'dual use,' 'civilian versus military' and 'aerospace' – are problematic. Because these concepts are more political, more charged and much less neutral than usually assumed, their historiographical utility is limited. Employing the concept of 'dual use' runs the risk of reproducing a simplistic binary opposition. Doing so easily perpetuates the illusion that a distinction between a civilian and a military side of technology can be made. The practical question is then whether the term's analytical validity outweighs its inbuilt ideological agenda, but there can be little doubt that it easily obscures as much as it illuminates.

Ironically, a fourth, if hardly less loaded neologism proves more helpful when trying to understand the martial dimension of spaceflight: Eisenhower's notorious 'military-industrial complex,' first introduced in his 1961 farewell speech. Alarmed by what he considered a possible threat to democracy, the outgoing US president warned against an unprecedented 'conjunction of an immense military establishment and a large arms industry' pursued by a scientific-technological elite in search of political hegemony. Public reactions were limited at first, yet beginning with the Vietnam War, the idea of a military-industrial complex began to resonate with a public fearing an uncontrolled and increasingly impenetrable mix of power, money and technology to which, some were quick to point out, academia should have been added.[17] Whether as a societal diagnosis historically accurate or not, what renders the term historiographically useful is precisely its open and critical character. Rather than blurring spheres of interest and power constellations, bringing the military-industrial complex into play helps to see otherwise hidden technopolitical entanglements between states, lobby groups and the defense industry.

Finally, a fifth conceptual duality – 'militarization versus weaponization' – must be discussed. Neither term is uncontested either, yet distinguishing between the militarization of outer space on the one hand and its weaponization on the other is key to the present enterprise. Whereas 'militarization' has been subject to considerable scholarship, not the least in the context of a conceptual history of terms such as 'military,' 'militia' and 'militarism' reaching back to the eighteenth century, 'weaponization' is the younger and seemingly more straightforward term.[18] An 'instrument of any kind used in warfare or in combat to attack and overcome an enemy,' the *Oxford English Dictionary* defines 'weapon,' referring the *explanandum* to its combatant context.[19] But do space weapons exist if no war has hitherto been waged in space? Despite a wide variety of past efforts to place weapons in low-earth orbit – including considerations to equip future spacefarers with 'spin stabilized micro guns,' 'gas weapons for close-in fighting' and 'spring propelled spherical projectiles' in addition to more concrete efforts such as arming the Soviet Almaz space station with a 23-millimeter cannon – to date no direct confrontation has been fought in space.[20] Accordingly, observers and critics have concluded time and again that space has not become a theater of war as no weapon systems have effectively been deployed or used there.[21]

For three reasons such a long-term consensus seems to have been breaking apart in recent years. First, much-discussed events such as the ASM-135 anti-satellite weapon (ASAT), tested by the United States Air Force in September 1985, or China's unannounced 2007 elimination of one of its derelict weather satellites, creating more than 2,000 pieces of space debris, have demonstrated that spacefaring nations are perfectly capable of deliberately destroying objects in low-earth orbit.[22] Second, not a single military operation is fought any longer without the support of satellite technology. The Global Positioning System (GPS), developed by the US Department of Defense during the 1970s and 1980s and used with overwhelming success in the 1991 Persian Gulf War, made commentators proclaim the 'world's first "space war",' thus substantiating pre-SDI forecasts of future warfare's reliance on space-based assets (see Figure 1.1 above). And third, because extraterrestrial infrastructure has now become so crucial to conventional warfare on earth, its protection from potential attacks is increasingly considered a massive security requirement.[23]

For about three decades social and cultural historians have been pleading for a broader and more encompassing understanding of 'militarization,' for instance, as 'the contradictory and tense social process in which civil society organizes itself for the production of violence.' In a similar vein, Michael S. Sherry has described militarization as 'the process by which war and national security became consuming anxieties and provided the memories, models, and metaphors that shaped broad areas of national life.'[24] Both 'production of violence' and 'memories, models, and metaphors' are key here as they resonate with the concept of astroculture underlying this book and its preceding companion volumes, *Imagining Outer Space* and *Limiting Outer Space*. With a view to broadening the historiographical focus on the interests, activities and experience of the historical actors themselves, historians of astroculture seek to understand the lure, potential and meaning space-crazed societies hoped to find in the heavens.[25]

Highlighting the imaginary as a driving force behind twentieth-century space exploration leads to an analytical approach that bypasses conventional narratives which posit the 'conquest of the stars' as an inevitable outcome of the Cold War. More than a symbolic front or a political function, spaceflight and space thought became essential to wider socio-cultural self-understanding, drawing from both fact and fiction. Scrutinizing astrocultural imaginaries and ideologies of technoscientific progress in a global perspective suggests that the militarization of space is neither the "evil" twin-brother of space utopias nor a mere epiphenomenon of the Cold War. Rather, from the outset it was part and parcel of the production of twentieth-century space itself.

II 'High ground' and the Cold War in orbital space

Warmongers and pacifists alike believed in the promise of space. Fifteen years and a world war after the 1930 *Gartenlaube* scenario, a 29-year-old radar technician, training instructor and flight lieutenant stationed outside Coventry, England, participated in an essay competition organized by the Royal Air Force. His manuscript on 'The Rocket and the Future of Warfare' radiated such detailed expert knowledge and prophetic clarity that the journal's jury awarded it first prize in every category offered. In its diagnosis, tone and outlook, the article showed surprising parallels to Oberth's, despite an ongoing technological revolution evidenced by the world's first ballistic missile launched in 1944 – the infamous 'vengeance' weapon V-2/A4 – as well as the first atomic bomb, dropped above Hiroshima just a few months earlier.[26]

The author agreed that long-range rockets promised to be a decisive, if not the ultimate, weapon in any future war and one against which no defense existed. He pondered Oberth's optimistic belief that deterrence as a strategic concept would make armed conflicts impossible, yet rejected the idea vehemently. Quite to the contrary, this author extrapolated, the combination of rocket and atomic power, now technically conceivable, would lead to an altogether new type of 'radiation war,' to be fought by 'automatic and guided,' that is, computerized weapons. A third parallel, equally hidden yet central to both scenarios, is particularly noteworthy in the present context. According to Oberth, all rocket technology would eventually culminate in the development of the *Weltraumschiff* or spaceship as the 'ultimate goal,' to arrive 'as surely as a prophesized solar eclipse.'[27] Writing fifteen years later, the author of 'The Rocket and the Future of Warfare' shared such cosmic enthusiasm. Faced with civilization's newly acquired ability to self-destruct, the long-range rocket offered at least a small glimpse of hope. From there, 'interplanetary rockets' aka 'true "spaceships", travelling in trajectories outside the atmosphere' would evolve, leading to the establishment of space stations as an 'attractive solution to the problem of world surveillance' and opening unforeseen vistas 'before which the imagination falters.' The rocket may have been the future of warfare, yet it could also serve as the foundation of quite a different and indeed utopian future – one prophesied to play out among the stars.[28]

The author of this prizewinning treatise was, of course, none other than the young Arthur C. Clarke (1917–2008), the so-hailed prophet of the Space Age, soon to turn into one of the most perspicacious spaceflight gurus and influential space *personae* of the twentieth century.[29] A year prior to the article's publication, in February 1945, Clarke had suggested in a letter to the editor of *Wireless World* to use the extant V-2s for stationing an 'artificial satellite' in a fixed earth orbit to give 'television and micro-wave coverage to the entire planet.' At that time the almost 3,200 German V-2 missile attacks on Antwerp, London, Norwich and many other West European cities were still rampant.[30] Despite having been on the receiving end of the Nazi terror weapons, Clarke nevertheless turned to military applications in his tireless efforts to make spaceflight a reality.

Advanced in this early piece almost en passant as one of the few optimistic side effects of an otherwise gloomy outlook, fantasies of world surveillance and military dominance from above figured as the central trope to justify any space militarization endeavors throughout the 1980s. Effectively updating a military tenet that can be traced back to ancient China – in his famous *The*

Art of War, Sun Tzu or Master Sun had advised military leaders to seek elevated territory whence to exert control and thus force the enemy to attack from a lower position – the increasing utilization of the third dimension for combat purposes made military strategists call for 'space superiority' or 'space supremacy,' obviously after, but also long before the Second World War.[31]

Former NASA archivist Lee Saegesser has compiled a comprehensive collection of almost fifty 'high ground'-references by politicians, military figures and journalists, including a handful from the Soviet Union, covering the 1890s to the 1980s.[32] It is unsurprising that two thirds of these references stem from the post-Sputnik 1950s and the early 1960s. What is surprising, however, is that, once again, Douhet and Oberth modernized the age-old doctrine in the 1920s by applying it to aerial warfare and armed space stations.[33] Triggered by fictional scenarios and expert interventions alike, particularly during the early Cold War, pleas for space supremacy abounded. Ranging from the 1950 science-fiction classic *Destination Moon* ('The first country that can use the moon for the launching of missiles will control the earth. That, gentlemen, is the most important military fact of this century') to an oft-cited 1952 *Collier's* editorial ('The first nation to do this will control the earth [...]. Whoever is the first to build a station in space can prevent any other nation from doing likewise') politicians soon followed suit. 'Control of space means control of the world,' senator Lyndon B. Johnson proclaimed in January 1958. John F. Kennedy reiterated the dictum during his 1960 election campaign: 'If the Soviets control space they can control earth.'[34] That this proved to be a largely American debate, with constant repetition turning the trope into a truism, did not go unnoticed. 'Hardly a week pauses without some American brass-hat discussing the use of Spaceships for spying on the enemy or claiming that he who controls the Moon will have the Earth at his mercy,' a *Daily Mail* commentator grumbled in 1959: 'What a pity that the cosmic horizons now unfolding should be thus darkened! What a tragedy that the first Space-vehicles had to be terrible weapons of war!'[35]

The lines between astroculture and power politics had been blurry before, but with the onset of the Cold War outer space gained unprecedented attractiveness and relevance. It was the battlefield central to all future warfare. In view of the evolving global confrontation, techno-fantasies previously dismissed as fringe were now imbued with a new sense of political urgency. Both man and machine must adapt to meet the demands of future extraterrestrial warfare.[36] 'Rocket-fever has taken hold,' the British *Picture Post* diagnosed in November 1952, providing its readers with an overview of ongoing guided missile research in the United States, West Germany and the United Kingdom's 'rocket playground' at Woomera, Australia.[37] One of the first instances to draw concrete consequences to secure the high ground was astronomer and science-fiction author Robert S. Richardson (1902–81). In a 1948 article titled 'Rocket Blitz from the Moon,' published in the mass-market *Collier's* magazine and lavishly illustrated by noted space artist Chesley Bonestell (1888–1986), Richardson added a new figure to the equation by suggesting to use the moon as a launch base for atomic bombs. 'Operation Knickerbocker,' as a German translation of the article published in the conservative Swiss weekly *Die Weltwoche* christened this doomsday scenario, presented an outlook similar to Oberth's 1930 nocturnal New York City raid. Eighteen years later, the rockets came straight from the moon.[38]

Other concrete proposals to gain strategic advantage by militarizing outer space are better known than many of the early, half-realistic flights of fancy, impactful as they were. They include, of course, Wernher von Braun's campaign for a manned space station (1946–56) that would serve as reconnaissance platform and orbiting battle station for achieving 'space superiority' over the Soviet Union, similarly propagated in *Collier's* and illustrated by Bonestell a few years after the 'Rocket Blitz.' Project Horizon, a feasibility study undertaken by the US Army in 1959, foresaw the construction of a 12-man lunar outpost reminiscent of a Wild West fort.[39] Some of these projects appear much less far-fetched if one takes into account that both superpowers conducted altogether nine nuclear tests in outer space from 1958 to 1961, at altitudes between 150 and 540 kilometers, before the 1963 Partial Test Ban Treaty (PTBT) banned all such tests in space, the atmosphere and underwater (but not underground). All the while, the pioneering US photo reconnaissance program CORONA and its successor systems GAMBIT and HEXAGON used satellites to keep a constant eye on Soviet and Chinese activities from an altitude of about 150 kilometers. Likewise, the Soviet Union began operating their own photo reconnaissance satellites in 1962. According to the CIA, over the course of CORONA's lifetime (1960–72) 121 missions produced roughly 800,000 images. The first mission alone led to the identification of 64 airfields and 26 new surface-to-air missile (SAM) sites on Soviet territory. The US government did not acknowledge that it used satellite systems for reconnaissance purposes until 1978. CORONA was declassified only in 1995, after the end of the Cold War.[40]

Yet the program's success demonstrated the power of satellite technology for defense and security. By establishing a space infrastructure, control could be exerted over the Cold War enemy via photographic surveillance. In 1978 the Stockholm International Peace Research Institute (SIPRI) estimated that 1,484 satellites or 60 percent served military purposes. Others believed this figure to be as high as 75 percent of a total of 1,810 launches since 1957.[41] Instrumental for multiple military purposes including communication, navigation, meteorology, reconnaissance and surveillance, satellites became important supportive combat tools that could possibly be transformed into weapons themselves. The development of ASATs designed to incapacitate satellites began in the late 1950s. Both superpowers experimented with conventional rockets, either hitting the target directly or damaging it from the blast and heat of nearby explosions; they also turned to energy weapons and lasers.

Shrouded in the same secrecy as CORONA and many other military programs, such newfangled weapon technologies intrigued and alarmed the public. Commentators anticipated 'satellite wars' in near-earth orbit and constantly invoked the enemy's efforts in developing ultra-sophisticated space weapons. European graphic artists such as Klaus Bürgle (1926–2015) and Tony Roberts (1950–) both envisioned satellites battling each other in breathtaking laser duels in 1967 and 1978, respectively (Figures 1.3 and 1.4). While motif and setting, with a portion of earth in the background, are similar, there is a slight, yet telling difference between the two scenarios. Anticipating a universally used technology, Bürgle's 1967 satellites bore no identification marks, while in the British equivalent, publicized eleven years later in *Radio Times*, they could easily be recognized as belonging to the two superpowers. Picturing a Soviet spaceship destroying its enemy, the article echoed increased Cold War tensions and fears of a military buildup. Another five years later, after Reagan's 'Star Wars' speech, the popular German scientific magazine *GEO* used exactly the same image to illustrate an article, 'Battlefield Outer Space,' warning of the dangers of SDI. Technoscientific imaginaries had become so intertwined with political realities that it proved impossible for the public to distinguish fiction from fact.[42]

Figure 1.3 In 1967 renowned West German graphic artist Klaus Bürgle (1926–2015) pictured a future space war scenario featuring satellites autonomously attacking each other with laser weapons.

Source: Klaus Bürgle, *Satellite War/Zerstörung durch Laserstrahlen*, tempera, 33 × 46.3 cm, ca. 1967. Courtesy of Klaus Bürgle.

Figure 1.4 In October 1978 the much-read (and still publishing) British weekly *Radio Times* pictured the destruction of an American spacecraft by a Soviet hunter-killer satellite, equipped with a new generation of laser weapons.
Source: *Radio Times* (21–27 October 1978), cover image. Illustration by Tony Roberts. Courtesy of *Radio Times*/Immediate Media Company.

Given widespread fears that the skies could become a full-scale battlefield, public pressure did much to spur a series of diplomatic and legal agreements during the 1960s. The 1967 Outer Space Treaty – formally the 'Treaty on Principles Governing the Activities of States in the Exploration and Use of

Outer Space, including the Moon and Other Celestial Bodies' – banned weapons of mass destruction in outer space. It also prohibited testing weapons, conducting military maneuvers and establishing the kind of bases envisioned, for instance, in the aforementioned Project Horizon.[43] Distinguishing between passive and aggressive military purposes, and declaring satellites to be 'no weapons themselves' but useful only 'to enhance military systems below' – that is, on earth – the international community deliberately turned a blind eye to the fighting capacity of space-based technology and reaffirmed the dual-use fiction prevailing in Western societies.[44] Rather than curtailing the accrual of militarization efforts, the impact of this body of laws proved primarily symbolic. Clinging to traditional doctrines of earth-bound confrontation without outlawing military activities per se, international legal debates conceptualized space as an extension of terrestrial warfare, despite all sweeping astrofuturist rhetorics à la 'battlefield of the future.'[45]

Finally, the Cold War requires further explanation, not the least as it figures prominently in this book's title. As indicated above, that 'Forty-Year Confrontation between East and West' plays a different part in the overall argument than readers might expect.[46] First and foremost, it is noteworthy that outer space is largely absent from the already extensive, yet ever-booming Cold War literature. Recent studies on the arms race or the nuclear crisis complex lack a space dimension and have little to offer on satellites, satellite reconnaissance or even SDI.[47] Those Cold War historians who "see" space have often reduced the third dimension's increasing significance to the bilateral Space Race. Most works assume a causal nexus between the development of spaceflight and the Manichean postwar world order, with outer space reduced to a surrogate theater of war transposed into low-earth orbit.[48] But neither the Space Age nor the development of spaceflight are mere consequences of the Cold War, and reiterating high-ground, dual-use, East/West binaries only reproduces contemporary clichés meant to sell and sanctify the military-industrial complex's reach into the heavens. Evoking a Cold War context is a necessary, yet insufficient framework to explain the making of outer space and its advancing and ever-intensifying militarization, both in terms of *Realpolitik* and the *imaginaire*.

Hence, the present book – possibly provocatively for some, self-evident, if not banal to others – attempts to decouple outer space and the Cold War without denying the latter's all too obvious repercussions. As an alternative it suggests to foreground the production of space itself and to pay particular attention to the spatiographical dimension of these processes. Surprisingly enough, few observers have commented on the fact *this* kind of conquest has largely been confined to low-earth orbit, that is, to altitudes of 500–1,200 kilometers above earth, with communication satellites in much higher geostationary orbit the exception to the rule. The militarization of outer space was thus notably geocentric, in stark contrast to those space wars of science fiction and science fantasy imagined in unspecified galaxies far, far away. In 1962 Eurospace's vice-president Michael N. Golovine (1903–65) suggested to replace 'aerospace' with 'orbital space' as a geographically more accurate alternative. While the term did not prevail, it is indeed in orbital space where the vast majority of realities and imaginaries under scrutiny in this volume are located and were played out.[49] Accordingly, visual depictions of warfare in space – realistic, artistic or both – almost invariably include a segment of planet Earth in the background, to serve not only as a token of scale but also as a point of reference (see Figures 1.1, 1.3 and 1.4 above, and 1.5 and 1.6 below). Space wars, these graphics implicitly remind their viewers, were essentially earth wars in disguise, extending Sun Tzu's classic high-ground tenet into low-earth orbit.

III Militant astroculture and fights of fancy

Fascination for anything space and military goes in both directions – space within the military, but also the military within space. 'The history of future-war fiction [offers] an account of the fears and expectations of a given historical moment,' literary scholar David Seed has argued.[50] *Spacewar!*, one of the very first video games, proclaimed the imminence of, simply put, space war in its own name, underlined by an exclamation mark. Conceived at MIT in early 1962, less than a year after the first human had orbited earth and set in an unspecified part of the universe, two players had to shoot down each other's spacecraft while trying to elude the gravity emanating from a bright, pulsating star in the middle of an otherwise black computer screen.[51] The program was a collaborative project of computer researchers or 'hackers.' Adaptations were soon disseminated to other labs and research facilities in the United States and the United Kingdom. Before the release of the arcade adaptation *Space Wars* in 1977, *Spacewar!* was not available to a mass audience. But it was not entirely obscure either. On 26 January 1969, legendary CBS anchorman Walter Cronkite (1916–2009) presented the game on his television show *The 21st Century*. 'No, the Pentagon is not designing interplanetary warships. This "space war" is an exercise programmers use to relax and to learn what their computer can do,' Cronkite was quick to dismiss any military applicability.[52] And in October 1972, *Rolling Stone* sponsored the First Intergalactic Spacewar Olympics with two dozen opponents. Stewart Brand (1938–), best known as editor of the *Whole Earth Catalog* (1968–72), reported at length.[53]

Space had fueled new types of media before, for instance, George Méliès's *Le Voyage dans la lune* (1902) in the case of film, and pulp magazines and science-fiction comic books in the 1920s. According to one of the game's co-authors, it was the excessive consumption of cheap space novels and 'cinematic junk food' which had 'established the mind-set that eventually led to *Spacewar!*' Another team member felt so offended by the 'random stars' in the background that he encoded the entire night sky and added a program to ensure that the stars on the computer screen were displayed in the correct positions to each other and their actual brightness.[54] From the 1900s to the 1920s to the 1960s, astroculture served as the subject of choice for film, pulp fiction or computer games, respectively, as each of these new media arose.

The body of 'militant astroculture,' that is, astroculture which renders space imaginaries into battlefield scenarios and dwells on weapons, warfare and violence in space, is vast.[55] From H. G. Wells's 1897 science-fiction novel *The War of the Worlds*, Orson Welles's notorious 1938 radio adaptation and the 1953 film version by Byron Haskin to lesser known B-movies such as *War of the Satellites* (1958), the Japanese *Battle in Outer Space* (1959) or the Italian-Spanish *Terrore nello spazio* (1965), combat with aliens or warfare set in distant worlds has been a popular subject for a wide variety of feature films.[56] Here, fantasies of imperial expansion propagated by films such as *Destination Moon* (1950), *Conquest of Space* (1955) or even *Around the World in Eighty Days* (1956, with Méliès and rocket launch footage in an onscreen prologue) collided with threats of alien invasion, be they peaceful (*It Came from Outer Space*, 1953), belligerent (*Invaders from Mars*, 1953), educational (*The Day the Earth Stood Still*, 1951) or apocalyptic (*Earth vs. the Flying Saucers*, 1956). After Kenneth Arnold's legendary 1947 encounter, one wave of flying saucer sightings after the other shook the world, not just the United States. These and numerous other movies would both profit from and contribute to the ensuing and enduring UFO myth.[57] Whether considered sympathetic or hostile, extraterrestrials undermined human claims for universal hegemony. They served to dramatize the notion that the exploration of space was tantamount to encountering hostile adversaries which had to be defeated, no matter the cost.

Raumpatrouille (1966), the first West German science-fiction series on television now widely considered canonical, was framed as a clash between two cultures, one terrestrial, the other alien. The first of seven episodes was entitled 'Attack from Outer Space,' whereas the mini-series' finale 'Invasion' 'confronts us yet again with the space enemy,' as a 1965 press release declared.[58] The most popular hero of German astroculture, however, is arguably Perry Rhodan who has been incessantly defending the universe against 'degenerated' and 'sneaky' foes since his first appearance in 1961. By 1978 the novel series had sold more than 300 million copies; to date approximately two billion copies have been issued.[59] Frequently criticized for propagating "fascist" mindsets, both *Raumpatrouille* and Perry Rhodan demonstrate that the stakes in militant astroculture correspond with those of classical mythology, with individual heroes fighting for the survival of a collective, be it a civilization or the entire galaxy.

From Méliès's *Voyage dans la lune* (1902) to Heinlein's *Starship Troopers* (1997), human 'exploration' resorted to violence whenever expansive ambitions were met with resistance. Astroculture's most popular comic heroes including Buck Rogers, Flash Gordon, Nick der Weltraumfahrer and Dan

Dare act as modern knights, usually busy protecting humankind from super-villains or fighting extraterrestrial monsters.[60] Admittedly, both friend and foe use high-tech laser weapons and travel in space, yet the actual combat more often than not resembles classical forms of terrestrial warfare with swords, spears or – at most – guns. Violence usually means archaic, physical violence, either in the form of dueling pairs or massive formations battling each other.

Another case in point is the hugely popular and still ongoing *James Bond* franchise. Whether supervillains disrupting a projected rocket launch (*Dr. No*, 1962), snatching both American and Soviet spacecraft (*You Only Live Twice*, 1967) or deploying a killer satellite to blackmail the superpowers (*Diamonds Are Forever*, 1971), space wars drive and motivate many of the earlier installments. *Moonraker* (1979) introduced the Space Shuttle as the future work-horse of American space exploration, capable of serving as a military supply vehicle and transporting Space Marines. Originally meant to be released alongside the Space Shuttle's first launch, the film debuted on schedule despite the postponement of the spacecraft's maiden flight until 1981. Ridiculed by critics as 'most outlandish,' *Moonraker* dramatized pre-SDI concerns about the military purpose of the shuttle.[61]

The militarization of astroculture peaked in the 1970s. 'Science Fiction is Hollywood's next big business,' West German weekly *Der Spiegel* observed in a lengthy discussion of blockbusters such as Lucas's *Star Wars* (1977) and Steven Spielberg's *Close Encounters of the Third Kind* (1977).[62] Entitled 'escape into space,' the 14-page article took umbrage that the entertainment industry's most recent offerings effectively centered around death, war and violence in space. For the BBC, that kind of science fiction owed more to fact than fantasy. 'This isn't just "Buck Rogers" science fiction. It's real. It's here. It's today,' host Tom Mangold explained in a two-part television special 'The Real War in Space,' broadcast in 1978. Military strategists George Keegan (1921–93) and John Hackett (1910–97), author of a widely read fictional study *The Third World War*, published in 1978, agreed that the battles of the future would be waged beyond earth.[63]

Signaling a reversal from futuristic to dystopian tropes, post-Apollo space imaginaries stood in sharp contrast to concurrent developments in earth orbit. For many, the 1975 Apollo-Soyuz Test Project (ASTP) – featuring the first docking of a US and a Soviet spacecraft – symbolized the end of the Space Race. Even though it generated much less public enthusiasm than the 1969–72 lunar landings, the United Nations General Assembly hailed the mission as an important step towards multilateralism and pronounced a new era of peaceful cooperation.[64]

Icelandic pop artist Erró (1932–) gave the heavenly ASTP encounter a different twist. In his 1975 collage *Meeting in Space*, he montaged a sectional drawing of the docking spacecraft onto Rubens's seventeenth-century oil painting *The Tiger, Lion and Leopard Hunt* in front of a blue sky and with a segment of clouded earth in the upper right, thus juxtaposing age-old tropes of colonial conquest and imperial domination (Figure 1.5). Joyful Cold War

Figure I.5 For his 1975 *Meeting in Space* painting, Icelandic pop artist Erró (pseudonym of Guðmundur Guðmundsson, 1932–), juxtaposed a sectional drawing of the first US/USSR space docking with scenes of violent confrontation, both among men and with wild animals, taken from Rubens's 1616 oil painting *The Tiger, Lion and Leopard Hunt.*

Source: Erró, *Meeting in Space* (1975), from the series 'Space 1974–77 (Homage to Robert McCall),' 93 × 68 cm, private collection, Paris. Courtesy of VG Bild-Kunst.

warriors, floating towards one another with open arms and exchanging handshakes in earth orbit, could not overcome the eternal struggle between man and nature underlying the veneer of all technological civilization. His Space Age version of *homo homini lupus* exposed man's foray into the stars as a continued struggle for power, survival and domination. Far from being a mere comment on a "real" meeting in space, Erró's *Meeting in Space* exemplifies the ways in which 1970s astroculture was particularly obsessed with violence.

With the rapid waning of the worldwide Apollo frenzy and human spaceflight coming to a temporary end in the United States, the optimism of the Space Age gave way to a sense of disillusion. During the post-Apollo period, outer space lost much of its popular allure and cultural significance. Commentators diagnosed a sense of 'general space fatigue' and saw widespread belief in infinite human expansion replaced by the discovery of inner space.[65] Yet at the same time its military-technical relevance massively increased. Fears of a Soviet arms buildup spread throughout the Western imagination, spilled over into the stars and contributed to the frequently evoked insecurities of the so-called crisis decade. An integral facet of all space imaginaries from the very onset, 1970s militant astroculture prefigured, prepared and preceded the most comprehensive program to weaponize space to date.

IV Star Wars wars

In his March 1983 'Star Wars' speech, briefly referred to at the outset, Ronald Reagan declared the skies of crucial importance for the West's future, if not worldwide hegemony. 'What if free people,' he disguised a potential global war in a single rhetorical question, 'could live secure in the knowledge that their security did not rest upon the threat of instant U.S. retaliation to deter a Soviet attack, that we could intercept and destroy strategic ballistic missiles before they reached our own soil or that of our allies?'[66] By the time of his announcement, Cold War tensions had been on the rise again. With NATO's disputed Double-Track Decision, the Soviet invasion in Afghanistan in late December 1979, the downing of Korean Airlines Flight 007 over Siberia in September 1983 and NATO's large-scale Able Archer 83 exercise, fears of military confrontation and an ensuing Armageddon were widespread.[67] Reagan's grandiose promise to render nuclear weapons 'impotent and obsolete' eventually resulted in the Strategic Defense Initiative (SDI), a large-scale space- and ground-based defense system designed to protect the United States from a nuclear strike.

Reagan connected his venture into the stars with the promise of a world liberated from the threat of nuclear war. While he did not explicitly discuss space, it was nonetheless contained in the 'Star Wars' speech, as an absent-present site of eschatological hope and *lieu de l'avenir* whence salvation would descend. With the help of a missile-based defense shield against intercontinental ballistic missiles (ICBM) placed in earth orbit, Reagan asseverated, the 'evil empire' in the East would be contained. Rather than relying on the notorious concept of Mutual Assured Destruction (MAD) and the threat of world extinction, fantastic technology would serve to restore a bygone future and offer 'a new hope for our children in the twenty-first century.'[68]

Even though the speech contained more suggestive speculation than concrete details, Reagan's proposal had a greater and longer lasting impact than any other proclamation of this kind. Coming from a former actor and self-declared science-fiction aficionado, it inaugurated a long-lasting international controversy, echoing far beyond the 1986 Reykjavík summit with Soviet party leader Mikhail S. Gorbachev and not abating for half a decade. 'Star Wars' led to a type of propaganda war never before fought in space. Long in the making, militant astroculture and superpower politics coalesced in an unprecedented militarization of outer space.[69]

Many of Reagan's allies, in particular in Western Europe, were less convinced by his vision. Some doubted the project's technical feasibility, whereas others feared having the wars of the future fought above their heads. European critics pointed to the continent's exposed position in the middle of a bipolar Cold War world. 'If both sides had shields,' wrote British social historian E. P. Thompson (1924–93), 'Europe would be a no man's land, with laser-zapped nukes falling on their heads.'[70] Thompson and many other European intellectuals maintained that abandoning deterrence theory as the key principle of global stability would expose Europe to the superpowers' belligerent gamble for hegemony. Traditionally espousing a different approach towards the so-called conquest of space, Reagan's initiative evoked severe rifts within Europe's space community, at least rhetorically strictly opposed to any efforts of embattling the heavens.[71]

Drawing a direct line from the reactions caused inside the Soviet Union to the collapse of its empire in Eastern Europe at the end of the decade, politicians and historians have declared SDI a major factor in ending the Cold War.[72] They have also pointed out that SDI was not a *space* program proper as it did not aim to further the appropriation, exploration or exploitation of the spatial surroundings of planet Earth. For instance, as a national security initiative, space policy veteran John Logsdon has recently reminded us, SDI only 'incidentally involved stationing defensive systems in orbit.'[73] While technically certainly correct, Logsdon's 'incidentally' is not. SDI's locale in outer space, as imaginary, audacious and futuristic as it seemed in the early 1980s, is by no means trivial but indeed fundamental to understanding the program's societal repercussions, both domestically and internationally. SDI caused worldwide fears because of a 'strange mingling of science and culture, expertise and popularisation' which was only feasible in outer space and happens to be a characteristic of all astroculture. 'From its very inception,' political scientist Edward Reiss has rightly observed, 'SDI was both a high-tech programme *and* a popular idea, a mélange of physics, psycho-politics and metaphysics [...]. Its secret strength lay in its emotive connotations, its capacity to motivate people by mobilising culture, discourse, emotion and fantasy.'[74] It is precisely its peripheral locale that explains the curious mixture of fascination and deterrence which the program caused on the part of a public weary of an arms race extended into the heavens.

The project's naming process is illustrative in this regard. As soon as Reagan's speech had been aired live on television, critics derided the idea of installing space-based weapons systems in order to render nuclear warfare obsolete as fantastic and far-fetched.[75] 'Science fiction becomes fact,' the *Times* of London headlined. But it was one of Reagan's major political opponents, Senator Edward M. Kennedy (1932–2009), who coined the vision's name when speaking of 'reckless "Star Wars" schemes,' in reference to the original space opera trilogy by the same title.[76] One reason for the swift attribution was that the military potential of the newly inaugurated Space Shuttle had indeed been discussed before, both in and outside the United States, for instance, in conjunction with *Star Wars*, the movie, and was pictured accordingly in other popular films such as the aforementioned *Moonraker*.[77] A great, if unexpected, success at the box office, *Star Wars* was omnipresent in the late 1970s. Subject to a worldwide marketing campaign, the third (or by its later counting, sixth) installment, *Return of the Jedi*, came to movie theaters two months after Reagan's speech, on 25 May 1983, and captured the public imagination like no other astrocultural product of the early 1980s. While some praised Reagan for the fact that 'Star Wars,' the defense project, defictionalized *Star Wars*, the movie, as an exotic dream world, most other observers vehemently disagreed.[78] Two years later, George Lucas (1944–), the saga's producer and main stakeholder, was entirely fed up. Sufficiently indignant about the instant and lasting rhetorical equation between his and Reagan's 'Star Wars,' Lucas decided to go to court with a view to protecting the *Star Wars* trademark from further copyright infringement by television commercials. Unsurprisingly, success was limited.[79]

To this day, Reagan's 'Address to the Nation on Defense and National Security' is better known as his 'Star Wars' speech, whereas the renewed ballistic missile defense program was only named 'Strategic Defense Initiative' (SDI) a year later, in January 1984. The ironic 'Star Wars'/*Star Wars* equation exemplifies not only a central characteristic of all astroculture, that is, the constant oscillation between technology and fiction. It also illustrates that public references to *Star Trek*, *Star Wars*, *Gravity* or 'Hollywood' in general do more than simply acknowledge the existence of a fictional realm where such matters are being discussed in parallel. Such imaginaries have themselves political "real-world" effects that continue to the present.[80] A quick look at the scores of political cartoons generated by the SDI debate reveals that the vast majority played with this outer space/*Star Wars* theme to deride Reagan's scenario as outlandish, lofty and up in the air, inspired by cheap science fiction.[81]

Commentators, critics and cartoonists did not quite realize how right they were. The source of Reagan's technological inspiration has frequently been credited to Senator Malcolm Wallop (1933–2001), a conservative politician from Wyoming, Edward Teller (1908–2003), the so-called father of the hydrogen bomb, and retired Lieutenant General Daniel O. Graham (1925–95). Wallop had shared his enthusiasm for space-based laser weapons with Reagan as early as 1979, Teller for his part offered detailed suggestions during the run-up to Reagan's speech, and Graham had given advice in the 1976 and 1980 election campaigns.[82] The year prior to the speech, Graham published *High Frontier: A New National Strategy*, a high-gloss manifesto disguised as a 'study' full of imaginative illustrations depicting 'military high performance spaceplanes' forming part of a 'Global Ballistic Missile Defense Engagement.' An aircraft carrier-turned-spaceship adorned the pamphlet's cover, thus transposing one of the most potent symbols of American military power into earth orbit (Figure 1.6). None other than noted science-fiction author Robert A. Heinlein (1907–88) contributed the introduction to the paperback version of this bellicose brochure, published after Reagan's notorious speech and dedicated to the president. The High Frontier concept, Heinlein flat-out declared, is 'the best news [...] since VJ Day.'[83]

In recent years a second, more radical line of interpretation has gained ground. In an article originally published in *Le Monde Diplomatique*, American science-fiction author Norman Spinrad (1940–) laid out that SDI's original spin doctor had been neither Wallop, Teller nor Graham, but rather Jerry Pournelle (1933–2017), another science-fiction author and right-leaning space believer. Disillusioned by the stagnation of the US spaceflight program after Apollo, Pournelle realized that only the military would be able to finance space infrastructures, including large orbital space stations, considered necessary for all future expansion plans. Either militarize space, the rationale went, or see all enlargement dreams implode. To Pournelle and others, SDI was the only way to relaunch the otherwise faltering conquest of space. Once Reagan had taken office, Pournelle and his friend Larry Niven (1938–) lobbied the idea to the White House together with Graham, Heinlein and others. According to Spinrad, Pournelle even helped craft Reagan's 'Star Wars' speech, with its catchy formulations including the above-quoted promise to 'make America strong again.' Apparently flattered, Pournelle later repudiated authorship but did not deny his impact.[84] If accurate, such a genealogy would explain why 'Star Wars' looked like science fiction – because it was. Neither incidental nor peripheral to space, SDI's resonance had everything to do with astroculture and its imaginaries.

As often, reactions in Europe were delayed, yet once unleashed came with a vengeance. Reagan's speech caught politicians by surprise and constituted the prelude to what observers termed the *Raketenwinter* – missile winter – of 1983/84.[85] Following NATO's highly controversial Double-Track Decision from December 1979, this was the heyday of the peace movement. Protests against nuclear armament and the militarization of space soon blended together. Large demonstrations were held in Rome, Amsterdam,

Figure 1.6 An aircraft carrier-turned-spaceship adorned the cover of Daniel O. Graham's controversial 1982 *High Frontier* manifesto.
Source: Daniel O. Graham, *High Frontier: A New National Strategy*, Washington, DC: The Heritage Foundation, 1982, cover image. Courtesy of the artist, Sharon Higgins.

Brussels, Bonn, Florence, London, Athens, Zurich and Copenhagen. At the second UNISPACE conference in Vienna in August 1982, thousands of delegates representing almost a hundred nations discussed reconnaissance satellites, anti-satellite weapons and the ongoing militarization of space. UN Secretary-General Javier Pérez de Cuéllar (1920–2020) set the tone when condemning and 'vigorously opposing' the escalating militarization of outer space during his opening speech: 'The forces of reason and peace [must] join together to counter what could be a frightening escalation of the arms race.' What had been planned as a forum for scientific-technical exchange morphed into a tribunal on the extension of the arms race into space and the fear of attacks from the skies.[86]

Public intellectuals and scientists including the aforementioned E. P. Thompson in the United Kingdom as well as Hans-Peter Dürr (1929–2014) in West Germany and actors such as Yves Montand (1921–91) in France stepped in and spoke up. According to Thompson, it was only after Reagan's 1984 reelection and the new priority given to SDI in his inaugural speech that Europeans, hitherto preoccupied with 'more terrestrial issues of oncoming cruise and Pershing missiles,' finally realized that the military stakes were much higher. *Star Wars: Science-Fiction Fantasy or Serious Probability?*, a booklet Thompson edited, asked the British public this very question. On its cover a laser beam surgically dismantled the American flag, one white star after another (Figure 1.7). An earlier pamphlet was subtitled *Self-Destruct Incorporated* – short for SDI – to pillory the military-industrial complex's attack on the freedom of the skies.[87] Montand warned his compatriots on television that Europe could find itself 'totally disarmed, naked' if SDI's triple challenge – military, political, techno-economic – were not met properly while both superpowers kept arming and shielding themselves.[88] Chanting 'Den Himmel von Raketen frei/Christen gegen SDI' ('Keep rockets off the sky/Christians against SDI'), German protesters organized large assemblies not only in Mainz, Hanover and Hamburg, but also in Jena, in the DDR's southwest. An anti-SDI petition launched on the fortieth anniversary of the Hiroshima bombing was signed by more than fifteen thousand adversaries, amongst them ex-chancellor Willy Brandt, Olympic gold-medalist Michael Gross, future Nobel Prize-recipient Günter Grass and rock star Udo Lindenberg.[89]

Over the course of a little more than half a century, space imaginaries had come full circle, from the 1930s missile raid scenarios to the 1980s 'Star Wars' defense shields. Long considered a sanctuary and the place of all utopia, then largely pushed aside as irrelevant for the future of humankind, space changed again at the beginning of the 1980s. From a place of longing and yearning, outer space or rather its perceived Gestalt transformed into a threat of impending Armageddon. Accordingly, SDI's numerous critics, both in Europe and the United States, were from the outset on the defensive. As much as they feared global extinction, they also worried that the brutal arms race on earth might spill into the skies, 'previously free of weapons' as one of many manifestos had it.[90]

One of Thompson's compatriots dissected with particular clarity what was going on in the skies. Forty years after the publication of his prize-winning

Figure 1.7 During the early 1980s, European intellectuals and *hommes des lettres* vehemently protested President Ronald Reagan's Strategic Defense Initiative. The cover of E. P. Thompson's booklet *Star Wars: Science-Fiction Fantasy or Serious Probability?* featured a laser beam surgically dismantling the American flag, one white star after another.

Source: E. P. Thompson, ed., *Star Wars: Science-Fiction Fantasy or Serious Probability?*, Harmondsworth: Penguin, 1985, cover image. Courtesy of Penguin Books.

essay in the *Royal Air Force Quarterly*, Arthur Clarke, still one of the foremost promulgators of space thought, realized that most of the space utopias he had never tired of advocating throughout his career suddenly looked stale, if not naïvely optimistic.[91] Rather than offering infinite travel destinations, enticing colonization opportunities and possible alien encounters, the heavens of the late 1970s and early 1980s were cluttered with orbiting fortresses, surgical laser weapons and battling killer satellites (see Figures 1.3, 1.4 and 1.6 above).[92] Even though the vast majority of these violent dystopias never materialized and the ferocious debates around them abated over the second half of the 1980s, they did not conclude with the end of the Cold War either. Changing the name but not the intention, post-SDI programs such as Ballistic Missile Defense (BMD), National Missile Defense (NMD) and Terminal High Altitude Area Defense (THAAD) have remained part of the superpower's arsenal, as present-day 'Space Force' phantasms remind us.[93]

V Militarizing outer space

Throughout the Space Age, the battles of the future were foreseen to be waged – and decided – beyond the confines of planet Earth. Long hailed as the site of heavenly utopias and otherworldly salvation, outer space figured as much a sanctuary as a threat, serving as the ultimate battlefield and providing the locale of interplanetary, if not galactic, warfare on a hitherto unforeseen scale. As should have become clear by now, the relation between militant astroculture and actual warfare is intricate. Hitherto no wars have been waged in space, and the differences between space assets and military imaginaries are undeniable. Ironically, as with civilian spaceflight, it is the earthward gaze that is decisive. Instead of securing hegemony over the Milky Way, earth itself has been the true object of all military expansion into space. Rather than taking warfare to the stars, low-earth orbit's militarization since the late 1950s has enhanced terrestrial warfare and reshaped war on earth. *Militarizing Outer Space* does not exoticize all things space but seeks to understand the complex dynamics between fantastic schemes, technoscientific feasibilities and military rationales. Examining and historicizing multifarious endeavors, both real and imagined, of arming the skies, conquering the heavens and waging space wars, the volume zooms in on this interplay between security and fantasy, technopolitics and knowledge. Concentrating on weapons, warfare and violence, it explores the violent and often dystopian dimensions of the Space Age and exposes the "dark" side of global astroculture.

As the third and final book in the *European Astroculture* trilogy, *Militarizing Outer Space* shares some of the overall objectives with the two preceding volumes, *Imagining Outer Space* and *Limiting Outer Space*. These include a cultural-interpretative approach, a commitment to combining a multiplicity of disciplinary perspectives into a coherent whole, and the intention to decenter the historiography of outer space, spaceflight and space exploration by pushing its geographical focus beyond the borders of the two Cold War superpowers. It is for this reason that the majority of chapters are situated in the United Kingdom, France, and West and East Germany, thus highlighting a particularly strong West European component; all other authors were strongly encouraged to work from a transnational perspective.

Arranged both thematically and chronologically, the book's fourteen chapters – including this introduction and the epilogue – are grouped in four sections: 'Embattling the Heavens,' 'Waging Future Wars,' 'Armoring Minds and Bodies' and 'Mounting Combat Infrastructures.' Contributions to the first part, 'Embattling the Heavens,' analyze the fundamental technopolitical developments underlying the violent and martial dimensions of twentieth-century space thought and astroculture. Dwelling on the historical origins of both the Atomic Age and the Space Age, they address the transfer and circulation of high-tech knowledge within and beyond Western societies in a world that felt threatened by all-out confrontation and apocalyptic doom. Chapters include a succinct discussion of the differences between weaponization and militarization, and the different logics of space fiction versus space fact (Michael J. Neufeld); an account of the development of the first ICBMs in the 1950s as a technopolitical prerequisite to the American and Soviet civilian spaceflight programs that soon followed (Christopher Gainor); and a critique of the oft-repeated stereotype that the European space effort emerged in the early Cold War as a distinctly civilian, non-military supranational enterprise. While the West European space community vehemently opposed any possible space weaponization (and does so to this very day), influential EEC member states such as the United Kingdom and France actively engaged in the development of space technologies which could also be used for military purposes (Michael Sheehan).

Contributions to the second section, 'Waging Future Wars,' explore different aspects of the 'battlefield of the future' topos in film, literature and Cold War politics. This part includes an analysis of military and belligerent aspects of twentieth-century space science-fiction movies, from Georges Méliès's *Voyage dans la lune* (1902) to Aleksey German's *Es ist nicht leicht, ein Gott zu sein* (*Trudno byt' bogom/Hard to be a God*, 1989), with special emphasis on tropes of alien invasion, apocalyptic disaster and social disintegration (Natalija Majsova); a close reading of C. S. Lewis's 'cosmic trilogy' (1938–45) that reveals an anti-rationalist critique of the 'conquest of space' underlying its rationale of reconciling faith and erudition (Oliver Dunnett); and a chapter on the ways in which clashes between East and West German engineers over astroculture, rocketry and the politics of 'dual use' contributed to the political division of Cold War Europe (Daniel Brandau).

The militarization of outer space corresponded with the militarization of inner space. Accordingly, the book's third section, 'Armoring Minds and Bodies,' turns to those inner spaces, from war imaginaries to their impact on bodies, subjectivities and the self, be it in medicine, poetry, literature or civilian research programs. Chapters analyze projects to redesign human nature for adaptation to the hostile space environment, culminating in the invention of the Cyborg in 1960 (Patrick Kilian); the form and function of space suits in twentieth-century astroculture, in particular Robert A. Heinlein's 1959 *Starship Troopers* novel, adapted for the screen by Dutch director Paul Verhoeven in 1997, which propagated physical protection and subjective transformation alike (Philipp Theisohn); and the surprising parallels between government-sponsored space observation programs and privately funded psychical research projects, sometimes labelled 'pseudo-scientific,' both developed and maintained as military surveillance technologies thought to penetrate the 'inner space' of the Cold War mindset (Anthony Enns).

'Mounting Combat Infrastructures,' the volume's fourth and last section, investigates the development and use of space hardware and discusses accompanying military rationales. Focusing on the 1970s and beyond, this part includes a chapter on the architecture of American, Soviet and European mission control rooms, essential elements of space operations initially built around the kernel of war rooms and the command and control paradigm of ballistic missile defense (Regina Peldszus); a contribution on the series of Salyut and Almaz space stations launched by the Soviet Union in the 1970s, military enterprises camouflaged in a civilian program, as an effort to reestablish and reaffirm Russian presence in earth orbit after having lost the race to the moon (Cathleen Lewis); and a comparative analysis of the military origins of the American satellite-based Global Positioning System (GPS) and its younger European and Chinese counterparts, the global navigation satellite systems Galileo (GNSS) and BeiDou (BDS) (Paul E. Ceruzzi). Finally, a comprehensive epilogue relates this volume's key findings to the larger analytical, conceptual and historiographical issues at stake when historicizing astroculture from a European perspective. Taking this particular moment in time, now that the *European Astroculture* trilogy has been completed with the publication of the present volume, as an occasion to pause, step back and take inventory, it identifies and sketches a set of research perspectives with a view to globalizing the study of global astroculture in the light of renewed political tensions and belligerent rhetorics (Alexander C. T. Geppert).

As discussed at the opening of this chapter, alert observers confronted with the quest to establish a special US Space Force instantly linked the moment to Ronald Reagan's 'Star Wars' program 35 years prior. 'Historically, that has always been the problem with missile defense,' the *New York Times* commented: 'The technology has always seemed more promising on paper than in reality. And the Pentagon has never kept up with Hollywood's imagining.'[94] While both the Pentagon and Hollywood have certainly shaped and perpetuated enduring space war imaginaries, the fourteen chapters in this book substantiate that the allure of ruling outer space – and, in effect, earth – has neither been limited to the early twenty-first century nor to the United States. Continuously oscillating between heavenly utopias and ultimate battlefields, between space realities and imaginaries, and fueled by fantasies of control, conquest and colonization, astroculture has proven instrumental in fathoming forms, functions and futures of warfare. From such a perspective, this book joins its two preceding volumes in taking outer space and its cultural history seriously. Located at the intersection of the imagination, technopolitics and knowledge, astroculture is not only a much richer and still largely untapped field of historiographical enquiry than hitherto recognized but key to the making of our planetized present.

Notes

1. 'Space Imperialism,' *Daily Mail* (5 January 1959), 1. For comments, criticism and suggestions we are grateful to Daniel Brandau, Ralf Bülow, Paul Ceruzzi, James David, Dwayne Day, Rainer Eisfeld, Anna Kathryn Kendrick, Michael Neufeld, Karlheinz Rohrwild and Bernd Weisbrod. Once again, Ruth Haake's congenial assistance has been as critical as it is deeply appreciated. All Internet sources were last accessed on 15 July 2020.
2. Christina Wilkie, 'Trump Floats the Idea of Creating a "Space Force" to Fight Wars in Space,' *CNBC* (13 March 2018), https://cnb.cx/36CMQQc; for the official announcement three months later, see Katie Rogers, 'Trump Wants "Space Force" to Be Added to Military,' *New York Times* (19 June 2018), A6. See also David Montgomery, 'Trump's Excellent Space Force Adventure,' *The Washington Post Magazine* (3 December 2019), https://wapo.st/2M1FzS9.
3. Ronald Reagan, 'Address to the Nation on Defense and National Security,' 23 March 1983, Ronald Reagan Presidential Library, available at https://www.reaganlibrary.gov/research/speeches/32383d. For a video, see https://youtu.be/ApTnYwh5KvE; see also *Daedalus* 114.3 (Summer 1985), 369–71.
4. G. Edward Pendray, 'Next Stop the Moon,' *Collier's* 21.36 (7 September 1946), 11–13, 75, here 12.
5. Hermann Oberth, 'Die 3 Gesichter der Rakete,' *Die Gartenlaube* 43 (23 October 1930), 887–91, here 891. There are unconfirmed hints that the article was ghost-written either by prolific spaceflight advocate Willy Ley (1906–69) or rocket engineer Rudolf Nebel (1894–1978).
6. For a comprehensive compilation of privately held source material, see Karlheinz Rohrwild, *Die Raumfahrt und die Militärs*, unpublished manuscript, Feucht: Hermann-Oberth-Raumfahrt-Museum, 2015, 3.
7. See Wilhelm Hoeppner-Flatow, 'Kriegswaffen der Zukunft,' *Kreuz-Zeitung* 85.283 (9 October 1932), 7; Peter J. Bowler, *A History of the Future: Prophets of Progress from H. G. Wells to Isaac Asimov*, Cambridge: Cambridge University Press, 2018, 151–65, here 151.
8. Hermann Oberth, *Die Rakete zu den Planetenräumen*, Munich: Oldenbourg, 1923 [*The Rocket into Planetary Space*, Munich: Oldenbourg, 2014]; idem, 'Das Raketengeschoß,' in idem, *Wege zur Raumschiffahrt*, Munich: Oldenbourg, 1929 [*Ways to Spaceflight*, Washington, DC: NASA, 1972], 202–5, here 202: '[...] ich glaube, es würde sicher nicht so leicht zu Kriegen kommen, wenn die Betreffenden wüssten: "Der erste, den es trifft, bist Du selbst".'
9. John MacCormac, 'War with Rockets Pictured by Oberth,' *New York Times* (31 January 1931), 8. Similarly, former Saxonian Major General Artur Baumgarten-Crusius (1858–1932) conceptualized the rocket of the future as a superweapon

which would bring an end to all warfare; see idem, *Die Rakete als Weltfriedenstaube*, Leipzig: Verband der Raketen-Forscher und -Förderer/Roßberg'sche Buchdruckerei, 1931.

10. 'Raketenflug! Aufruf' [1930], Deutsches Technikmuseum Berlin, Archiv, Kleine Erwerbungen 1671, 1: 'derartige Vorteile, daß mit einem Schlage [Deutschlands] frühere Weltgeltung wiederhergestellt wird.' On the military appeal of rocketry, see also Rudolf Nebel, 'Raketen schützen die Grenzen,' *B.Z. am Mittag* 55.82 (9 April 1931), 3. On the Weimar Republic space fad, see Michael J. Neufeld, 'Weimar Culture and Futuristic Technology: The Rocketry and Spaceflight Fad in Germany, 1923–1933,' *Technology and Culture* 31.4 (October 1990), 725–52; Alexander C. T. Geppert, 'Space *Personae*: Cosmopolitan Networks of Peripheral Knowledge, 1927–1957,' *Journal of Modern European History* 6.2 (2008), 262–86; and idem and Tilmann Siebeneichner, '*Lieux de l'Avenir*: Zur Lokalgeschichte des Weltraumdenkens,' *Technikgeschichte* 84.4 (2017), 285–304.
11. Roderick Morison, 'By Rocket to the Planets,' *Daily Mail* (4 November 1936). See, for instance, 'Rockets Filled with Powder and Oil to Ignite Airplanes,' *Science and Invention* 9.2 (June 1921), 117; 'Death-Dealing Rocket,' ibid. (November 1924), 656; Gawain Edwards [= G. Edward Pendray], 'Will the Rocket Replace Artillery?,' ibid. (November 1930), 600, 632; 'Death Rockets Rain Fiery Metal,' *Popular Science Monthly* (November 1924), 40; and David Lasser, 'The Rocket in the Next War?,' *Everyday Science and Mechanics* (March 1932), 326–7.
12. The most influential manifesto is Giulio Douhet's *Il dominio dell'aria: saggio sull'arte della guerra aerea* [1921], 2nd edn, Rome: Istituto nazionale fascista di cultura, 1927. See also David Edgerton, *England and the Aeroplane: An Essay on a Militant and Technological Nation*, Basingstoke: Macmillan, 1991.
13. Helmuth Trischler and Hans Weinberger, 'Engineering Europe: Big Technologies and Military Systems in the Making of Europe,' *History and Technology* 21.1 (Spring 2005), 49–83, here 69.
14. Roger Handberg, *Seeking New World Vistas: The Militarization of Space*, Westport: Praeger, 2000, 53; idem, 'Military Space Policy: Debating the Future,' *Astropolitics* 2.1 (Spring 2004), 79–89, here 82. For an extensive discussion of what she terms the 'plasticity' of dual use, see the contribution by Regina Peldszus, Chapter 11 in this volume.
15. See Yanek Mieczkowski, *Eisenhower's Sputnik Moment: The Race for Space and World Prestige*, Ithaca: Cornell University Press, 2013, 166–71; Kenneth Osgood, *Total Cold War: Eisenhower's Secret Propaganda Battle at Home and Abroad*, Lawrence: Kansas University Press, 2006; and, in particular, the contribution by Christopher Gainor, Chapter 3 in this volume.
16. See, for instance, Kevin Madders, *A New Force at a New Frontier: Europe's Development in the Space Field in the Light of Its Main Actors, Policies, Laws, and Activities from Its Beginnings up to the Present*, Cambridge: Cambridge University Press, 1996, 44; and Joan Johnson-Freese, *Space as a Strategic Asset*, New York: Columbia University Press, 2007, 27–50. For a critical and more nuanced perspective on the European space effort, see Michael Sheehan's contribution, Chapter 4 in this volume.
17. Dwight D. Eisenhower, 'Farewell Address,' 17 January 1961, available at https://www.eisenhowerlibrary.gov/sites/default/files/all-about-ike/speeches/wav-files/farewell-address.mp3. For a detailed analysis, see Alex Roland, *The*

Military-Industrial Complex, Washington, DC: American Historical Association, 2001; and idem, 'The Military-Industrial Complex: Lobby and Trope,' in Andrew J. Bacevich, ed., *The Long War: A New History of U.S. National Security Policy since World War II*, New York: Columbia University Press, 2007, 335–70.

18. Karl P. Mueller, 'Totem and Taboo: Depolarizing the Space Weaponization Debate,' *Astropolitics* 1.1 (Spring 2003), 4–28. For a comprehensive conceptual history of 'militarism,' see Werner Conze, Reinhard Stumpf and Michael Geyer, 'Militarismus,' in Otto Brunner, Werner Conze and Reinhart Koselleck, eds, *Geschichtliche Grundbegriffe: Historisches Lexikon zur politisch-sozialen Sprache in Deutschland*, vol. 4, Stuttgart: Klett-Cotta, 1978, 1–47.
19. 'Weapon,' *Oxford English Dictionary*, 2nd edn, vol. 20, Oxford: Clarendon Press, 44–5. The definition suggested by Karl D. Hebert – 'any asset, Earth-based or space-based, designed to attack targets in space (Earth-to-space and space-to-space)' – does not go beyond stating that space weapons are weapons used in space; see idem, 'Regulation of Space Weapons: Ensuring Stability and Continued Use of Outer Space,' *Astropolitics* 12.1 (March 2014), 1–26, here 3. The crucial difference between 'militarization' and 'weaponization' is also the focus of Michael Neufeld's contribution, Chapter 2 in this volume.
20. See, for instance, US Army Weapons Command, *Meanderings of a Weapon Oriented Mind When Applied in a Vacuum such as on the Moon*, Rock Island: US Army Weapons Command, 1965. For the Almaz R-23M Kartech cannon, see Cathleen Lewis's contribution, Chapter 12 in this volume.
21. For instance, Paul B. Stares, *The Militarization of Space: U.S. Policy, 1945–1984*, Ithaca: Cornell University Press, 1985, 17; and Matthew Mowthorpe, *The Militarization and Weaponization of Space*, Lanham: Lexington Books, 2004, 3. Military historian Sean Kalic implicitly implores his colleagues to follow their historical actants' parlance. According to him, US Presidents Eisenhower, Kennedy and Johnson referred to the 'militarization of space' when relating to the use of space for non-aggressive military purposes, whereas 'weaponization' stood for space-based systems that actually threatened space as an open frontier; see idem, *US Presidents and the Militarization of Space, 1946–1967*, College Station: Texas A&M University Press, 2012, 5–6. We would contend that such a counterintuitive distinction between aggressive versus non-aggressive creates more conceptual confusion than it clarifies.
22. The United States and the Soviet Union began negotiating an ASAT treaty in the late 1970s, but it was never concluded. Today, other space powers would have to be included as well.
23. Handberg, *Seeking New World Vistas*, 3. On the history of GPS, see Greg Milner, *Pinpoint: How GPS Is Changing Technology, Culture, and Our Minds*, New York: Norton, 2016; Paul E. Ceruzzi, *GPS*, Cambridge, MA: MIT Press, 2018; and, in particular, Ceruzzi's contribution to this volume, Chapter 13.
24. Michael Geyer, 'The Militarization of Europe, 1914–1945,' in John R. Gillis, ed., *Militarization of the Western World*, New Brunswick: Rutgers University Press, 1989, 65–102, here 79; Michael S. Sherry, *In the Shadow of War: The United States since the 1930s*, New Haven: Yale University Press, 1995, xi.
25. For a definition and discussion of 'astroculture' as underlying all three volumes in the *European Astroculture* trilogy, see Alexander C. T. Geppert, 'European Astrofuturism, Cosmic Provincialism: Historicizing the Space Age,' in idem, ed., *Imagining Outer Space: European Astroculture in the Twentieth Century*, 2nd edn,

London: Palgrave Macmillan, 2018, 3–28, here 9 (= *European Astroculture*, vol. 1). See also his epilogue, Chapter 14 in the present volume.

26. The Nazi term *Vergeltungswaffe* is conventionally mistranslated as 'vengeance weapon'; 'reprisal weapon' would be more accurate.
27. Oberth, 'Die 3 Gesichter der Rakete,' 891.
28. For a detailed discussion of East and West Germany's differing stances toward rocketry, see Daniel Brandau's contribution, Chapter 7 in this volume; see also his recent monograph *Raketenträume: Raumfahrt- und Technikenthusiasmus in Deutschland 1923–1963*, Paderborn: Schöningh, 2019.
29. Arthur C. Clarke, 'The Rocket and the Future of Warfare,' *Royal Air Force Quarterly* 17.2 (March 1946), 61–9, here 61, 65, 66; John Reddy, 'Arthur Clarke: Prophet of the Space Age,' *Reader's Digest* (November 1969), 74–8. To his biographer, Neil McAleer, this episode is worth precisely two sentences, in addition to the two factual errors it contains. As shown above, the concept of mutually assured destruction existed long before 1946, and Clarke was not an advocate but a critic; see Neil McAleer, *Visionary: The Odyssey of Sir Arthur C. Clarke*, Baltimore: The Clarke Project, 2010, 45. See in this context also Robert Poole, 'The Myth of Progress: *2001: A Space Odyssey*,' in Alexander C. T. Geppert, ed., *Limiting Outer Space: Astroculture after Apollo*, London: Palgrave Macmillan, 2018, 283–302, here 291–4 (= *European Astroculture*, vol. 2); and Oliver Dunnett's contribution, Chapter 6 in this volume.
30. Not more than two thirds actually reached their target. Arthur C. Clarke, 'Peacetime Uses for V2,' *Wireless World* 51.2 (February 1945), 58. It was this brief letter to the editor that Clarke would later credit as the first time he had ever mentioned his idea of a synchronous communication satellite, not the usually cited article published in the same journal in October of that year ('Extra-Terrestrial Relays: Can Rocket Stations Give World-Wide Radio Coverage?,' ibid. 51.10 [October 1945], 305–8); see Arthur C. Clarke to Eugene Emme, 2 August 1970, Smithsonian National Air and Space Museum Archives, Arthur C. Clarke Collection (hereafter NASMA/ACCC), 007/01.
31. Sun-Tzu (Sunzi), *The Art of War* [1772], New York: Penguin, 2003, 61: 'He who occupies/ High Yang ground/ And ensures/ His line of supplies/ Will fight/ To advantage.'
32. Lee D. Saegesser, *Space: The High Ground*, unpublished manuscript, Washington, DC: NASA History Office, ca. 1983, NASA Historical Reference Collection, NASA History Division, NASA Headquarters Washington, DC, 17086.
33. Douhet, *Dominio dell'aria*, 37; Oberth, *Wege zur Raumschiffahrt*, 352, 355. Note that the Italian original of Douhet's oft-quoted doctrine ('Chi possegga il dominio dell'aria e disponga di una adeguata forza offensiva, mentre da un lato preserva tutto il proprio territorio ed il proprio mare dalle offese aeree nemiche e toglie all'avversario la possibilità di qualsiasi azione ausiliaria aerea [concorso degli aerei alle operazioni di terra e di mare] dall'altro si trova in grado di esercitare sul nemico azioni offensive di un ordine di grandezza terrificante, contro le quali all'avversario non resta alcun modo di reagire') is far less catchy, if by no means less drastic than the standard English translation ('A nation which has command of the air is in a position to protect its own territory from enemy aerial attack and even to put a halt to the enemy's auxiliary actions in support of his land and sea operations, leaving him powerless to do much of anything'; see, for instance, Saegesser, *Space*, n.p.) and does not bear any reference to the nation either.
34. *Destination Moon*, directed by Irving Pichel, USA 1950 (George Pal Productions); Joseph Kaplan, Oscar Schachter, Fred Freeman and Cornelius Ryan, 'What Are

We Waiting For?,' *Collier's* (22 March 1952), 23; John F. Kennedy, 'Countdown for Survival: "If the Soviets Control Space... They Can Control Earth",' *Missiles and Rockets* (10 October 1960), 12–13; and Saegesser, *Space*, n.p.

35. 'Space Imperialism.'
36. See Patrick Kilian's contribution, Chapter 8 in this volume.
37. Trevor Philpott, 'Guided Robots Go To War,' *Picture Post* 57.8 (22 November 1952), 15–19, here 15, 19.
38. See Robert S. Richardson, 'Rocket Blitz from the Moon,' *Collier's* 122 (23 October 1948), 24–5, 44–5, here 25: 'Certainly, when space travel comes into being, the first nation to gain control of the moon will be able to control the earth'; and idem, 'Militärbasis auf dem Mond?,' *Die Weltwoche* 17.812 (3 June 1949), 9.
39. On von Braun's iconic space station, see Michael J. Neufeld, '"Space Superiority": Wernher von Braun's Campaign for a Nuclear-Armed Space Station, 1946–1956,' *Space Policy* 22.1 (February 2006), 57–62. For Project Horizon, consult Frederick I. Ordway, Mitchell R. Sharpe and Ronald C. Wakeford, 'Project Horizon: An Early Study of a Lunar Outpost,' *Acta Astronautica* 17.10 (October 1988), 1105–21; and Anthony M. Springer, 'Securing the High Ground: The Army's Quest for the Moon,' *Quest: The History of Spaceflight Quarterly* 7.2 (Summer/Fall 1999), 342–8.
40. The CIA's own history is Kevin C. Ruffner, ed., *CORONA: America's First Satellite Program*, Washington, DC: CIA, 1995, xiii. For more scholarly treatments, see Curtis Peebles, *The Corona Project: America's First Spy Satellites*, Annapolis: Naval Institute Press, 1997; Dwayne A. Day, John M. Logsdon and Brian Latell, eds, *Eye in the Sky: The Story of the Corona Spy Satellites*, Washington, DC: Smithsonian Institution Press, 1998; and, in particular, John Cloud, 'Imaging the World in a Barrel: CORONA and the Clandestine Convergence of the Earth Sciences,' *Social Studies of Science* 31.2 (April 2001), 231–51. See also Michael Neufeld's and Anthony Enns's respective contribution, Chapters 2 and 10 in this volume.
41. 1957–77: USSR 902; USA 563; USA for United Kingdom 5; USA for France 3; USA for NATO 4; France 5; China 2. See Jasani Bhupendra, *Outer Space: Battlefield of the Future?*, London: Taylor & Francis, 1978, inserted erratum page; Desmond King-Hele et al., eds, *The RAE Table of Earth Satellites 1957–1986: Compiled at the Royal Aircraft Establishment, Farnborough, Hants, England*, 3rd edn, New York: Stockton Press, 1987, vii, 9; and Hubert Feigl, 'Militärisch nutzbare Satelliten: Ihre Bedeutung für Sicherheit und Rüstungskontrolle,' in Karl Kaiser and Stephan Frhr. von Wenck, eds, *Weltraum und internationale Politik*, Munich: Oldenbourg 1987, 189–207, here 189.
42. John Hall, 'Space Wars,' *Radio Times* (21–27 October 1978), 92–101, here 92; Rüdiger Proske, 'Schlachtfeld Weltraum: Die neuen Fronten werden am Himmel aufgebaut,' *Geo Special 'Weltraum'* 8.3 (1983), 42–51.
43. See 'Treaty on Principles Governing the Activities of States in the Exploration and Use of Outer Space, including the Moon and Other Celestial Bodies,' 19 December 1966, United Nations Office for Outer Space Affairs, available at http://www.unoosa.org/oosa/en/ourwork/spacelaw/treaties/introouterspacetreaty.html. See in this context Luca Follis, 'The Province and Heritage of Humankind: Space Law's Imaginary of Outer Space, 1967–79,' in Geppert, *Limiting Outer Space*, 183–205.

44. Gerald Steinberg, cited in Detlev Wolter, *Common Security in Outer Space and International Law*, Geneva: UNIDIR, 2006, 26; see also Everett C. Dolman, *Astropolitik: Classical Geopolitics in the Space Age*, London: Routledge, 2002, 8.
45. See Everett C. Dolman, 'A Debate About Weapons in Space: For US Military Transformation and Weapons in Space,' *SAIS Review of International Affairs* 26.1 (Spring 2006), 163–75.
46. Ignatius F. Clarke, *Voices Prophesying War: Future Wars, 1763–3749*, 2nd edn, Oxford: Oxford University Press, 1992, 174.
47. See, for instance, Bernd Greiner, 'Zwischenbilanzen zum Kalten Krieg,' *Mittelweg 36* 16.3 (June/July 2007), 51–8, and his multi-volume *Studien zum Kalten Krieg*, 7 vols, Hamburg: Hamburger Edition, 2006–13; John Lewis Gaddis, *The Cold War: A New History*, New York: Penguin, 2005; Annette Vowinckel, Marcus M. Payk and Thomas Lindenberger, eds, *Cold War Cultures: Perspectives on Eastern and Western European Societies*, Oxford: Berghahn, 2012; Christof Becker-Schaum, Philipp Gassert, Martin Klimke and Wilfried Mausbach, eds, *The Nuclear Crisis: The Arms Race, Cold War Anxiety, and the German Peace Movement of the 1980s*, Oxford: Berghahn, 2016; Eckart Conze, Martin Klimke and Jeremy Varon, eds, *Nuclear Threats, Nuclear Fear and the Cold War of the 1980s*, Cambridge: Cambridge University Press, 2016; and Matthew Grant and Benjamin Ziemann, *Understanding the Imaginary War: Culture, Thought and Nuclear Conflict, 1945–90*, Manchester: Manchester University Press, 2016. The exception to the rule is Naomi Oreskes and John Krige, eds, *Science and Technology in the Global Cold War*, Cambridge, MA: MIT Press, 2014.
48. See only Tom Wolfe, *The Right Stuff*, New York: Farrar, Straus & Giroux, 1979; Karsten Werth, *Ersatzkrieg im Weltraum: Das US-Raumfahrtprogramm in der Öffentlichkeit der 1960er Jahre*, Frankfurt am Main: Campus, 2006; and Jeremy Black, *The Cold War: A Military History*, London: Bloomsbury, 2015, 98–103. To give but three further examples of such a blind spot: The *Cambridge History of Warfare* does not mention space war or the possibility of a future war decided in space, not even in an epilogue entitled 'The Future of Western Warfare'; see Geoffrey Parker, in idem, ed., *The Cambridge History of Warfare*, Cambridge: Cambridge University Press, 2005, 413–32. "Dean of British strategic studies" Lawrence Freedman, in his magisterial *The Future of War*, devotes a single paragraph each to outer space in a Cold War context and to SDI; see idem, *The Future of War: A History*, New York: PublicAffairs, 2017, 89–90, 97. And a recent militarization anthology has little to offer on space but contemporary speculation and indexes SDI under 'Star Wars'; see David H. Price, 'Militarizing Space,' in Roberto J. González, Hugh Gusterson and Gustaaf Houtman, eds, *Militarization: A Reader*, Durham, NC: Duke University Press, 2019, 316–19.
49. Michael N. Golovine, *Conflict in Space: A Pattern of War in a New Dimension*, London: Temple Press, 1962, xiii; 'A Space War Prophet in Disguise,' *Times* (14 April 1962), 5.
50. David Seed, 'Introduction,' in idem, ed., *Future Wars: The Anticipation and the Fears*, Liverpool: Liverpool University Press, 2012, 1–8, here 2. See also the contributions to related edited collections such as George E. Slusser and Eric S. Rabkin, eds, *Fights of Fancy: Armed Conflict in Science Fiction and Fantasy*, Athens: University of Georgia Press, 1993, and Grant and Ziemann, *Understanding the Imaginary War*.

51. For a demonstration of *Spacewar!*, see https://youtu.be/eePWlLKm_Bg. The best analysis is Christian Ulrik Andersen, '*Monopoly* and the Logic of Sensation in *Spacewar!*,' in Olga Goriunova, ed., *Fun and Software: Exploring Pleasure, Paradox, and Pain in Computing*, London: Bloomsbury, 2014, 197–212, here 202–8. See also Natalija Majsova, 'Outer Space and Cyberspace: An Outline of Where and How to Think of Outer Space in Video Games,' *Teorija in praksa* 51.1 (February 2014), 106–22. We have taken 'fights of fancy' from Slusser and Rabkin, *Fights of Fancy*.
52. For an oral history of the program's distribution at various American and English universities throughout the 1960s and the Cronkite quote, see Devin Monnens and Martin Goldberg, 'Space Odyssey: The Long Journey of *Spacewar!* from MIT to Computer Labs Around the World,' *Kinephanos* 6 (June 2015), 124–47, here 130–1, 140–1. It is unclear when and where the game was played for the first time on the European continent and whether it preceded popular space-themed video games such as *Space Invaders* (1978) and *Asteroids* (1979).
53. Stewart Brand, 'Spacewar: Fanatic Life and Symbolic Death Among the Computer Bums,' *Rolling Stone* 123 (7 December 1972), 50–6.
54. For an autobiographical account by one of the game's creators, see J. M. Graetz, 'The Origin of *Spacewar!*,' *Creative Computing Video and Arcade Game* (1983), 78–84, here 78 and 82–3.
55. The use of the term 'militant' is inspired by geographer Felix Driver who, based on a 1924 Joseph Conrad essay, has charted a 'fabulous,' a 'militant' and a 'triumphant' epoch in the history of exploration and geographical knowledge; see his *Geography Militant: Cultures of Exploration and Empire*, Oxford: Blackwell, 2001, 3–4. As a qualifier 'militant' seems better suited than 'martial,' 'military,' 'militaristic,' 'combative' or even 'belligerent' as it highlights the dominant themes of that particular form of astroculture without necessarily advocating the militarization or weaponization of space; neither is militant astroculture generated from, within or by the military. But see Steffen Hantke, 'Military Culture,' in Rob Latham, ed., *The Oxford Handbook of Science Fiction*, Oxford: Oxford University Press, 2014, 329–39.
56. H. G. Wells, 'The War of the Worlds,' *Pearson's Magazine* 3/4.16–24 (April–December 1897), 363–73, 487–96, 587–96, 108–19, 221–32, 329–39, 447–56, 558–68, 736–45; idem, *The War of the Worlds*, London: William Heinemann, 1898. On Orson Welles's notorious radio show, often exaggerated in its effects, see only Hadley Cantril, *The Invasion from Mars: A Study in the Psychology of Panic*, Princeton: Princeton University Press, 1940; Peter J. Beck, *The War of the Worlds from H. G. Wells to Orson Welles, Jeff Wayne, Steven Spielberg and beyond*, London: Bloomsbury, 2016; and, in particular, Natalija Majsova's contribution to the present volume, Chapter 5.
57. In lieu of a much larger historical (but not historiographical) body of literature, see Alexander C. T. Geppert, 'Extraterrestrial Encounters: UFOs, Science and the Quest for Transcendence, 1947–1972,' *History and Technology* 28.3 (September 2012), 335–62.
58. 'Das letzte Abenteuer unserer Science-Fiction-Serie konfrontiert uns noch einmal mit dem Weltraumfeind'; see Bavaria Presse-Information 'Raumpatrouille: Die phantastischen Abenteuer des Raumschiffs Orion: Invasion,' Westdeutscher Rundfunk, Historisches Archiv, Cologne, unverzeichneter Bestand Bavaria (1966), April 1965.

59. 'Science Fiction: Flucht ins Weltall,' *Der Spiegel* 32.6 (6 February 1978), 158–71, here 161. On Perry Rhodan, see, for instance, Klaus Bollhöfener, Klaus Farin and Dierk Spreen, eds, *Spurensuche im All: Perry Rhodan Studies*, Berlin: Thomas Tilsner, 2003; and Niels Werber, 'Selbstbeschreibungen des Politischen – in Serie: Perry Rhodan 1961–2018,' *Kulturwissenschaftliche Zeitschrift* 3.1 (2018), 75–98.
60. See in this context Guillaume de Syon, 'Balloons on the Moon: Visions of Space Travel in Francophone Comic Strips,' in Geppert, *Imagining Outer Space*, 187–207.
61. *Dr. No*, directed by Terence Young, UK 1962 (United Artists); *You Only Live Twice*, directed by Lewis Gilbert, UK 1967 (United Artists); *Diamonds Are Forever*, directed by Guy Hamilton, UK 1971 (United Artists); and *Moonraker*, directed by Lewis Gilbert, USA/UK/FR 1979 (United Artists). On the *James Bond* franchise, see Andrè J. Millard, *Equipping James Bond: Guns, Gadgets, and Technological Enthusiasm*, Baltimore: Johns Hopkins University Press, 2018; see in this context also Tilmann Siebeneichner, 'Spacelab: Peace, Progress and European Politics in Outer Space, 1973–85,' in Geppert, *Limiting Outer Space*, 259–82.
62. 'Science Fiction: Flucht ins Weltall.'
63. *Radio Times* (21–27 October 1978), 55; Hall, 'Space Wars'; John Hackett, *The Third World War: August 1985*, New York: Macmillan, 1978, 202–6.
64. United Nations, General Assembly, 'Resolution on International Cooperation in the Peaceful Use of Outer Space,' 18 November 1975, Historical Archives of the European Union/European Space Agency, 4609.
65. For greater detail on 1970s disillusionment, consult the contributions to Geppert, *Limiting Outer Space*. On 'space fatigue' in particular, see idem, 'The Post-Apollo Paradox: Envisioning Limits During the Planetized 1970s,' in ibid., 3–28.
66. Reagan, 'Address to the Nation.' On Reagan's SDI/'Star Wars' speech, see Jakob Schissler, 'SDI und die politische Kultur der USA,' in Bernd W. Kubbig, ed., *Die militärische Eroberung des Weltraums*, Frankfurt am Main: Suhrkamp, 1990, vol. 1, 168–97; Karsten Zimmermann, 'Reagans "Star Wars"-Rede,' in ibid., 56–93; and Edward Reiss, *The Strategic Defense Initiative*, Cambridge: Cambridge University Press, 1992, 37–47.
67. The so-called Double-Track Decision, adopted during a special meeting of NATO foreign and defense ministers on 12 December 1979, combined the threat of weapon modernization with the offer of mutual limitation: NATO would deploy additional nuclear arms unless the Soviet-led Warsaw Pact stopped the stationing of SS-20 missiles in Western Europe.
68. Reagan, 'Address to the Nation.' For the concept of *lieu de l'avenir*, see Geppert and Siebeneichner, '*Lieux de l'Avenir*.'
69. There is a stark contrast between the vast amount of literature directly generated by SDI in the mid- to late 1980s and its superficial historicization, especially from an international and non-US-centered perspective. Arguably the best study is still Edward Reiss's *Strategic Defense Initiative*, while Frances FitzGerald, *Way Out There in the Blue: Reagan, Star Wars and the End of the Cold War*, New York: Simon & Schuster, 2000 is often referenced as a standard work, despite its journalistic character. For more detailed information, see the following ten (!) bibliographies in chronological order: Caroline D. Harnly and David A. Tyckoson, eds, *Space Weapons*, Phoenix: Oryx Press, 1985; Gerhard Weber,

Militarisierung des Weltraums: Krieg der Sterne, SDI. Auswahlbibliographie, 2 vols, Jena: Universitätsbibliothek, 1986/1988; Robert M. Lawrence, *Strategic Defense Initiative: Bibliography and Research Guide*, Boulder: Westview Press, 1987; Alva W. Stewart, *The Strategic Defense Initiative: A Brief Bibliography*, Monticello: Vance Bibliographies, 1987; Hans Günter Brauch and Rainer Fischbach, eds, *Military Use of Outer Space: A Research Bibliography*, Stuttgart: AG Friedensforschung und Europäische Sicherheitspolitik (AFES), Institut für Politikwissenschaft, Universität Stuttgart, 1986; eidem, *Militärische Nutzung des Weltraums: Eine Bibliographie*, Berlin: Arno Spitz, 1988; Volker Schiller, *Strategic Defense Initiative (SDI): Auswahlbibliographie (1979–1987)*, Bonn: Deutscher Bundestag, 1988; Russell R. Tobias, ed., *America in Space: An Annotated Bibliography*, Pasadena: Salem Press, 1991; Thomas Hübner, ed., *Raumfahrt-Bibliographie: Ein Verzeichnis nichttechnischer deutschsprachiger Literatur von 1923 bis 1997*, Hörstel: Raumfahrt-Info-Dienst, 1998; as well as Wilson W. S. Wong and James Gordon Fergusson, *Military Space Power: A Guide to the Issues*, Santa Barbara: Praeger, 2010. Gary E. McCuen, ed., *Militarizing Space*, Hudson: Gary E. McCuen Publications, 1989, is a textbook featuring a wide variety of primary sources. For an excellent overview of recent scholarship, see first and foremost David C. Arnold, 'Space and War,' in Dennis Showalter et al., eds, *Military History: Oxford Bibliographies Online*, Oxford: Oxford University Press, 2017, http://www.oxfordbibliographies.com/view/document/obo-9780199791279/obo-9780199791279-0168.xml (last modified 28 February 2017).

70. E. P. Thompson, 'The Real Meaning of Star Wars,' *Nation* 240 (9 March 1985), 257, 273–5, here 274.
71. On Space Europe, see Walter A. McDougall, 'Space-Age Europe: Gaullism, Euro-Gaullism, and the American Dilemma,' *Technology and Culture* 26.2 (April 1985), 179–203, here 189–90, 197; and Helmuth Trischler, 'Contesting Europe in Space,' in Martin Kohlrausch and Helmuth Trischler, *Building Europe on Expertise: Innovators, Organizers, Networkers*, Basingstoke: Palgrave Macmillan, 2014, 243–75, here 271–2.
72. See only John Lewis Gaddis, *The United States and the End of the Cold War: Implications, Reconsiderations, Provocations*, New York: Oxford University Press, 1992, 43–4; Mira Duric, *The Strategic Defence Initiative: US Policy and the Soviet Union*, Aldershot: Ashgate, 2003, 53–7; and Michael J. Neufeld, *Spaceflight: A Concise History*, Cambridge, MA: MIT Press, 2018, 68–70, in addition to many others.
73. John M. Logsdon, *Ronald Reagan and the Space Frontier*, London: Palgrave Macmillan, 2019, 2.
74. Reiss, *Strategic Defense Initiative*, 153, 163.
75. Reagan, 'Address to the Nation.'
76. Pearce Wright, 'Science Fiction Becomes Fact,' *Times* (25 March 1983), 6; '"Star Wars": How the Term Arose,' *New York Times* (25 September 1985), A10.
77. Charles Mohr, 'It's a Long Way to Star Wars,' *New York Times* (27 June 1982), E9; *Moonraker*.
78. See, for instance, Harry Waldman, *The Dictionary of SDI*, Wilmington: Scholarly Resources, 1988, ix: '"Star Wars" – a phrase that once conjured up the exotic dream world of H. G. Wells, Isaac Asimov, or Stephen Spielberg – is science fiction no longer. Over the last four years, since the media gave the popular tag to Ronald Reagan's new American space priority, the term has entered into our everyday consciousness.'

79. Lawrence Feinberg, 'Lucasfilm's "Star Wars" Lawsuit,' *Washington Post* (11 November 1985), C3.
80. For one example of evoking such references without granting them "real-world" effectiveness, see James Clay Moltz, *Crowded Orbits: Conflict and Cooperation in Space*, New York: Columbia University Press, 2014, 1–5, 171. At the time of writing in the summer and fall of 2019 there were numerous similar science-fiction references in the press, wondering whether the current US president had watched too many Hollywood blockbusters.
81. Numerous cartoons can be found in McCuen, *Militarizing Space*, and Edward Tabor Linenthal, *Symbolic Defense: The Cultural Significance of the Strategic Defense Initiative*, Urbana: University of Illinois Press, 1989; see also Rebecca S. Bjork, *The Strategic Defense Initiative: Symbolic Containment of the Nuclear Threat*, Albany: SUNY Press, 1992. Unfortunately, all three books limit themselves to a US perspective.
82. William J. Broad, 'Reagan's "Star Wars" Bid: Many Ideas Converging,' *New York Times* (4 March 1985), 1, 8; H. Bruce Franklin, *War Stars: The Superweapon and the American Imagination* [1988], 2nd edn, Amherst: University of Massachusetts Press, 2008, 200–5; FitzGerald, *Way Out There in the Blue*, 121–31.
83. 'VJ Day' stands for victory over Japan. Daniel O. Graham, *High Frontier: A New National Strategy*, Washington, DC: The Heritage Foundation, 1982; idem, *High Frontier: A Strategy for National Survival*, New York: Tom Doherty, 1983, 1. On the history of this "study", see Christopher Lee, *War in Space*, London: Hamish Hamilton, 1986, 116–22; and Donald R. Baucom, *The Origins of SDI, 1944–1983*, Lawrence: University Press of Kansas, 1992, 141–70. On its author, consult 'Daniel O. Graham, General, Dies at 70,' *Washington Times* (3 January 1996); and on Heinlein, see Philipp Theisohn's contribution, Chapter 9 in this volume.
84. See Norman Spinrad, 'Quand "La Guerre des étoiles" devient réalité,' *Le Monde Diplomatique* 46.544 (July 1999), 28; and idem, 'Too High the Moon: From Jules Verne to Star Wars,' *Le Monde Diplomatique* (July 1999), available at https://mondediplo.com/1999/07/14star. For Pournelle's response, see https://www.jerrypournelle.com/debates/nasa-sdi.html. For a more detailed analysis of the relation of science-fiction authors to government and military agencies, see David Seed, 'The Strategic Defense Initiative: A Utopian Fantasy,' in idem, ed., *Future Wars: The Anticipations and the Fears*, Liverpool: Liverpool University Press, 2012, 180–200; and in particular, Peter J. Westwick, 'From the Club of Rome to Star Wars: The Era of Limits, Space Colonization and the Origins of SDI,' in Geppert, *Limiting Outer Space*, 283–302, here 291–4.
85. Larry Pressler, *Star Wars: The Strategic Defense Initiative Debates in Congress*, New York: Praeger, 1986, 144; 'Angst vor einer ungewissen Zukunft,' *Der Spiegel* 46 (14 November 1983), 17–19, here 18. See also Gustav Seibt, 'Bielefeld im Raketenwinter 1983/84,' in Sonja Asal and Stephan Schalk, eds, *Was war Bielefeld? Eine ideengeschichtliche Nachfrage*, Göttingen: Wallstein, 2009, 171–8. On European reactions, consult John Wilkinson, T. B. Millar and Marie-France Garaud, 'Foreign Perspectives on the SDI,' *Daedalus* 114.3 (Summer 1985), 297–313; Michael Lucas, 'SDI and Europe,' *World Policy Journal* 3.2 (Spring 1986), 219–49; Ivo H. Daalder, *The SDI Challenge to Europe*, Cambridge, MA: Ballinger, 1987; Hans Günter Brauch, ed., *Star Wars and European Defence: Implications for Europe. Perceptions and Assessments*, Basingstoke: Macmillan, 1987; David S. Yost, 'Western Europe and the U.S. Strategic Defense Initiative,'

Journal of International Affairs 41.2 (Summer 1988), 269–323; Robert C. Hughes, *SDI: A View from Europe*, Washington, DC: National Defense University Press, 1990; Mireille Couston and Louis Pilandon, *L'Europe puissance spatiale*, Brussels: Bruylant, 1991; Reiss, *Strategic Defense Initiative*, 125–36; and Sean N. Kalic, 'Reagan's SDI Announcement and the European Reaction: Diplomacy in the Last Decade of the Cold War,' in Leopoldo Nuti, ed., *The Crisis of Détente in Europe: From Helsinki to Gorbachev, 1975–1985*, London: Routledge, 2009, 99–110.

86. Franklin, *War Stars*, 197; Dieter O. A. Wolf, Hubertus M. Hoose and Manfred A. Dauses, *Die Militarisierung des Weltraums: Rüstungswettlauf in der vierten Dimension*, Koblenz: Bernard und Graefe, 1983, 7; *Report of the Second United Nations Conference on the Exploration and Peaceful Uses of Outer Space, Vienna, 9–21 August 1982* (A/CONF.101/10 and Corr. 1 and 2), Vienna: United Nations, 1982, 114. On arms races before, during and since the Cold War, see the contributions in Thomas G. Mahnken, Joseph Maiolo and David Stevenson, eds, *Arms Races in International Politics: From the Nineteenth to the Twenty-First Century*, Oxford: Oxford University Press, 2016.
87. E. P. Thompson, 'Why Is Star Wars?,' in idem, ed., *Star Wars: Science-Fiction Fantasy or Serious Probability?*, Harmondsworth: Penguin, 1985, 9–27, here 27; idem and Ben Thompson, *Star Wars: Self-Destruct Incorporated*, London: Merlin, 1985; Eckart Spoo, 'Naturwissenschaftler warnen vor der Militarisierung des Weltraums,' *Frankfurter Rundschau* (9 July 1984); and Hans-Peter Dürr, 'Die forschungspolitischen Auswirkungen von SDI,' *Gewerkschaftliche Monatshefte* 12 (1985), 725–37.
88. Jacques Amalric, 'Yves Montand face à la guerre,' *Le Monde* (18 April 1985); Michael Dobbs, 'French Debate Role of Atomic Force,' *Washington Post* (24 April 1985), A27.
89. 'Wissenschaftler warnen vor SDI-Programm,' *Frankfurter Allgemeine Zeitung* (7 August 1985), 2; Wayne C. Thompson, 'West Germany and SDI,' in Steven W. Guerrier and Wayne C. Thompson, eds, *Perspectives on Strategic Defense*, Boulder: Westview Press, 1987, 151–83, here 161; 'Anhang: Kongressprogramm,' in Gunnar Lindström, ed., *Bewaffnung des Weltraums: Ursachen – Gefahren – Folgen*, Berlin: Dietrich Reimer, 1986, 189–90; Harald Kunze, ed., *Friedenskampf gegen USA-Weltraumrüstung: Protokoll des interdisziplinären wissenschaftlichen Kolloquiums vom 31.1.1986 in Jena zum Thema 'Die Strategie der USA zur Militarisierung des Weltraums und der Kampf der Friedenskräfte,'* Jena: Friedrich-Schiller-Universität, 1986.
90. 'Jetzt soll das Wettrüsten in den Weltraum getragen werden, der bisher von Waffen frei war. Das würde mit Sicherheit die Chancen für die Abrüstung auf der Erde verringern.' Quoted after 'Aufruf der Tagung über die bewaffnete militärische Nutzung des Weltraums am 15.09.1985 in Hannover,' flyer by Rainer Braun, 'Weltraumwaffen: Wir haben die Argumente geprüft,' Krefeld, ca. 1985: Krefelder Initiative; Alexander Geppert's personal archive.
91. Arthur C. Clarke, 'Star Wars and Star Peace,' *Interdisciplinary Science Review* 12.3 (September 1987), 272–7 (also published as idem, 'Star Wars, Star Peace,' *The Daily Telegraph* [3 January 1987]). This piece was originally delivered in New Delhi on 13 November 1986, as the nineteenth Jawaharlal Nehru Memorial lecture; the original manuscript can be found in NASMA/ACCC, 143/02.

92. See, for example, Willy Lützenkirchen and Egmont R. Koch, 'Der Himmel hängt voller Spione: Die Militärs erobern den Weltraum,' *X – Unsere Welt heute* 4.8 (August 1972), 56–61; and Hall, 'Space Wars.'
93. Reuben Steff, *Strategic Thinking, Deterrence, and the US Ballistic Missile Defense Project: From Truman to Obama*, Farnham: Ashgate, 2013; Columba Peoples, *Justifying Ballistic Missile Defense: Technology, Security and Culture*, Cambridge: Cambridge University Press, 2010.
94. David E. Sanger and William J. Broad, 'At Pentagon, Trump Shares His Ambitions to Lift Missile Defenses,' *New York Times* (18 January 2019), A9.

PART I

Embattling the Heavens

CHAPTER 2

Cold War – But No War – in Space

Michael J. Neufeld

Space war has been a fixture of astroculture since the blossoming of science fiction in the late nineteenth century. Battles with aliens, space fighters, ray guns and laser weapons have been depicted in novels, comic books, movies and computer games, and this genre got a new lease on life with the release of the *Star Wars* motion picture in 1977. Yet in the more than seventy years since the end of the Second World War, when outer space was first penetrated by the V-2 ballistic missile, no hostile military action between two powers has ever taken place outside the atmosphere. Weapons, including nuclear warheads, have been tested in space and nations have destroyed their own spacecraft in anti-satellite (ASAT) systems tests. The Cold War between the United States, the Soviet Union and their allies drove the expenditure of trillions of dollars on military space systems. The end of that contest around 1990 did not significantly change the trajectory either. Still, no shots – or lasers – have been fired in engagements between space powers.[1]

During the Cold War, space near the earth militarized but did not weaponize. Multiple national security satellite systems were put into space, but no weapons were permanently stationed in orbit or on the moon. The great-power consensus behind that process, which has had only a partial basis in international law and has sometimes looked like it might collapse, has remained in place until today because military satellite systems have stabilized, rather than destabilized, world order. While nuclear deterrence was the fundamental reason why the Cold War became, in the words of historian

Michael J. Neufeld (✉)
Smithsonian National Air and Space Museum, Washington, DC, USA
e-mail: NeufeldM@si.edu

Alexander C. T. Geppert et al. (eds), *Militarizing Outer Space*
European Astroculture, vol. 3
https://doi.org/10.1057/978-1-349-95851-1_2

John Lewis Gaddis, 'the long peace' (at least in terms of great-power war, not the devastating proxy wars in the so-called Third World), reconnaissance and early warning spacecraft made a nuclear war much less likely.[2] Nuclear arms control and eventual reduction were only possible because the superpowers could use 'national technical means of verification,' in the deliberately vague language of US-Soviet treaties, to determine how many delivery systems the other side had and what their capability was. Navigation and geodetic satellites were launched to make nuclear targeting much more accurate, and became critical to precision conventional strikes on earth after the Cold War was over, yet they are now essential to civilian life through vehicle and handheld navigation systems. In short and on balance, the militarization of near-earth space has been a positive force for global stability and the global economy, notwithstanding repeated threats to destabilize the regime with space weaponry. One more aspect is equally striking: the gulf between space fiction and space reality in the military realm only widened during and after the Cold War. Space war makes for popular entertainment, but so far, at least, it has made very little military or political sense.

I Militarizing outer space, 1943–62

Cold War military activity in outer space can be divided into three periods. First came an era in which spaceflight and satellite technology was in the process of invention (1943–62), and no international consensus existed about what was the proper role of the military in space. The superpower nuclear-arms race threatened to extend into earth orbit. Second was a period (1963–83) in which the two sides accepted a de facto regime of stability. No weapons were deployed in space, although some were tested. Finally came a brief period at the end of the Cold War (1983–89) in which President Ronald Reagan's Strategic Defense Initiative (SDI) threatened to collapse that regime. But the crisis was short-lived because the Cold War ended, effectively restoring the status quo.

In the first phase, it is important to note that near-earth space was a military realm from the moment a human device first entered it. If we take the now widely accepted definition of its lower boundary as 100 km (62.1 miles), then a German V-2 missile passed that line sometime in 1943, and routinely travelled through space during attacks on Allied cities beginning in September 1944. After the war, captured or reproduced V-2s became the starting point of the American and Soviet ballistic missile programs and were also deployed to gather scientific data on the upper atmosphere and near space useful for both military and civilian purposes.[3] Only the military services of the great powers had the capability to launch anything into space, whether a sounding rocket, a satellite or a deep-space probe, well past the formation of the US civilian National Aeronautics and Space Administration (NASA) in 1958. The agency depended on military rockets for much of its early history.

Thus the militarization of space was not a process that began after the Soviet Union launched Sputnik in 1957 on an intercontinental ballistic missile (ICBM). Rather, near-earth space partly "civilianized" after 1958, as non-military space agencies and corporations began to launch payloads. Military and intelligence services still controlled the majority of everything sent into space. The militarization of space was not an intrusion upon a civilian realm, rather it was an expansion of national security systems in a space that had been military from the outset.

The first serious discussions of military space operations began at the end of the Second World War. Already on 15 May 1945, just two weeks into his captivity, Wernher von Braun (1912–77) handed two British interrogators a document, 'Survey of Development of Liquid Rockets in Germany and Their Future Prospects.' Among his breathtaking predications was a piloted space vehicle or station: 'The whole of the earth's surface could be continuously observed from such a rocket. The crew could be equipped with very powerful telescopes' and observe 'ships, icebergs, troop movements, constructional work, etc.' He also mentioned a space mirror, lightly constructed and 'kilometers' in diameter, which could focus sunlight to modify the weather or destroy things on earth. Both ideas were taken from the 1920s books of Hermann Oberth (1894–1989), who von Braun considered to be his inspiration and mentor. Oberth was one of the few interwar space theoreticians who had seriously examined military uses of space travel; he also mentioned long-range missile attacks with poison-gas warheads in 1929.[4] But von Braun's comments aside, most late-Second World War and postwar discussion focused on the missile warfare introduced by the Germans. The far-sighted Commanding General of the US Army Air Forces, General Henry H. 'Hap' Arnold (1886–1950), discussed a future 'manless' air force and after Hiroshima noted the possibility of nuclear-armed ICBM attacks from space, as did several other postwar experts.[5]

A few definitions are in order here. Ballistic missiles of the V-2's range and longer travel through outer space, but as the launch point and target are both at the surface, I wish to exclude them from my definition of space war and military space systems, otherwise they would take up too much of the narrative. Practically speaking, it is easiest to focus on military and national security systems in earth orbit or beyond, as they stay in space on a longer or quasi-permanent basis. Any weapon stationed in orbit, even ones designed to attack ICBMs or ground facilities, can be included in the definition of space warfare, as can any ground-based military systems designed to attack space-based assets. But as noted previously, the capability to wage space war has never come to fruition beyond a few limited, ground-based ASAT systems. Virtually all military and national security spacecraft are passive, that is, without any offensive capability.

In the United States immediately after the Second World War, the Navy and Air Force funded a feasibility study of a satellite and what military and civilian uses it might have. The Air Force and its think tank, Research and Development (RAND), noted its potential for reconnaissance. But the post-war satellite projects quickly died in the budget cutbacks of the late 1940s, and long-range missile programs were greatly reduced. Interest revived after the 1949 Soviet atomic bomb test and especially after the North Korean invasion of the South in June 1950, which stoked fears of a Soviet attack on Western Europe. RAND began new reconnaissance satellite studies in 1951, leading to the Project Feedback report of 1954, the first formal proposal for what such a vehicle might look like.[6] A year earlier, President Dwight Eisenhower (1890–1969) had come into office. He was deeply concerned by the impenetrability of the Soviet Union and its potential capability to pull off a Pearl Harbor-style surprise attack with nuclear weapons. One of America's central strategic problems was a Soviet Union very difficult to infiltrate with human spies and conventional technology, making Soviet military capabilities hard to estimate. The vast USSR landmass was largely inaccessible except through dangerous and illegal aircraft overflights.[7]

Reconnaissance thus dominated secret US discussions of the military uses of outer space in the 1950s. As is well known from the work of Walter McDougall, Cargill Hall and Dwayne Day, in 1954 Eisenhower formed a secret advisory group on the surprise-attack threat called the Technological Capabilities Panel (TCP). It recommended a stopgap, high-altitude reconnaissance aircraft (the U-2), a reconnaissance satellite and a scientific satellite for the International Geophysical Year (1957–58) to establish a 'freedom of space' precedent. Lawyers had already argued that national airspace ended with the atmosphere and that outer space was like the high seas, an international commons, although no one could predict whether the Soviets would accept that. Eisenhower also approved crash program status for the Atlas ICBM project in 1954, based on breakthroughs in thermonuclear warheads light enough to be launched by more reasonably sized missiles.[8] The Soviet strategic position, on the other hand, was dominated by what it saw as a threatening encirclement by US bases and allies, as well as by a growing threat of direct attack from North America. Hence dictator Josef Stalin (1878–1953) wanted to create nuclear weapons, long-range aircraft and missiles to attack the United States and its allies. Military satellites were not a priority, although space advocates inside the Soviet Union were well aware of the possibilities.[9]

In Western Europe, discussions of military space technology were equally theoretical. When the spaceflight movement revived in the early 1950s, with the creation of the International Astronautical Congress and Federation, British and German advocates made utopian proposals that spaceflight should be carried out by a civilian international organization such as the United Nations. But as von Braun pointed out to Arthur 'Val' Cleaver (1917–77) of the British Interplanetary Society (BIS), 'we should stop bewailing the fact that our beloved space travel idea is being pulled into the capacious maw of the military,' because only it had the money.[10] Even that dictum did not apply in Western Europe, as the United Kingdom and France, the only two nations trying to remain great powers, could not afford substantial rocket programs, let alone space activities. The United Kingdom built its own nuclear-armed bomber force, but postponed long-range missile development until the late 1950s. France invested more, but only began to accelerate its rocket projects when Charles de Gaulle returned to power in 1958.[11] The net result was that until after Sputnik, satellites with national security purposes were primarily an American concern. But US military satellite projects remained underfunded before 1957 because the president was a fiscal conservative who worried that the existing military buildup was already a threat to the American system of government.[12]

In the US public realm, but quite separate from the many images of space warfare and alien attacks in the flourishing subculture of popular science fiction, there were new discussions of military spaceflight in the early 1950s. Notably, when the first in a series of *Collier's* magazines focusing on space came out in March 1952, its main article, written by Wernher von Braun, advocated an all-purpose space station that could observe and dominate the Soviet Union. It could include nuclear missiles, in a sub-station orbiting ahead of the main station, which would exercise control over targeting. In other places, von Braun contemplated using them for preventive, nuclear first strikes to rob the Soviet Union of its space capability. Von Braun and his friends later whitewashed his station advocacy as part of a master plan for peaceful human spaceflight, but he was probably the first person to advocate 'space superiority' through control of orbital space as a means to ensure American victory in the Cold War. After Sputnik, von Braun, his Army superior General Bruce Medaris (1902–90) and others made statements that orbital space and even the moon was the new 'high ground' that had to be controlled, otherwise the world faced Soviet domination.[13]

Behind the scenes, US Air Force (USAF) officers had sponsored a series of studies of military space planes, perhaps armed, that reflected their view that space was the natural extension of their turf. Yet the Eisenhower administration was resistant to proposals for deploying weapons in space that might trigger an arms race, threatening the assets they thought were really important: reconnaissance vehicles. In 1956 the USAF formalized the spy satellite project as Weapons System WS-117L. But the only space program with authorization to build hardware was Vanguard, the US Navy project to launch a satellite for the International Geophysical Year. It was the 'stalking horse' to establish the 'freedom of space' principle for later reconnaissance missions, while simultaneously garnering international prestige for the United States in the Cold War. Or so it was hoped.[14]

The Soviets themselves had established a scientific satellite project in response to the American public announcement in late July 1955, which led to a very large vehicle that was launched as Sputnik 3 in 1958. The first, minimal Sputnik, however, was a late 1956 initiative of Sergey P. Korolyov's design bureau, when fears mounted that the United States might get to orbit first. It was launched, to world acclaim, on 4 October 1957. Sputnik 2, carrying an ill-fated dog, was even more of an improvisation, thrown together within a month after the unexpected international acclaim for the first. The Sputniks made an extraordinary impression on American elites and the public, but also on Europe and the emerging nations in the Global South. Soviet claims as to the superiority of socialism became more credible in the latter. The satellites also legitimized previous ICBM test announcements, making the nuclear threat to America and its allies seem much more real.[15]

The Soviet Army's missile troops, which Soviet Premier Nikita Khrushchev made in 1959 into a separate military service, the Strategic Rocket Forces, had launched these missions. The USSR Academy of Sciences' prominent role in public announcements and international meetings was window-dressing in a space program that was controlled by the military and its industrial design bureaus. The Academy's only substantive role was in the creation of scientific experiments.[16]

The category of civilian spaceflight thus was effectively invented by the United States in the course of creating NASA out of the National Advisory Committee for Aeronautics in 1958. Eisenhower, Senate Majority Leader Lyndon Johnson (1908–73) and other members of the political elite were motivated by both domestic and foreign policy considerations. They were annoyed by the interservice rivalry that grew out of the existing, often bitter competition for roles in ballistic missiles, particularly between the Army and the Air Force.[17] Both services had presented proposals for human spaceflight and made claims to run the whole space program. But equally critical was the global image the US government wanted to project in the Cold War. Space accomplishments that countered Soviet firsts could reassure allies and influence the new nations in the Global South rapidly being created by the devolution of European empires. In the face of relentless Soviet propaganda about America's militarism and imperialism, "peaceful" and "scientific" space exploration looked better when it was carried out by a civilian agency. NASA was in fact embedded in the national security establishment, with deep connections to the Central Intelligence Agency (CIA) and the military services. For the sake of its image, it had to obscure those links as far as was feasible.[18]

Although Eisenhower restrained and channeled the space ambitions of the armed services, the Space Race sparked by the Sputniks dramatically accelerated spending on national-security space systems, notably reconnaissance. The Air Force WS-117L project spun off the Satellite and Missile Observation System (SAMOS), which encompassed both reconnaissance and signals intelligence payloads, and Missile Defense Alarm System (MIDAS), the earliest experiment in creating a warning satellite network to scan the Soviet Union for launches. While SAMOS got all the early publicity regarding space reconnaissance, the program that was really important, CORONA, was conducted in total secrecy. It was to be a stopgap system using film-return capsules, until the SAMOS TV or film read-out spacecraft worked, which they never did. Officially, the Eisenhower administration cancelled the WS-117L firm-return project in early 1958, while reconstituting it as a super-secret, joint CIA-USAF program on the model of the U-2. Launches were hidden under the appellation Discoverer, which supposedly were carrying defense science payloads. Twelve straight failures were only tolerated by the US political system because of the perceived urgency of the project. Finally, in August 1960, Discoverer 13 made the first successful return of a capsule from orbit; a couple of weeks later Discoverer 14 returned the first photos of the Soviet Union from space. It came less than four months after the embarrassing shoot-down of Francis Gary Powers's aircraft ended U-2 overflights of the Soviet Union. CORONA went on to become the mainstay of US overhead reconnaissance

Figure 2.1 President Dwight Eisenhower (1890–1969) shows off the Discoverer 13 capsule in an August 1960 press conference, likely at the White House. It was the first man-made object recovered from orbit. With him are (from left) Secretary of the Air Force Dudley Sharp, Secretary of Defense Thomas Gates, Air Force Chief of Staff Thomas White, White House Press Secretary James Hagerty (in back), Colonel Charles Mathison and White House Appointments Secretary Thomas Stephens. All are pretending it was for a defense science program, when it was the first successful test of the CORONA film-return system. The capsule now sits in the National Air and Space Museum's main hall in Washington, DC.
Source: Courtesy of National Archives and Records Administration.

in the 1960s, supplemented by higher-resolution systems and by signals intelligence satellites, the first of which was launched by the US Navy even before Discoverer 13. In September 1961 President John F. Kennedy created the National Reconnaissance Office (NRO), jointly staffed by the CIA and the Defense Department. Even the name was a secret. Effectively the United States now had three major space programs: civilian (mostly NASA), military (mostly USAF) and intelligence (NRO) (Figure 2.1).[19]

The Soviets countered US military missions in space with a barrage of propaganda about American aggressiveness. US reconnaissance satellites were a particular sore point, and Khrushchev and his spokespeople refused to publicly acknowledge the 'freedom of space' principle. There was a lot of hypocrisy in that stand. Korolyov had received approval in principle in 1956 to pursue a reconnaissance satellite, although it was in part his own stalking horse for a human spacecraft: he imagined the same large capsule could be used to recover either the entire camera system (not just the film) or a cosmonaut. After the Space Race began, the Soviet leadership approved both the Zenit for reconnaissance and the Vostok for humans. The effort to develop Vostok in a military-industrial establishment strained by the ballistic missile race resulted in delays to Zenit. The first successful flights were in spring and summer 1962. Half a year earlier, in October 1961, the Soviet government had authorized a major expansion of military satellite programs in view of a perceived American militarization of space (Figure 2.2).[20]

Soviet attempts in the United Nations to have reconnaissance satellites declared illegal under international law forced the United States into a diplomatic counteroffensive. Its representatives argued for the legitimacy of 'peaceful' (rather than 'nonmilitary') space activity, implying the inclusion of passive military systems. President Kennedy and his advisers believed, with good reason, that overhead photography of the Soviet Union was critical to US national security. This imagery certainly improved nuclear targeting, but it also greatly reduced the gnawing uncertainty about Soviet capability, which had led to fear-mongering assertions of a 'bomber gap' and a 'missile gap' during the mid- to late 1950s.[21]

If the ongoing diplomatic battle over space was not disturbing enough, the United States and Soviet Union each conducted three space nuclear tests in 1962, the largest and most spectacular being the United States' Starfish Prime, a 1.4 megaton explosion 400 km (250 miles) over the Pacific Ocean. It produced an electromagnetic pulse that damaged electrical and telephone equipment in Honolulu, Hawaii, almost 1,450 km (900 miles) away. The United States had earlier conducted other high-altitude and space nuclear tests in 1958, testing concepts for missile defense, including the possibility of creating an artificial radiation belt to damage incoming warheads. Trapped energetic particles from Starfish Prime actually disabled several satellites. That same year, Defense Secretary Robert McNamara (1916–2009) authorized the

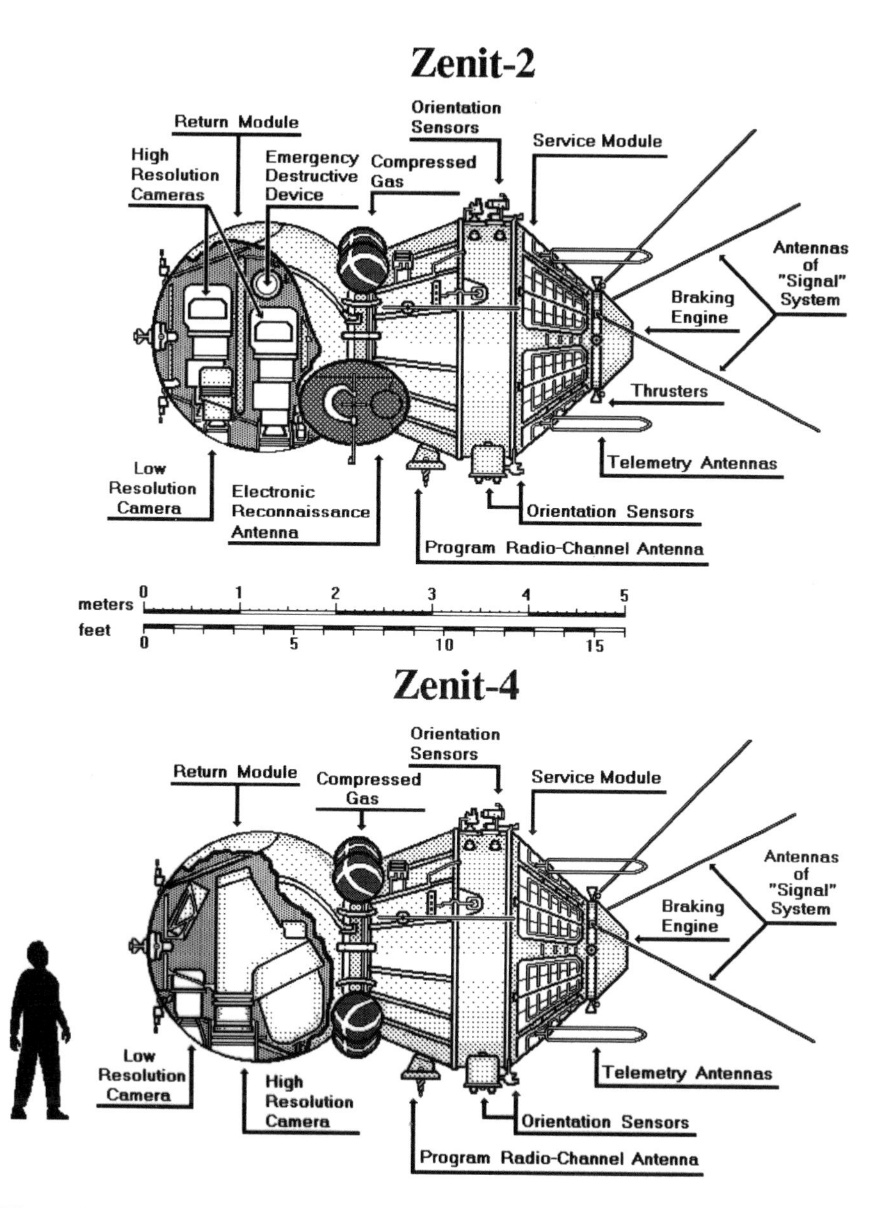

Figure 2.2 Zenit 2 and Zenit 4 were the initial and high-resolution versions of the first Soviet reconnaissance satellite. Based on the Vostok human spacecraft, Zenit returned its entire camera system, not just the film, in its spherical re-entry module. Zenit 2 was first launched in spring 1962 and Zenit 4 in late 1963. In 1970 they were replaced by more advanced spacecraft.

Source: Peter A. Gorin, 'Zenit: The First Soviet Photo-Reconnaissance Satellite,' *Journal of the British Interplanetary Society* 50.11 (November 1997), 441–8, here 442. Courtesy of British Interplanetary Society.

Army and the Air Force each to set up a limited anti-satellite capability with their own nuclear-armed missiles on two different Pacific islands. It was to be a possible emergency response to any Soviet orbital bombs.[22] There was every reason to believe that any future war would include nuclear combat in space. Thus, the de facto arrangement that soon emerged – space could be militarized but not weaponized – was still nowhere to be seen in 1962.

II The era of relative space stability, 1963–83

The rhythm of space nuclear testing and arguments over military systems reflected the overall state of US-Soviet tension, with periods of maximum stress during the Berlin confrontations of 1958 and 1961–62, and the frightening Cuban Missile Crisis of October 1962. The sobering effect of that latter near-death experience had much to do with why a new space consensus emerged so quickly in 1963. Notably, the Limited Test Ban Treaty of August 1963, signed by the United States, the Soviet Union and the United Kingdom, banned nuclear explosions in outer space, as well as in the atmosphere or under the sea. In October, the two superpowers agreed not to station nuclear weapons in orbit. And the Soviet Union silently dropped its verbal offensive against reconnaissance satellites, now that it had a capability as well. The diplomatic process culminated in the United Nations Outer Space Treaty, signed in 1967, which banned the stationing of 'weapons of mass destruction' in orbit or on other heavenly bodies.[23]

The treaty did not forbid stationing other weapons in space. The de facto agreement not to do so prevailed because of both sides' growing perception that they were better off not starting such an arms race. That did not mean that there were no challenges in this period. Beginning in 1967, the Soviet Union began testing a co-orbital ASAT, in which one orbiting spacecraft maneuvered past a previously launched target, then exploded at a safe distance. While this caused some disturbance in American military circles, many tests were failures, and the United States did not assess it as a significant threat. The United States retained its nuclear-armed interceptors on Pacific islands until 1969 for the Army system and 1975 for the Air Force, but they were then abandoned without any demand for a quid pro quo. Later in the 1970s the USAF began developing a smaller ASAT missile to be launched from a climbing F-15 fighter, a response to renewed Soviet ASAT testing after a long lull. But all systems were only capable of destroying satellites in low-earth orbit (LEO) and, given their minimal and experimental nature, were no more than minor exceptions to the informal regime. Another irritant was Soviet testing of what the United States called a Fractional-Orbital Bombardment System, which could launch a nuclear warhead into a partial orbit to attack the United States from the south, getting around its mostly north-facing warning systems. But it too never seemed like a fundamental threat and, by my definition, should not count as a space weapon as effectively it is a variant ICBM.[24]

The fundamental characteristic of the two decades of relative stability was the deployment of a large number of military space systems by the two superpowers in near-total secrecy, with very little public discussion. These served many purposes, but the four most important were intelligence, early warning, navigation and communications. In each case the United States was two to ten years ahead, but the Soviets mirrored every American system eventually, sustaining strategic stability. The Soviet Space Forces, part of the Strategic Rocket Forces, became the largest launch service in the world – orbiting over fifteen hundred satellites under the generic 'Kosmos' label by the end of 1983. The first Western European military spacecraft appeared around 1970: communications satellites for the United Kingdom and the North Atlantic Treaty Organization (NATO), followed by the French in 1984, but the dominant European space activities were explicitly non-military and cooperative, leading to the formation of the European Space Agency (ESA) in 1975. Beginning in 1970, the Chinese People's Liberation Army began launching satellites too, although they had little national security capability before the 1990s. Thus the superpower military/space duopoly was essentially intact until the end of the Cold War.[25]

Intelligence satellites included photography and signals intelligence, soon joined by radar reconnaissance of surface ships and by radar imaging through clouds. In the 1970s both sides began flying bigger film-return photoreconnaissance spacecraft, followed by the first digital-image-return spacecraft by the United States in 1976 and the Soviet Union in 1982. These had much longer lifetimes in orbit, as they did not have to be abandoned or deorbited when they ran out of film. A related technology for the United States was the Defense Meteorological Satellite system, which began as a program to get more usable CORONA imagery; too much film had been taken of overcast targets. Soon both sides were using weather-satellite images and data produced by both military and civilian constellations to provide global forecasting for their militaries. The Soviets had only one system, as effectively they had only one space program – military (Figure 2.3).[26]

A peculiar dead end of this period, but one that cost billions of dollars and rubles, was the human-operated reconnaissance station, a small, special-purpose spacecraft quite unlike von Braun's gigantic, multi-use, rotating station concepts. In 1963 the Kennedy administration cancelled the Air Force's Dyna-Soar space plane, but authorized the Manned Orbiting Laboratory (MOL), essentially the cover name for a super-high-resolution optical camera operated by military astronauts. The Richard Nixon administration cancelled it in 1969, before it ever flew, as much cheaper robotic systems were in prospect. The Soviets developed a parallel Almaz system and actually flew two stations as Salyuts in the mid-1970s, using the same name as the quasi-civilian stations they pretended to be. Results were not that impressive. One oddity was the incorporation of a cannon to ward off American attackers, probably a response to loose talk about the military applications of

Figure 2.3 In September 2011, the National Reconnaissance Office unveiled the gigantic HEXAGON reconnaissance satellite as part of the organization's fiftieth anniversary party at the National Air and Space Museum's Steven F. Udvar-Hazy Center in Chantilly, Virginia. The vehicle's main imaging system was the KH-9 broad-area-search and mapping camera. HEXAGON carried four film-return capsules significantly larger than the CORONA ones. Operating from 1976 to 1984, it was the last US spy satellite to carry film.
Source: Courtesy of Smithsonian National Air and Space Museum.

the US Space Shuttle authorized in 1972. One possibility discussed was using it to recover or inspect Soviet satellites.[27]

As reconnaissance systems improved, they became a significant factor in the production of relative geopolitical stability and even détente in the 1970s. Historian Gaddis, in his investigation of the causes of the 'long peace,' notes the inherently stabilizing character of a bipolar system between two comparatively evenly balanced blocs on opposite sides of the world that did not much compete economically. He should have put more emphasis on nuclear deterrence. But he also notes the growing global transparency created by satellite technology from the 1960s on, which he calls the 'reconnaissance revolution.' Indeed it is impossible to imagine the major arms control treaties the superpowers concluded in the 1970s without intelligence satellites. It was not that the threat of nuclear war had vanished; there were frightening moments created by the Israeli-Arab war of 1973, by false early-warning alarms and by Soviet fear of the aggressive talk of the Ronald Reagan administration in the early 1980s. The displacement of US-Soviet confrontations to a series of proxy wars, notably in Vietnam, Angola, Ethiopia and Afghanistan, devastated those countries and cost the lives of millions. So the Cold War was hardly a benign 'long peace,' as Gaddis himself acknowledges, but a global catastrophe was avoided and military satellites played no small part.[28]

Early warning satellites themselves contributed to strategic stability, by increasing warning time and decreasing the threat of an accidental nuclear war created by false alarms from radar systems. Radar provided as little as five to fifteen minutes notice of attack. Coming out of the MIDAS program, the US Air Force developed 'Program 461' (the Kennedy administration decided in 1962 to make all national security launches anonymous and all programs hidden behind bland numbers). In 1966 that project launched the first test satellites for a system in polar orbit about 3,000 km (2,000 miles) high, using infrared sensors to look for rocket plumes from launches. But that is not the one that was actually deployed. At about that time, the first successful experimental and commercial geosynchronous and geostationary communication satellites in 24-hour orbits demonstrated the technical maturity of systems stationed there. The Defense Support Program, as the new American system was obscurely named, launched its first-generation satellites between 1970 and 1973. The active constellation, once it was constructed by 1973, always had one spacecraft watching Eurasia from over the Indian Ocean, plus one each staring at the Atlantic and Pacific for submarine-launched ballistic missiles, plus an on-orbit spare. Subsidiary instruments looked for space nuclear explosions.[29]

The Soviets began experimenting with their first early-warning satellite in 1972. They put their primary constellation in highly elliptical, 12-hour orbits with apogees over the northern hemisphere, the same as their Molniya communications satellites. Spacecraft were needed in at least four different orbital planes so that one was always viewing North America. These were supplemented by geostationary satellites after 1985. The Oko system (Old Russian for 'eye') had numerous satellite failures and did not become fully operational until 1982. The Soviets never depended on it as much as the Americans did. Oko also caused a frightening false alarm in 1983 when sunlight reflected off clouds produced a warning of an American missile attack. An alert operator dismissed it as a technical problem. Thus missile early-warning satellites were far from foolproof, but once thoroughly tested they became, on balance, factors for making nuclear war less likely, because they reduced fear of surprise attack and added warning time and knowledge about the other side's capabilities.[30]

Navigation satellites, an application scarcely imagined before the Space Race, began in the wake of Sputnik. Two engineers at the Johns Hopkins University's Applied Physics Laboratory outside Baltimore, a major US Navy contractor, noticed that tracking the Doppler shift in Sputnik's radio transmission as it moved toward or away from the observer, which was used to determine its orbit, could be turned around to fix a position on earth, if the orbit was well known. That led to the Navy's Transit system, the primary purpose of which was to provide locations to ships at sea, particularly ballistic-missile submarines. Without Transit, there was little chance that the inertial guidance systems in the latter's missiles would have a launch position precise enough to hit their targets in a nuclear war. The first successful launch was in 1960 and the system became operational in 1964. Transit was sometimes used by the other services for positioning and civilian users even adopted it for surveying and other applications.[31]

Beginning in the late 1960s, a sometimes tense collaboration between the Navy and the Air Force led to a much more accurate system based on another principle, very precise time delivered by satellite atomic clocks. This became the Air-Force-operated Global Positioning System (GPS), which began launching satellites in 1978 and came into limited operations in the early 1980s. The Soviets imitated Transit and followed that with the first launch of the GPS-clone GLONASS system in 1982. These mature navigation systems really came into their own only after the end of the Cold War and made little direct contribution to the era of relative stability in spaceflight. Transit's role was especially ambiguous, as it primarily increased submarine-launched missile accuracy, as did GPS, although never enough that the Soviets felt their nuclear forces were threatened by a first strike from those weapons (unlike US land-based ICBMs). But the increasing dependency of both sides on new satellite systems tended to reinforce the unwritten consensus not to deploy active military space systems – when tension between the superpowers did not override that consensus, as it did after President Ronald Reagan's 1983 'Star Wars' speech.[32]

The final major category of new military space systems during the 1963–83 period was that of communications. American discussions go back to the RAND report of 1946, and there was extensive experimentation in the early years of the Space Race, using both NASA and commercial satellites. After first thinking that global military communications would be more cheaply and easily done by leasing commercial satellite circuits, Defense Secretary McNamara authorized the Initial Defense Satellite Communications System in 1964. First launches came in 1966. Commercial leasing did not provide the high level of security and flexibility required for many military missions, although it continues down to the present as a supplement. The first dedicated military satellites were in near-geosynchronous orbits that drifted around the earth, but the second-generation system, first launched in 1971, was in geostationary orbit. They were supplemented by a wide variety of other satellites, some dedicated to the Navy or to tactical use, leading to a forest of acronyms. As befits their unitary space program, the Soviets deployed Molniya satellites for both military and civilian use in highly elliptical orbits beginning in 1965 and small, low-earth-orbit Strela satellites for tactical, store-and-dump communications starting in 1964. Beginning in the late 1970s the Soviet Union added several large geostationary systems. As noted previously, the United Kingdom and NATO put their first geostationary communications satellites up around 1970, at first simply buying American satellite technology and launch services. Overall, these systems, much like navigation satellites, empowered global military operations but also fortified the credibility of nuclear deterrence, a foundation stone of the 'long peace' and global stalemate. Increasing dependency on space communications further reinforced the systemic stability of the military space consensus – assuming again that neither superpower made a move to destabilize it – as happened in 1983.

III Stability threatened and stability restored, 1983–89

President Reagan's 23 March 1983 national address on ballistic missile defense – soon dubbed the 'Star Wars' speech, from the feature film series that launched in 1977 – was a milestone moment. It signaled a prospective overthrow of the de facto consensus on space weapons by contemplating active anti-missile and anti-satellite systems in orbit, such as laser battle stations. It provoked an enormous uproar, notably in American and European media, popular culture and policy circles, reinforced by the Soviet propaganda about the US threat to global peace through advanced weapons development. Its popular label itself was a nod to the prevalence of space war in contemporary science fiction and astroculture, and to public fear that such a war might actually happen.

Yet in hindsight, 'Star Wars' seems ephemeral as a political and military phenomenon. By 1986 the president was already discussing arms control treaties with the new Soviet leader, Mikhail Gorbachev (1931–). By 1988 the Cold War appeared over, as was confirmed by the collapse of Eastern European communist governments in 1989 and of the Soviet Union itself in 1991. As a result, Reagan's grandiose Strategic Defense Initiative (SDI) dwindled to a minimal, ground-based ICBM defense in the 1990s. The military space reality of 1993 was little different from that of 1983, except that Russian systems were in decay due to the economic collapse that followed the Soviet Union's breakup.[33]

Although Reagan's speech was sudden and unexpected, even among Washington defense intellectuals, as it stemmed from the president's personal, somewhat fantastical wish to abolish the threat of nuclear annihilation, it came against a background of years of rising tension. The overall state of US-Soviet relations was, as always, primary, with military space systems playing only a subsidiary role. The détente of the early and mid-1970s slowly fell apart, partly as a result of Soviet and Cuban intervention in Angola and Ethiopia, followed by the Soviet invasion of Afghanistan at the end of 1979. The Strategic Rocket Forces pursued a relentless ICBM buildup and modernization, moderated only by the détente arms control treaties. American defense officials worried that deterrence could be destabilized by a growing Soviet first-strike threat; not surprisingly, the other side felt the same way about US capabilities. The Soviets resumed their co-orbital ASAT tests in 1976 and that same year authorized a large directed-energy (laser or particle-beam) weapons development program that increasingly agitated conservative US officers and defense intellectuals. Reagan's election in 1980 signaled a new, hard-line approach to the Cold War, but it had already revived under President Jimmy Carter, especially because of Afghanistan. The Reagan administration immediately accelerated the development of the F-15-launched ASAT missile after it came into office in 1981. Yet the 'Star Wars' speech was an inflection point in military space history, as it threatened the emergence of a whole new orbital regime.[34]

The early public controversy over SDI was dominated by the most futuristic solutions promoted by Hungarian-American physicist Edward Teller (1908–2003) and other hard-line Cold Warriors from the nuclear-weapons establishment, notably the nuclear-pumped X-ray laser station, which would annihilate itself in a nuclear explosion while sending out laser beams to destroy Soviet ICBMs and warheads in flight. To provide a complete defense, however, would require many layers, including expanded space-based early-warning systems, ground-based anti-ballistic missiles (ABMs), and perhaps boost-phase intercept systems, either missiles or directed energy weapons, to destroy ballistic missiles early in flight, when their rocket plumes made them most identifiable. In reality, the Strategic Defense Initiative Organization (SDIO) and other US military services and agencies quickly retreated from exotic ideas like the X-ray laser, and realized that a perfect defense had always been impossible. By the later 1980s SDIO settled on small, maneuverable, non-explosive impactors (first nicknamed 'smart rocks,' and when further downsized and improved, 'brilliant pebbles'), based on orbiting satellites and ground-based missiles (Figure 2.4).[35]

The Soviet reaction was mixed. Pavel Podvig argues that far from unnerving the Soviet military, 'Star Wars' empowered its defense establishment to try to accelerate funding for its directed-energy, ASAT and ABM programs. The first flight of the new Energia superbooster in 1987, which was also designed to carry the Buran Space Shuttle, launched a test version of the Skif-DM laser battle station. Gorbachev and the party leadership imposed strict controls on testing due to the growing rapprochement with the Reagan administration, but the vehicle failed to orbit anyway. Peter Westwick, on the other hand, has shown that the Soviet political leadership was 'obsessed' by SDI and by the alleged 'space strike' potential of US battle stations against ground targets, something the American side found hard to understand. Yet by 1987 Gorbachev had de-linked SDI from arms control, perhaps because the Soviet military argued that it could easily be defeated by countermeasures. It appears that Soviet elites were far from united about what SDI meant, and that balance changed quickly over time. In any case, Gorbachev's domestic and foreign initiatives soon overtook both sides' weapons developments, resulting in nothing new being deployed before the Cold War petered out.[36]

The SDI debate did draw Western European elites and publics into a controversy over space militarization unlike anything that had occurred earlier. The space nuclear tests of 1958 and 1962 had inevitably produced media

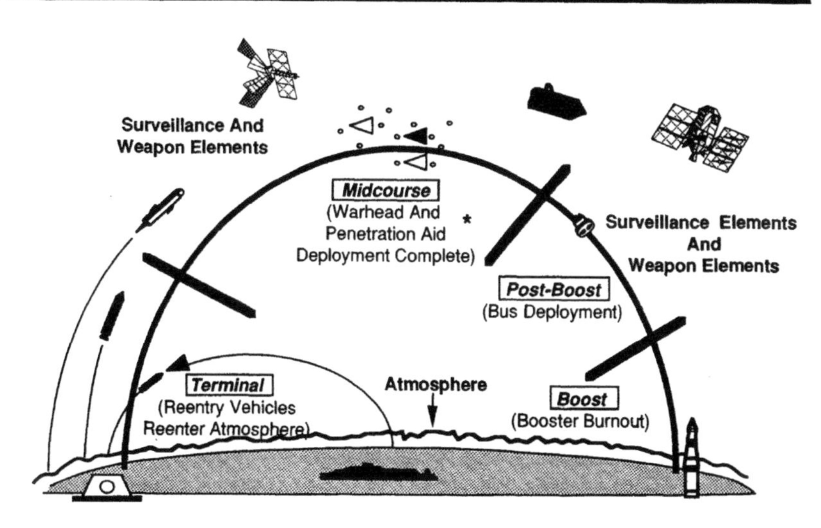

Figure 2.4 This illustration from March 1989 shows the later, more modest version of the Strategic Defense System (SDS), part of President Reagan's Strategic Defense Initiative (SDI). It depicts missile defense based on kinetic impactors rather than the earlier, more exotic proposals for X-ray lasers in orbit.
Source: Courtesy of Ballistic Missile Defense Organization.

commentary and peace protests, although scholarship on this topic is minimal. But Bernd Mütter's analysis of how outer space was treated in West Germany's media, particularly television, shows that discussion of the military uses plummeted after Sputnik and did not revive until the SDI debate. Before then it was almost a taboo topic, as West German journalists only wanted to present spaceflight in the context of science and exploration. Their reporting covered what NASA and European civilian space agencies were doing. The journalists were profoundly uncomfortable with any other dimension, reflecting a deep, post-Second World War pacifism in their audience. But 'Star Wars' provoked several to denounce the American initiative on television. The applicability of Mütter's study to other Western European countries is unknown, but it would not be surprising if there was also a studied ignorance of, or reluctance to talk about, military space applications until the 'Star Wars' controversy of the mid-1980s.[37]

Certainly, the debate agitated Europeans who distrusted the Reagan administration or the United States generally, notably on the left. The president's announcement also caught military and political elites connected to the United States and NATO by surprise and disturbed British and French leaders regarding the future viability of their limited nuclear deterrents. But the nations closest to the United States wanted to receive part of the billions Reagan planned to spend on SDI, to benefit its industry and keep up technologically. The United Kingdom signed a memorandum of understanding for participation in late 1985 and West Germany in early 1986. The fear of technological eclipse drove France, at that time not part of the NATO military structure and led by socialist President François Mitterrand (1916–96), to launch an independent initiative, EUREKA, to foster mostly civilian research at home and cooperating Western European countries. In the end, the United States did not spend much money on space weapons and missile defense in Europe, but that was hard to anticipate as the Reagan administration's rhetoric was grand and virtually no one foresaw the imminent collapse of the Soviet bloc.[38]

In the midst of the SDI controversy, the Soviet Union announced it was suspending ASAT tests in 1983, to bolster its arguments for a treaty against space weaponization. In 1985 the US Congress suspended flights of the F-15-launched missile. Although SDI dragged on into the early 1990s and was not without a technological legacy in the United States in the areas of rocket and spacecraft development, electronics and ground-based missile defense, the net impact of the furor on the de facto space regime was effectively zero. The only difference is that the United States was now a hyperpower, greatly superior to every other nation in military and space capability, while Russian budget crises led to declines in the competence and capability of its aerospace industry and military space infrastructure.

IV Military space in a post-Cold War world

In place of the bipolar military space world of the 1980s arose a multipolar one in which China, notably, became the new challenge to US hegemony. Its destruction of one of its own defunct weather satellites in 2007 in an ASAT test created a cloud of orbital debris and further inflamed American military leaders and policy-makers who were already suspicious of the People's Republic. Under the conservative George W. Bush administration, air-force-connected advocates for 'space control' through US domination of near-earth space with weapons grew in influence and volubility.[39]

Their reasoning, and that of their Chinese counterparts, was driven by the realization that American forces had become completely dependent upon military space infrastructure, in part because it also greatly increased battlefield effectiveness. In the Persian Gulf War of 1991, navigation, early warning, reconnaissance, communications and weather satellites, first launched in the strategic competition with the Soviet Union, were critical to the decisive victory over a regional power, Iraq. The next year, Air Force Chief of Staff General Merrill A. McPeak (1936–) called that conflict the 'first space war,' not in the sense used in this essay, but rather to emphasize the centrality of space resources to the ground, air and anti-missile battle.[40] So-called drone warfare after 2001 is a further expression of that capability, as large remotely piloted reconnaissance and attack aircraft like the Predator are typically flown through communication satellites and depend on GPS for navigation. The United States can now not afford to do without such an infrastructure, and other large and medium-sized powers need to ally themselves with the United States or develop their own capability if they do not wish to be completely outclassed. It is no wonder that the Chinese and the resurgent Russians continue to work on ASATs and other military space systems. Moreover, all sides have limited anti-missile capability that can be used to shoot down low-earth-orbit satellites.

Yet in the more than a quarter century since the end of the Cold War, outer space still remains without orbiting weapons. The possibility of igniting a new arms race and destabilizing the environment for all space infrastructure, civilian and military, are powerful disincentives to the United States, Russia, China or anyone else violating the de facto regime. Moreover, the militarization of space from geostationary orbit on down has on the whole been a force for global stability, even if it may now contribute to local instability, notably in US attacks in the Middle East in the 'war on terror.' Military space systems made arms control treaties possible and nuclear war less likely during the Cold War and after. The rise of global transparency has only accelerated since the mid-1980s, as commercial imaging satellites have made high-resolution capability widely available even to human-rights groups and journalists. Threats to weaponize space will continue, dependent on the state of great-power competition, but there seems to be every hope that stability in that realm will continue as our dependency on space assets grows. Uncomfortable as the conclusion may be for some, it is apparent that, on the whole, the militarization of space has been, on balance, a positive force for peace.

Finally, I cannot help but note the complete disjuncture between space war in astroculture and the actual evolution of military space technology in the Cold War and after. On the one hand, we have the long-standing and ongoing popularity of space battles in the *Star Wars* and *Star Trek* movie franchises and in video games and novels and so forth. They make for drama and entertainment and sometimes for social comment. On the other, we have the development of a complex host of satellites for different military purposes, most of which are invisible to the public, either out of secrecy or media and public disinterest, leading to potential problems with civilian, democratic control. For much of the public, infrastructure is boring, even when it is civilian and open. One only notices its absence when it does not work. Thus it is not surprising that science fiction has largely operated in discourses disconnected from the dull but critical reality of the military in space since Sputnik.

Notes

1. A possible minor exception are claims that American military satellites have been "illuminated" by Soviet/Russian or Chinese lasers, temporarily blinding them. These claims remain unproven. See Jeffrey T. Richelson, *America's Space Sentinels: The History of the DSP and SBIRS Satellite Systems*, [1999], 2nd edn, Lawrence: University Press of Kansas, 2012, 76–7; and Dean B. Cheng, 'The Long March Upward: A Review of China's Space Program,' in Paul G. Gillespie and Grant T. Weller, eds, *Harnessing the Heavens: National Defense Through Space*, Chicago: Imprint Publications, 2008, 151–63, here 161. I want to acknowledge the influence of Alex Roland's paper, 'Cold War in Space,' given at the *Embattled Heavens* conference that is the origin of this book. All Internet sources were last accessed on 15 July 2020.
2. John Lewis Gaddis, 'The Long Peace: Elements of Stability in the Postwar International System,' *International Security* 10.4 (Spring 1986), 99–142; idem, *The Long Peace: Inquiries into the History of the Cold War*, New York: Oxford University Press, 1987; Odd Arne Westad, *The Global Cold War: Third World Interventions and the Making of Our Times*, Cambridge: Cambridge University Press, 2005.
3. Michael J. Neufeld, *The Rocket and the Reich: Peenemünde and the Coming of the Ballistic Missile Era*, New York: Free Press, 1995; David H. DeVorkin, *Science with a Vengeance: How the Military Created the US Space Sciences after World War II*, New York: Springer, 1992; Matthias Uhl, *Stalins V-2: Der Technologietransfer der deutschen Fernlenkwaffentechnik in die UdSSR und der Aufbau der sowjetischen Raketenindustrie 1945 bis 1959*, Bonn: Bernard & Graefe, 2001. Much has been made of the first successful A4 (later V-2) launch on 3 October 1942, also by me, but its apogee was about 80–90 km. It was a milestone for the program and for the history of rocketry, but the first missile with range long enough to ascend beyond 100 km altitude was probably fired on 14 April 1943. See the July 1943 document 'Versuchsschießen A4' pictured in Walter Dornberger, *Peenemünde: Die Geschichte der V-Waffen*, Frankfurt am Main: Ullstein, 1989, photo insert. A V-2 launched vertically from an island off Peenemünde on 14 September 1944 reached 176 km (102 miles); see Wernher von Braun to Arthur C. Clarke, 30 August 1951, US Space and Rocket Center Archives, Wernher von Braun Papers, file 400–2.
4. Wernher von Braun, 'Survey,' in Fritz Zwicky, *Report on Certain Phases of War Research in Germany*, vol. 1, Pasadena: Aerojet Engineering Corporation, 1945, 66–72; Hermann Oberth, *Die Rakete zu den Planetenräumen* [1923], Nuremberg: Uni-Verlag, 1960, 86–9; and idem, *Wege zur Raumschiffahrt* [1929], Bucharest: Kriterion, 1974, 199–200. The American rocket pioneer Robert Goddard (1882–1945) mentioned long-range rocket attacks in 1916 when he first wrote to the Smithsonian Institution looking for support for his experiments; see Goddard to 'President, Smithsonian Institution, 27 September 1916,' in Esther C. Goddard and G. Edward Pendray, eds, *The Papers of Robert H. Goddard*, vol. 1: *1898–1924*, New York: McGraw-Hill, 1970, 170–5, here 171.
5. Roger D. Launius, 'National Security, Space and the Course of Recent U.S. History,' in Gillespie and Weller, *Harnessing the Heavens*, 5–23, here 7; Paul B. Stares, *The Militarization of Space: U.S. Policy, 1945–1984*, Ithaca: Cornell

University Press, 1985, 23–9; Walter A. McDougall, ...*The Heavens and the Earth: A Political History of the Space Age*, New York, Basic, 1985, 88.

6. Ibid., 101–2; Jeffrey T. Richelson, *America's Secret Eyes in Space: The U.S. Keyhole Spy Satellite Program*, New York: Harper & Row, 1990, 2–8. The Douglas Aircraft Company first report for the air force (effectively RAND report no. 1), 'Preliminary Design of an Experimental World-Circling Spaceship,' 2 May 1946, can be downloaded at http://www.rand.org/pubs/special_memoranda/SM11827.html.
7. See McDougall, ...*The Heavens and the Earth*, 112–15; R. Cargill Hall, 'Postwar Strategic Reconnaissance and the Genesis of Corona'; and Dwayne A. Day, 'A Strategy for Reconnaissance: Dwight D. Eisenhower and Freedom of Space,' in idem, John M. Logsdon and Brian Latell, eds, *Eye in the Sky: The Story of the Corona Spy Satellites*, Washington, DC: Smithsonian Institution Press, 1998, 86–142, 260–79. See also Sean N. Kalic, *US Presidents and the Militarization of Space, 1946–1967*, College Station: Texas A&M University Press, 2012, 26–35.
8. McDougall, ...*The Heavens and the Earth*, 106–15; Hall, 'Postwar Strategic Reconnaissance'; Day, 'A Strategy for Reconnaissance,' 86–142; R. Cargill Hall, 'The Eisenhower Administration and the Cold War: Framing American Astronautics to Serve National Security,' *Prologue: Quarterly of the National Archives and Records Administration* 27.1 (Spring 1995), 58–72; Christopher Gainor, 'The Atlas and the Air Force: Reassessing the Beginnings of America's First Intercontinental Ballistic Missile,' *Technology and Culture* 54.2 (April 2013), 346–70; and his contribution, Chapter 3 in this volume.
9. Asif A. Siddiqi, *The Red Rocket's Glare: Spaceflight and the Soviet Imagination, 1857–1957*, Cambridge: Cambridge University Press, 2010, 241–314; idem, 'Soviet Space Power During the Cold War,' in Gillespie and Weller, *Harnessing the Heavens*, 135–50.
10. Alexander C. T. Geppert, 'Space *Personae*: Cosmopolitan Networks of Peripheral Knowledge, 1927–1957,' *Journal of Modern European History* 6.2 (2008), 262–86, here 280–3; Von Braun to Cleaver, 12 May 1951, US Space and Rocket Center, Wernher von Braun Papers, file 400–2.
11. See Stephen Robert Twigge, *The Early Development of Guided Weapons in the United Kingdom, 1940–1960*, Chur: Harwood, 1993; Olivier Huwart, *Du V2 à Véronique: La naissance des fusées françaises*, Paris: Marines éditions, 2004; and Doug Millard, 'A Grounding in Space: Were the 1970s a Period of Transition in Britain's Exploration of Outer Space?,' in Alexander C. T. Geppert, ed., *Limiting Outer Space: Astroculture after Apollo*, London: Palgrave Macmillan, 2018, 79–99 (= *European Astroculture*, vol. 2).
12. McDougall, ...*The Heavens and the Earth*, 126–34.
13. Michael J. Neufeld, '"Space Superiority": Wernher von Braun's Campaign for a Nuclear-Armed Space Station, 1946–1956,' *Space Policy* 22.1 (February 2006), 57–62; idem, 'The "Von Braun Paradigm" and NASA's Long-Term Planning for Human Spaceflight,' in Steven J. Dick, ed., *NASA's First 50 Years: Historical Perspectives*, Washington, DC: NASA, 2010, 325–47; William E. Burrows, 'Beyond the Blue Horizon: Lunar Missile Base Concepts in the Early Cold War,' in Gillespie and Weller, *Harnessing the Heavens*, 25–34. For a genealogy of the 'high ground' trope, see Alexander Geppert and Tilmann Siebeneichner's contribution, Chapter 1 in this volume.

14. Roy F. Houchin, 'Technology in Transition: Dyna-Soar and the Military Spaceplane,' in Gillespie and Weller, *Harnessing the Heavens*, 177–89, here 177–80; Hall, 'Postwar Strategic Reconnaissance,' 105–10; Day, 'A Strategy for Reconnaissance,' 129–37; Kalic, *US Presidents*, 26, 36–42.
15. Siddiqi, *Red Rocket's Glare*, 324–35.
16. Idem, 'Soviet Space Power.'
17. Michael J. Neufeld, 'The End of the Army Space Program: Interservice Rivalry and the Transfer of the Von Braun Group to NASA, 1958–1959,' *Journal of Military History* 69.3 (July 2005), 737–58.
18. McDougall, *...The Heavens and the Earth*, 164–76; James E. David, *Spies and Shuttles: NASA's Secret Relationships with the DoD and CIA*, Gainesville: University Press of Florida, 2015.
19. See most of the essays in Day, Logsdon and Latell, *Eye in the Sky;* and also Hall, 'Postwar Strategic Reconnaissance'; Day, 'A Strategy for Reconnaissance'; and Mark A. Erickson, 'Reconnaissance and Prestige: The Creation of a Trinitarian U.S. Space Program,' in Gillespie and Weller, *Harnessing the Heavens*, 49–63, here 56–8. CORONA mapping also penetrated the barrier between the secret and unclassified worlds by influencing global geodesy; see John Cloud, 'Imaging the World in a Barrel: CORONA and the Clandestine Convergence of the Earth Sciences,' *Social Studies of Science* 31.2 (April 2001), 231–51.
20. Peter A. Gorin, 'Zenit: The Soviet Response to CORONA,' in Day, Logsdon and Latell, *Eye in the Sky*, 157–70; Siddiqi, 'Soviet Space Power,' 140–1.
21. McDougall, *...The Heavens and the Earth*, 177–94, 344–51; Stares, *Militarization of Space*, 54–91.
22. Ibid., 76–82, 107–8; Jack Manno, *Arming the Heavens: The Hidden Military Agenda for Space, 1945–1995*, New York: Dodd, Mead, 1984, 54–60, 81–5; 'Starfish Prime,' https://en.wikipedia.org/wiki/Starfish_Prime; Gilbert King, 'Going Nuclear over the Pacific,' 15 August 2012, http://www.smithsonianmag.com/history/going-nuclear-over-the-pacific-24428997; and Neufeld, *Von Braun*, 327–8, 332.
23. McDougall, *...The Heavens and the Earth*, 344–8; Stares, *Militarization of Space*, 86–91, 101–2; for the UN treaty text, see http://www.unoosa.org/oosa/en/ourwork/spacelaw/treaties/introouterspacetreaty.html. See also Luca Follis, 'The Province and Heritage of Humankind: Space Law's Imaginary of Outer Space, 1967–79,' in Geppert, *Limiting Outer Space*, 183–205.
24. Stares, *Militarization of Space*, 80–1, 99–100, 136–45, 206–9; Siddiqi, 'Soviet Space Power,' 145.
25. Ibid., 143–6; Alasdair W. M. McLean, *Western European Military Space Policy*, Aldershot: Dartmouth, 1992, 88–93; Alexander C. T. Geppert, 'European Astrofuturism, Cosmic Provincialism: Historicizing the Space Age,' in idem, ed., *Imagining Outer Space: European Astroculture in the Twentieth Century*, 2nd edn, London: Palgrave Macmillan, 2018, 3–28, here 9–11 (= *European Astroculture*, vol. 1); Cheng, 'The Long March,' 154–5. For more on European attitudes, see Michael Sheehan's contribution, Chapter 4 in this volume.
26. Richelson, *America's Secret Eyes in Space*, 92–143; Siddiqi, 'Soviet Space Power,' 144–5. Seeing what the "other" side was doing took the superpowers into strange territory, including the employment of alleged psychics who might explore things seen in reconnaissance photographs. See Anthony Enns's contribution, Chapter 10 in this volume.

27. Stares, *Militarization of Space*, 97–9, 159–60; Dwayne A. Day, 'Behind the Blue: The Unknown U.S. Air Force Manned Space Program,' in Gillespie and Weller, *Harnessing the Heavens*, 83–93; Siddiqi, 'Soviet Space Power,' 147–8. For more on the Almaz stations, see Cathleen Lewis's contribution, Chapter 12 in this volume.
28. Gaddis, 'The Long Peace'; and idem, *The Long Peace*, 195–245, 291–302; Westad, *Global Cold War*.
29. Richelson, *America's Space Sentinels*.
30. Pavel Podvig, 'History and the Current Status of the Russian Early-Warning System,' *Science and Global Security* 10.1 (February 2002), 21–60; 'US-KM and US-KMO Constellations,' http://www.russianspaceweb.com/oko.html.
31. Richard D. Easton and Eric F. Frazier, *GPS Declassified: From Smart Bombs to Smartphones*, Lincoln: Potomac Books, 2013, 25–9, 42–55. See esp. Paul Ceruzzi's contribution, Chapter 13 in this volume.
32. Easton and Frazier, *GPS Declassified*, 57–107; Podvig, 'Russian Early-Warning System,' 31; Ceruzzi, Chapter 13 in this volume; https://en.wikipedia.org/wiki/Global_Positioning_System.
33. Almost the only general study of SDI, but one that is largely restricted to Reagan and his inner circle, is that of Frances FitzGerald, *Way Out There in the Blue: Reagan, Star Wars and the End of the Cold War*, New York: Simon & Schuster, 2000. Roger Handberg, *Seeking New World Vistas: The Militarization of Space*, Westport: Praeger, 2000, 63–85, gives an overview of SDI's impact on US military space policy. See also Alexander Geppert and Tilmann Siebeneichner's introduction, Chapter 1 in this volume.
34. Stares, *Militarization of Space*, 180–243; Westad, *Global Cold War*, 250–363; and Siddiqi, 'Soviet Space Power,' 145–6.
35. FitzGerald, *Way Out There*; William Broad, *Star Warriors: A Penetrating Look into the Lives of the Young Scientists Behind Our Space Age Weaponry*, New York: Simon & Schuster, 1985.
36. Pavel Podvig, 'Did Star Wars Help End the Cold War? Soviet Response to the SDI Program,' 17 March 2013, http://russianforces.org/podvig/2013/03/did_star_wars_help_end_the_col.shtml; Peter J. Westwick, '"Space-Strike Weapons" and the Soviet Response to SDI,' *Diplomatic History* 32.5 (November 2008), 955–79; and Siddiqi, 'Soviet Space Power,' 144–5. See also Westwick's 'From the Club of Rome to Star Wars: The Era of Limits, Space Colonization, and the Origins of SDI,' in Geppert, *Limiting Outer Space*, 283–302.
37. Manno, *Arming the Heavens*, 82–5; Bernd Mütter, '*Per Media Ad Astra?* Outer Space in West Germany's Media, 1957–1987,' in Geppert, *Imagining Outer Space*, 165–86.
38. McLean, *Western European Military Space Policy*, 40–54; Johannes M. Becker, 'EUREKA: Eine westeuropäische Antwort auf SDI,' in Bernd W. Kubbig, ed., *Die militärische Eroberung des Weltraums*, Frankfurt am Main: Suhrkamp, 1990, vol. 2, 616–43; idem, 'Die SDI-Rahmenvereinbarungen zwischen Bonn und Washington: Eine erste Bilanz,' in ibid., vol. 2, 644–719; Sean N. Kalic, 'Reagan's SDI Announcement and the European Reaction: Diplomacy in the Last Decade of the Cold War,' in Leopoldo Nuti, ed., *The Crisis of Détente in Europe: From Helsinki to Gorbachev, 1975–1985*, London: Routledge, 2009, 99–110; and Geoffrey Lee Williams and Alan Lee Williams, *The European Defence Initiative: Europe's Bid for Equality*, New York: St. Martin's Press, 1986, esp. chapter 13,

'European Defence and the Strategic Defence Initiative,' 200–12. See also Michael Sheehan's contribution, Chapter 4 in the present volume.

39. Cheng, 'The Long March,' 161; Everett C. Dolman, 'Astropolitics and *Astropolitik*: Strategy and Space Deployment,' in Gillespie and Weller, *Harnessing the Heavens*, 111–33, but also the comment by Roger D. Launius, 'National Security,' 12–18, in the same volume.
40. McPeak quoted in David N. Spires, 'Recurring Themes in U.S. Air Force Space History,' in Gillespie and Weller, *Harnessing the Heavens*, 95–109, here 100. See also Handberg, *Seeking New World Vistas*, 87–107.

CHAPTER 3

The Nuclear Roots of the Space Race

Christopher Gainor

Humankind's reach into space in the 1950s and 1960s has often been presented as being based on the work of space pioneers from various nations, and linked to the human desire to explore and colonize. This argument, which was advanced during that time by participants and promoters of space exploration, has obscured the fact that the beginnings of space exploration were based on powerful rockets that were originally built for another purpose – to carry weapons, especially nuclear weapons, over long distances as part of the Cold War confrontation between the United States and the Soviet Union. Historically, the military played a major role in moving space exploration from the realm of dreams and drawings to real spacecraft carrying humans and their creations into outer space.

This chapter examines how the creation of intercontinental ballistic missiles (ICBM), originally developed to carry nuclear weapons, opened the door to space exploration when these missiles were pressed into service as space launch vehicles. By implication, acknowledging the central role of ICBMs in the early Space Race of the Cold War period means we must reconsider the place of the military in the history of space exploration. Historians are taking into account new perspectives in space history, such as those expressed by critics of spaceflight or the environmental movement. These new views go beyond the traditional narrative developed by space exploration promoters that holds that spaceflight is just the latest expression of the human desire to explore.[1] This chapter also discusses how similar missiles and space launch vehicles have been developed by countries

Christopher Gainor (✉)
Sidney, BC, Canada
e-mail: cgainor@shaw.ca

Alexander C. T. Geppert et al. (eds), *Militarizing Outer Space*
European Astroculture, vol. 3
https://doi.org/10.1057/978-1-349-95851-1_3

in Europe and Asia beyond the two superpowers. It highlights new scholarship that focuses on the history of ICBMs from a military viewpoint, and that has upended long-held assumptions about ICBMs. These include beliefs that the Soviet Union had begun to develop these missiles when the Second World War came to an end and that the United States should also have done so, as well as the idea that the Soviet Union held the clear lead in ICBMs in the late 1950s and early 1960s. The first ICBMs quickly proved to be dual-use devices whose second use – as space launch vehicles – overshadowed their original military purpose, at least in the eyes of the public. While ICBMs were less prominent as part of nuclear weapons systems than they might have been, the misleading ideas about their development mentioned above have caused them to be seen more as vital parts of space programs than as weapons.

I Military needs and spaceflight dreams

To understand how missiles were viewed during the time of the Cold War, one must examine how space exploration arose earlier in the twentieth century. Early historical treatments of the history of spaceflight came from promoters and participants in the early US space program, notably people such as Wernher von Braun (1912–77; Figure 3.1), who led the German effort to create the first long-range ballistic rocket missile, the A4, and then held top positions in the US Army missile program and the National Aeronautics and Space Administration (NASA), and Willy Ley (1906–69), who wrote extensively about space exploration in his native Germany and later in the United States.[2]

The idea of space travel began to gain popularity in the nineteenth century due to fictional accounts such as those of Jules Verne (1828–1905), and enthusiasts in many countries began to develop methods of making these speculations a reality. Among them were two school teachers who developed theories that were applied to space exploration, the Russian Konstantin E. Tsiolkovsky (1857–1935) and the Transylvanian-German Hermann Oberth (1894–1989), along with American physicist Robert H. Goddard (1882–1945). Early in the twentieth century, Tsiolkovsky published his ideas in obscure Russian journals, and he was later credited for inspiring spaceflight enthusiasts in the early years of the Soviet Union. When Goddard declared in a 1920 paper that rockets could fly to the moon, the resulting publicity reverberated around the world. About that time, Oberth began publishing his ideas about space travel that inspired many people who joined rocket societies in many countries. The Verein für Raumschiffahrt, which was founded in 1927, drew von Braun and many other German space enthusiasts. During the height of the competition in space between the United States and the Soviet Union in the 1960s, the idea that these three theorists led the drive into space gained currency amongst spaceflight advocates and historians.[3]

Figure 3.1 Wernher von Braun (1912–77), the rocket pioneer who led the German Army rocket program that created the V-2 missile in the Second World War, is shown here celebrating in 1958 shortly after his US Army team launched the United States' first satellite, Explorer 1. Other members of the German rocket team worked in Soviet and French missile programs after the war.
Source: *Der Spiegel* 12.7 (12 February 1958), cover image. Courtesy of Spiegel-Verlag.

There were also several other spaceflight theorists and pioneers before the Second World War, notably Robert Esnault-Pelterie (1881–1957) of France, Herman Potočnik (1892–1929) of Slovenia, Eugen Sänger (1905–64) of Austria, Walter Hohmann (1880–1945) of Germany, and Friedrich Zander (1887–1933) and Yury Kondratyuk (1897–1942) in the Soviet Union. But the American and Soviet rocket and space programs arose after the Second World War mainly due to the knowledge gained by the German Army rocket program of the Nazi period that created the A4, which first flew in October 1942 and entered operational service in 1944 under the propaganda name V-2 (for *Vergeltungswaffe*, or Vengeance Weapon). That program had begun in 1932, when von Braun and others were recruited by the German Army rocket program and quickly found that they had to postpone or forget their dreams of space travel. Instead, they built rockets designed purely to project military power in the form of the world's first long-range ballistic missile armed with explosive warheads.[4] Many other new weapons and military technologies were developed during the Second World War, most importantly the atomic bomb. A gigantic effort by scientists and engineers supervised by the US Army Corps of Engineers culminated in the detonation of the first atomic or fission bomb in July 1945 and its use against Japan the following month.[5]

After the war, a number of countries began to develop long-range rockets for use as weapons and also to conduct scientific research high in earth's atmosphere and in outer space. The United States and the Soviet Union had been allies in the Second World War, but after the war their ideological differences and political confrontations in central Europe and elsewhere caused what has become known as the Cold War. The two superpowers built nuclear weapons and bomber aircraft to carry them in the early years of the Cold War, and in 1957 both began testing the first ICBMs that were capable of launching nuclear weapons to targets thousands of kilometers away. These missiles could also launch payloads into orbit around the earth and beyond, as the Soviet Union most dramatically demonstrated on 4 October 1957 when it placed the first artificial satellite, Sputnik 1, into earth orbit. Sputnik provided graphic proof that the Soviets possessed an ICBM, a fact that caused great controversy in American political circles, with Senate Majority Leader Lyndon B. Johnson (1908–73), a political adversary of President Dwight D. Eisenhower (1890–1969), comparing the Russian satellite to Japan's attack on Pearl Harbor in 1941.[6] Sputnik and the reaction to it strongly affected how historians viewed the development of space exploration for many years.[7]

The Soviet satellite marked the beginning of a competition in space between the United States and Soviet Russia that lasted until the end of the Cold War. Using the R-7 ICBM that was built so big that it was far more useful as a space launch vehicle than as a military missile, the Soviets also launched the first living being into space, the dog Laika on Sputnik 2, the first vehicle to go into orbit around the sun, the first probe to strike the moon and the first spacecraft to photograph the far side of the moon, amongst other space 'firsts.' The first human to fly into space, Yury Gagarin (1934–68), rode a modified R-7 into orbit in 1961. The string of Soviet space achievements culminating with Gagarin's flight sparked US President John F. Kennedy (1917–63) and the US Congress to proceed with a program to land American astronauts on the moon and recover America's status as the world's leading technological power. The 1960s competition in space between the United States and the Soviet Union led to the landing of twelve US Apollo astronauts on the moon between 1969 and 1972.[8]

The military background of the R-7 was well known and indeed helped fuel the concerns that underlay the political controversy in the United States following the launch of Sputnik. While many decisions about space programs were made by the top Soviet authorities, the program was run by the Ministry of Defense. This led to decisions such as the Soviet Vostok spacecraft that carried Gagarin and other early cosmonauts into space being designed and built in parallel with Zenit military reconnaissance satellites, and the creation of military space stations in the 1970s. But the extent of military control of the Soviet space program was kept under wraps until Russian archives were opened up in the late 1980s as the Cold War wound down.[9] The space successes at the beginning of the Space Race were presented by Soviet authorities as a scientific triumph, when in fact the political leaders of the time had been indifferent to space and the early space successes had been created by space enthusiasts working in the military-industrial complex.[10] The United States also relied on military ICBMs to launch most of its civilian and military satellites and spacecraft through the end of the twentieth century. In the months following the launch of Sputnik, Eisenhower and the US Congress decided to create NASA to run America's civilian space program, which would launch scientific and non-military remote sensing space vehicles, along with the US human space program. NASA had a much higher public profile than the secretive military space programs run out of the Pentagon. Especially with human space efforts such as the Space Shuttle, NASA found itself operating spacecraft and programs that served both military and non-military purposes.[11]

II ICBMs around the world

A fresh look at the United States Air Force's (USAF) treatment of bomber aircraft and ballistic missiles between 1945 and 1954 shows that the USAF and American leaders did not see a need for ICBMs during that time. While the US military was able to deliver two nuclear bombs to targets in Japan in August 1945 using B-29 bombers based on an island relatively close to Japan, many challenges stood in the way of creating a bomber force capable of dropping nuclear bombs on targets in the Soviet Union. New bomber aircraft with longer ranges needed to be developed to bridge the great distances between the United States and the Soviet Union, forward bases in Europe and Asia were established, aircraft crews required high standards of training, and aerial refueling needed to be advanced to extend the range of bomber aircraft. Moreover, the fission or atomic bombs developed in the Second World War and used against Japan were in very short supply until 1948, and they were difficult to arm. At the end of the Second World War, political leaders did not see a need to have a capability to deliver nuclear bombs to the Soviet Union, which had been America's ally in the war. But as Cold War confrontations between the two sides gained in urgency, notably when the Russians blockaded West Berlin in 1948 and then exploded their own nuclear bomb in 1949, American policy-makers joined military officials in recognizing the strategic value of being able to bomb the Soviet Union if war broke out between the two superpowers. President Harry S. Truman (1884–1972) had declined to place nuclear weapons under military control until the deepening Cold War caused him to change his mind in 1951.[12]

Although promoters of rockets in the late 1940s saw the potential of developing long-range ballistic missiles to carry nuclear weapons to distant targets, the V-2's drawbacks as a weapons system, including its poor accuracy and great expense, caused US military and scientific leaders to believe that a workable long-range nuclear-armed ballistic missile was still years off. Vannevar Bush (1890–1974), the American engineer who directed US military research and development efforts during the Second World War and the early postwar years, became a well-known critic of missiles. The great difficulties involved in building piloted aircraft to fly to targets thousands of kilometers away put the bigger problems involved in creating robotic missiles to accurately hit these distant targets into sharp focus. As well, the five years between the Second World War and the Korean War were a time of retrenchment in the US military, and the Air Force's early missile programs, including the forerunner program to Atlas, the MX-774, were affected by cutbacks.[13]

The first explosion of a thermonuclear bomb took place on the Pacific Island of Eniwetok on 1 November 1952, unleashing a force of more than 800 times the power of the fission bomb dropped on Hiroshima. This explosion was the culmination of a highly controversial development program authorized by President Truman on 31 January 1950 amidst political and scientific conflict between the physicists who had created the first nuclear weapon in the Second World War. Although the first thermonuclear bomb weighed 74 tons, physicists were confident that a new design would soon lead to thermonuclear bombs that weighed far less than the fission bombs that had been used in 1945 and built in the seven years that followed. The new lightweight design was proven in a test explosion on 1 March 1954.[14]

With the creation of lightweight thermonuclear bombs, two major problems with the idea of ICBMs were sharply reduced – the great size and lifting power needed for early nuclear weapons, and the stringent accuracy requirements to make effective use of fission weapons. Officials in the newly installed Eisenhower administration who knew of the potential of long-range missiles carrying thermonuclear weapons pressed for creation of an ICBM in 1953. When a committee of top experts headed by mathematician and physicist John von Neumann (1903–57) called in their February 1954 report for the development of an ICBM, the US Air Force moved ahead on Atlas a few weeks later. The report noted that the massive power of thermonuclear weapons meant that accuracy requirements for ICBMs could be relaxed, and a RAND Corporation report made available to the committee and government leaders reiterated the same point.[15]

At the time and later, Air Force leaders such as General Bernard Schriever (1910–2005; Figure 3.2), who directed the creation of Atlas, Titan, Thor and other ballistic missiles, stressed the importance of what he called the 'thermonuclear breakthrough' in reducing the size and accuracy requirements for ICBMs in leading the USAF to these powerful missiles. Schriever testified to Congress in 1956 that the limited explosive power of early nuclear weapons did not suggest that an ICBM would be 'a particularly useful military weapon' because it would require highly accurate guidance systems that were not available at the time and an extremely large missile. He repeated the same point in later testimony.[16] The impact of lightweight thermonuclear bombs can be shown in the size of ICBMs: A 1952 Convair design for the Atlas ICBM carrying a four-ton warhead weighed 300 tons, stood 49 meters tall and had seven engines. This design was much larger than any American ICBM ever built, including the Atlas, which when completed weighed 110 tons, stood 23 meters tall and used only three engines of the same type. Atlas was designed to carry a lightweight 1.5-ton thermonuclear warhead.[17] Evidence emerged after the end of the Cold War that the Soviet government gave the go-ahead to its first ICBM, the R-7, at roughly the same time in the spring of 1954 for the same reason the US Air Force did – the military potential of thermonuclear weapons carried by long-range missiles.[18]

Both Atlas and the R-7 began test flights in 1957, with the Soviet flights taking place in secret. An R-7 lofted Sputnik into orbit that year, while the USAF deliberately held Atlas away from launching satellites until it had proven itself in test flights in 1958. The failure of a much smaller US Vanguard rocket to put a satellite into orbit in December 1957 contributed to the sense of crisis in the United States that followed Sputnik, and the subsequent success by von Braun's group in placing the first US satellite into orbit in January 1958 fostered the misleading idea that von Braun and his team were at the center of US military missile programs (see Figure 3.1 above). Because the R-7 was designed at a time when the Soviet authorities anticipated a need to launch heavier warheads, it could launch larger payloads into space than Atlas or any other American rocket of the 1950s. The series of successful 'space firsts' in the late 1950s and early 1960s using the R-7 bolstered the misleading idea that the Soviets were ahead of the United States in missile technology.[19]

Figure 3.2 US Air Force General Bernard Schriever (1910–2005), who directed the effort that led to the creation and deployment of the Atlas and Titan ICBMs and the Thor intermediate-range ballistic missile (IRBM), with an Atlas ICBM inside an early missile silo. Schriever was focused on creating missiles to carry nuclear warheads and held back Atlas from use as a space launch vehicle.
Source: Courtesy of National Museum of the US Air Force.

The R-7 was an efficient space launch vehicle, and derivatives of the R-7 remain in use to the present day. But because of its great size, it was of limited use as an ICBM. Launch pads for the R-7 were so large and expensive that only six were built, which meant that only six of these missiles could be put on alert at any one time. The Atlas and Titan vehicles also had limitations as ICBMs, but their smaller size allowed the United States to field these ICBMs sooner and in greater quantity than the Soviet Union, a point that was driven home during the Cuban Missile Crisis of 1962. At that time, the United States could strike the Soviet Union with about 4,000 nuclear warheads, mostly from bomber aircraft, but including 179 ICBMs and more than 100 submarine-launched ballistic missiles. The Soviets could reply with only about 220 nuclear warheads, including just 20 from ICBMs. In spite of these facts, the idea that Soviet Russia had beaten the United States to a workable ICBM lasted through the end of the Cold War.[20]

The Atlas and the R-7 were the first ICBMs put into operational service, with the first Atlas ICBMs entering operational status in September 1959. In 1962 both the Soviets and Americans began deploying ICBMs using liquid fuels that ignited on contact and did not require the cryogenic oxidizers that the Atlas and R-7 did, marginally easing fueling issues. The R-7 was quickly removed from ICBM duty when the new missiles came on stream, and Atlas was stood down from ICBM use by 1965. Also in 1962, the USAF began deploying solid-fuel Minuteman ICBMs, and today advanced versions of the Minuteman ICBM remain on duty. Russia also developed solid-fuel ICBMs. They were far superior to their liquid-fuel predecessors as ICBMs because they could be kept on station for long periods of time and were far easier to maintain. Similar missiles were also built for submarine-launched ballistic missiles. Since that time the American and Russian militaries have maintained strategic nuclear forces that include nuclear weapons that can be delivered long distance by bomber aircraft, ICBMs and submarine-launched ballistic missiles.[21]

The persistence of incorrect ideas that arose in the Sputnik crisis about the superiority of Soviet missile forces helped maintain support for military programs in the United States through the Cold War, notably in the 'missile gap' controversy of the 1960 presidential election, and again in the 1980s, when conservatives aligned with President Ronald Reagan (1911–2004) attacked arms control treaties and promoted more defense spending and programs such as anti-ballistic missile schemes. By then, the fears that many Americans had expressed following Sputnik were updated with fears about Soviet ICBMs built and deployed in the 1960s and 1970s.[22] Aside from these controversies, rockets developed to be ICBMs gained greater prominence in their role as space launch vehicles. In his study of popular fear related to nuclear arms and nuclear energy, historian Spencer R. Weart makes few references to nuclear-armed missiles and instead examines the angst caused by nuclear bombs and nuclear reactors. He suggests that the decline of the anti-nuclear movement in the early 1960s took place because of the Limited Test Ban Treaty of 1963, symbolizing a relaxation of the Cold War, or because of a 'catharsis' caused by the Cuban Missile Crisis.[23] The early 1960s were also the time of the most popular fictional depictions of Cold War nuclear confrontations including *Fail Safe* (1962), a novel by Eugene Burdick and Harvey Wheeler that was also made into a movie, and the famous black comedy film, Stanley Kubrick's *Dr. Strangelove or: How I Learned to Stop Worrying and Love the Bomb* (1964), which focused on bomber aircraft as the delivery vehicles for nuclear bombs.[24]

By the 1960s both the United Kingdom and France were bringing Europe into the ranks of powers with nuclear weapons and space programs, and like the two superpowers, French and British space programs followed on their nuclear ambitions. In the 1950s Europe was still emerging from the destruction and disruption of the Second World War. While the launch of Sputnik heralded the possibilities of space exploration along with concerns about the military implications of ICBMs, these European countries had to make decisions about the benefits of their nuclear weapons and their space technologies versus their high cost.[25]

The United Kingdom began work on its own fission bombs soon after the end of the Second World War because of its own Cold War concerns about Soviet forces in Europe. The United Kingdom had cooperated with the United States and Canada during the war in the Manhattan Project, which developed the first nuclear bombs. But the United Kingdom felt obliged to develop nuclear weapons on its own after the American government declined to share information gained in the Manhattan Project. The United Kingdom exploded its first fission bomb in October 1952, not long before the United States exploded the first thermonuclear weapon. In 1954 the British government embarked on developing its own thermonuclear weapons. To deliver nuclear weapons, the United Kingdom had begun designing its V bombers in the late 1940s. In the spring of 1954, when the US government was beginning work on its Atlas ICBM, it suggested that the British build an intermediate-range ballistic missile (IRBM), a missile that would give the United Kingdom the ability to strike targets in the Soviet Union with thermonuclear weapons. The British soon began to develop its Blue Streak IRBM. Contrary to earlier indications, the Americans began building their own IRBMs. Questions about the cost of Blue Streak and space technologies such as reconnaissance satellites at a time when the country was still recovering from the war caused vigorous debate inside the British government. After accepting an American offer to use the Thor IRBM, which was developed by the USAF, the British government concluded in 1960 that Blue Streak would not meet its needs as a nuclear weapons delivery vehicle. After Thor was retired and the US government cancelled an air-launched missile that the British hoped to arm with its own nuclear weapons, the United Kingdom ceased to use land-based strategic missiles with its nuclear weapons and purchased American submarine-based strategic missiles, Polaris and later Trident.

The United Kingdom also chose to accept American offers of cooperation in its early satellite program as it decided to have only a modest space program.[26]

The British government was reluctant to totally stop work on Blue Streak because to do so would open the government to charges that the money spent on the missile had been a waste and that the United Kingdom's ability to conduct space research would be dependent on the United States. Cancellation could also be seen as a blow to the United Kingdom's position as a great power, and so it continued work on the missile and began discussions first with the French government and then the German and Italian governments about building a European satellite launcher. Blue Streak became the first stage of the Europa launch vehicle being developed by the European Launcher Development Organization (ELDO), which was established in 1962. Europa had a French second stage and a third stage from Germany, along with an Italian satellite test vehicle. But as the troubles for Europa mounted, and the United Kingdom's hopes to join the European Common Market faded in the face of France's veto, the British support for ELDO faltered in 1966, and the United Kingdom left the field in 1969.[27]

France had begun small-scale work on rockets in the late 1940s, helped by a limited number of rocket experts from the German Army's Second World War rocket program. In 1956 it launched a nuclear weapons program. France's military and technological ambitions grew greatly after Charles de Gaulle (1890–1970) became president in 1958. De Gaulle, who wished to re-establish France as a leading political, military and economic power, pushed for development of the French *force de frappe* with nuclear weapons mounted on missiles. In 1960 France began developing its own family of 'precious stones' rockets that led to French ballistic missiles and included the satellite launcher Diamant, starting to launch in 1965. The French government agreed in 1961 to join ELDO in part because it believed that it could gain knowledge from the British work on Blue Streak for both its own strategic missiles and its budding space ambitions. ELDO continued after the British withdrawal, but continued failures of the Europa rockets built by ELDO led to the organization and its rockets being wound up when the European Space Agency (ESA) was formed in 1975. The French government decided to build on its work on Europa, Diamant and military rockets, and with cooperation from other European countries, this work led to the European launch vehicle Ariane, which first flew in 1979. France was interested in ensuring that it had an independent capability of launching satellites into space, and it also saw space as an arena for economic growth. But its work on missiles also had a military dimension. Building on the 'precious stones' rockets, France used land-based strategic missiles during the Cold War, and its submarine-based strategic missiles were first deployed during the Cold War period.[28]

After the long period of growth for ICBM and submarine-launched ballistic missile forces in Russia, the United States, the United Kingdom and

France in the 1960s and 1970s, the nuclear disarmament movement revived in the 1980s. The resurgent peace movement was sparked by differences over intermediate-range missiles in Europe, the election of Ronald Reagan as US president and his subsequent effort to build an anti-ballistic missile defense shield known variously as the Strategic Defense Initiative (SDI) and 'Star Wars.'[29] While bombers took center stage in the public imagination and shorter-range missiles and defenses against ICBMs caused controversy, ICBMs seemed to be an accepted part of the nuclear landscape, except in the minds of Reagan and some of his advisors, who pushed aggressively for SDI and then proposed the elimination of these missiles at the Reykjavik summit in 1986.[30]

By then, new nuclear powers with space programs were emerging in Asia. China had its own unique role in the Cold War, first siding with the Soviet Union and then playing the United States and the Soviets against each other. It joined the ranks of nuclear powers with its first nuclear weapons test in 1964, and followed with the beginning of a space program with its first satellite launch in 1970. Although China's prominence in both fields faded during the period of the Cultural Revolution in the early and mid-1970s, China began to compete for a share of the commercial launch market in the 1990s. For many years, China's space and nuclear weapons programs operated under a high level of secrecy. China developed launch vehicles after building nuclear weapons and liquid-fuel missiles to carry them, including the DF-5A, an ICBM that is also used as a satellite launcher.[31]

Two other powers with space programs that include launch vehicles differ from the pattern seen with the United States, Russia, Europe and China, with clear relationships to nuclear arms. Japan has a strong space program with its own launch vehicles, but as the only country to undergo an attack with nuclear weapons, it has never possessed weapons of this type. However, Japan has developed space launch vehicles based on the USAF Thor missile, which was originally built to carry nuclear warheads.[32] India, which engineered nuclear weapons starting in the 1970s and then solid-fuel rockets to launch them, was not directly involved in the Cold War but has a history of confrontation with Pakistan. India has also built solid-fuel satellite launch vehicles, but its current generation of space launch vehicles, which were developed with assistance from Russia, use liquid fuels supplemented by solid-fuel rockets. It is marketing launch services using its new space launch vehicles. Because of political and military tensions between India, Pakistan and China, India has tried to play down the ties between its nuclear forces and its space program and emphasized the economic aspects of going into space.[33]

To make ICBMs effective as weapons, nuclear powers had to build infrastructures not only on earth but also in space. The United States and Russia began building and launching reconnaissance satellites that provided imagery of ICBMs and other military infrastructure belonging to adversaries. Other military satellites included mapping satellites to help with ICBM targeting, weather satellites to support military operations, geodetic research satellites that helped determine the earth's correct shape to refine ICBM targeting, early warning satellites to detect nuclear explosions, missile launches, radars, radio and television transmissions, dedicated communications and navigation satellites, amongst others. Other space weapons included anti-ballistic missile systems and anti-satellite systems. The United Kingdom, France and China have also developed military space programs, often to support command and control functions. The creation of ICBMs led to the creation of military space programs and the militarization of space, and in many cases the military assets that support ICBMs from orbit were placed there by ICBMs converted to space launch vehicles.[34]

Liquid-fuel ICBMs such as Atlas and Titan found a second purpose as space launch vehicles in the civilian NASA space program, along with other rockets originally developed for military uses such as Delta, and once they had been stood down as ICBMs in favor of solid-fuel missiles, they were used exclusively to launch payloads into space. These missiles are a great example of dual-use technology but far from the only one in military and civilian space programs. Weather, communications and other types of satellites are useful both in military and civilian space programs. NASA has always had a strong tradition of cooperation with the US military, including carrying military experiments and personnel aboard its spacecraft, and military personnel have played important roles in many NASA programs, including Apollo and the Space Shuttle. Indeed, many scholars argue that the US civilian space program itself was another arena in the Cold War, particularly in regard to Kennedy's decision to send astronauts to the moon.[35]

III Changing historiographical perceptions of ICBMs

Historians are re-examining the early history of the US Air Force and long-range missiles in the late 1940s and early 1950s. Just like Sputnik and many other Russian spacecraft were launched on the R-7 ICBM, many American satellites, space probes and astronauts were launched on top of Atlas and Titan rockets, which were originally developed as ICBMs for the USAF. In spite of their fame as rocket builders in the United States during these postwar years, von Braun and his group were not directly involved in the design or development of ICBMs. Instead, in the 1950s they built the Redstone medium-range ballistic missile and the Jupiter IRBM for the US Army. A derivative of the Redstone rocket launched America's first satellite, Explorer 1, in January 1958 (see Figure 3.1 above). Redstones also launched two US astronauts on suborbital flights in 1961. But along with the Delta launch vehicle, which grew out of the Air Force's Thor IRBM, Atlas and Titan became the primary launch vehicles of the US space program from 1958 through the balance of the twentieth century. While these rockets were originally built for the Air Force and some were used for military satellite launches, large numbers were launched for NASA, and some orbited payloads for space agencies from Canada, Europe and elsewhere, and for corporate customers from many countries. Atlas was America's first ICBM, and it was used to launch the first American astronauts to go into orbit (Figure 3.3), many early US satellites, and the first American space probes to go to the moon and many planets in the solar system.[36]

The USAF's decision in 1954 to proceed with Atlas was one of the key decisions leading to America's reach into space. Yet for many years, the only analytical works to examine this decision in detail were two 1960s papers by former air force historian Robert L. Perry and political scientist Edmund Beard's influential 1976 book, *Developing the ICBM: A Study in Bureaucratic Politics.*[37] These studies were written in the shadow of the political controversy in the United States that followed Sputnik, as evidenced by Beard's assertion that the Soviet Union had defeated the United States in the race to develop the first ICBMs, based on the idea that the orbiting of Sputnik had proven that the Soviet Union had a workable ICBM. In fact, a top American official admitted in 1961 that the United States was well ahead of the Soviet Union when it came to ICBMs.[38] While ICBMs have the power to place satellites into orbit, the fact that the Soviets had a rocket that could orbit satellites did not mean that they had a functional ICBM. The R-7's great size, which made it an effective space launch vehicle, also limited its military utility because of the gigantic launch complexes it required, a fact which made it nearly impossible to protect during launch preparations. As well, evidence emerged after the Cold War that R-7 had guidance problems, and its test flights failed to deliver dummy warheads to their targets because of problems with the re-entry vehicles.[39]

According to Beard, the United States could have developed its first ICBM considerably earlier than it did, but waited until 1954 to begin its ICBM program while the Soviet Union began work on its ICBM in 1946.[40] Both Perry and Beard criticized the US Air Force for not beginning serious development work on ICBMs before 1954, due to what Perry called the Air Force's 'complacency and self delusion,' and Beard termed its 'neglect and indifference' to ICBMs and a strong cultural bias in favor of bomber aircraft. Beard added: 'My opinion is that the United States could have developed an ICBM considerably earlier than it did but that such development was hindered by organizational structures and belief patterns that did not permit it.'[41] Both compared the USAF's supposed foot dragging on ICBMs to the US Navy's resistance to new gunnery methods at the beginning of the twentieth century, with Perry asserting that the USAF showed 'cultural resistance' to ICBMs.[42] Unstated but strongly implied in these texts was the idea that the immediate need for long-range ballistic missiles topped with nuclear weapons should have been blindingly obvious to US military leaders as soon as the first nuclear weapons were tested in 1945. Their beliefs about the early Soviet jump on developing ICBMs was knocked down by scholarship that emerged

Figure 3.3 A French Véronique scientific research rocket being prepared for launch. Starting with help from German rocket experts, the French government built Véronique rockets to improve their knowledge of rocket technology for use as military missiles, research rockets, and eventually in space launch vehicles such as Ariane. Véronique was famously used to launch the cat Félicette on a suborbital flight in 1963.
Source: Courtesy of Centre national d'études spatiales, 1962.

after Soviet archives opened in the late 1980s showing that the Soviet ICBM program also began in 1954, within a few weeks of the start of serious development work on Atlas.[43]

Early treatments of spaceflight history underemphasized the role of the USAF in the development of US rockets. Especially at the height of the Space Race, little was said about the history of Atlas, in part because of the deliberate low profile adopted by the Air Force. One popular book, *Atlas: The Story of a Missile*, appeared in 1960.[44] Aside from Perry's early paper, the first discussion of Atlas by historians was found in an official NASA history of Project Mercury, NASA's first human space program, and this book related the creation of Atlas to thermonuclear weapons.[45] Edmund Beard in *Developing the ICBM* dismissed the thermonuclear warhead advance as a 'false issue' because plans existed earlier to mount nuclear warheads on cruise missiles and because of incremental improvements to fission weapons in the late 1940s, an analysis that dismissed the major increase in power between the two classes of weapons.[46]

Despite its many deficiencies, Beard's *Developing the ICBM* remains a groundbreaking work in the academic study of missiles because it was the first book in this subject area to try to put these weapons into their social context and away from the all-too-common concentration on the artifact. Historians sensitive to the role of political forces in driving technology argued, as Beard did, that the arrival of the Eisenhower administration in 1953 caused the Air Force to proceed with ICBMs. These scholars included historian Walter A. McDougall, Rip Bulkeley, who wrote a book on early US space policy, former RAND analyst Carl H. Builder, and the distinguished historian of technology, Thomas P. Hughes, who suggested in *Rescuing Prometheus* that 'a conservative momentum, or inertia' involving both institutions and hardware set the pace of the Air Force's ICBM program.[47] While social, economic and political factors were indeed very important influences in the decisions leading to the creation of Atlas and other ICBMs, these works did not consider the wider political considerations affecting nuclear weapons in the late 1940s or the problems the USAF faced in making its bomber force effective. Military weapons programs during the Cold War usually had little trouble gaining support from US military leaders, politicians and defense contractors once a persuasive case for their utility as weapons was advanced. In the case of ICBMs, the creation of thermonuclear weapons persuaded decision-makers that they would be effective weapons.[48]

Not all historians of ICBMs followed the lead of Perry and Beard, whose analyses came to be treated as historical fact. Historians who wrote mainly on military topics and were more attentive to the military needs that triggered creation of these missiles, including Jacob Neufeld, John Clayton Lonnquest, Robert Frank Futrell and Stephen B. Johnson, pointed to the thermonuclear breakthrough as the most important factor that caused the USAF and the Eisenhower administration to proceed with Atlas and later ICBMs.[49]

As the Cold War came to an end with the fall of the Soviet Union in 1991, historians began to view the history of the Space Race in new and more critical ways. While one cause was the revelations coming from Russia about the true nature of the Soviet space program, the changing views of the early reach into space broadened considerably. For example, historian Walter A. McDougall in 1985 famously called Sputnik – the event that initiated the Space Race – a 'saltation, an evolutionary leap' for humankind. Just twelve years later, in 1997, McDougall disavowed his assessment and instead said Sputnik was an 'ephemeral episode in the larger history of the Cold War.'[50] In other words, Sputnik, which rode into space on an R-7 ICBM, was a product of the larger military and nuclear standoff of the Cold War years. Other historians and scholars have re-evaluated Sputnik and its meaning, with many questioning whether the American reaction to Sputnik extended beyond the political, media and military fields.[51] They have taken fresh looks at the Sputnik crisis in the United States, sometimes offering conflicting viewpoints. For example, in his analysis of how Eisenhower used fears of nuclear war to advance his agenda, Ira Chernus has shown that the Sputnik controversy was the result of 'nearly five years of frightening Cold War rhetoric' from the Eisenhower administration, while Yanek Mieczkowski, in his sympathetic treatment of Eisenhower's response to Sputnik, did not raise this issue and noted that Eisenhower was often criticized for passivity.[52]

The wider changes to spaceflight history were labeled by Roger Launius as the 'New Aerospace History,' which aimed to move beyond concentration on individual rockets or spacecraft to wider social, political and cultural issues relating to aircraft, missiles and space vehicles. Launius and other historians also criticized what became known as the 'Huntsville School' of histories of American space travel and rocketry, which overstated the contribution of von Braun and his German rocket experts to the development of American rocketry during the Space Race and therefore underplayed the importance of the development of the Air Force's work developing the Atlas, Titan and Thor missiles.[53]

Works such as Howard E. McCurdy's *Space and the American Imagination* published in 1997 examined how promoters of spaceflight exploited Cold War fears of nuclear Armageddon to boost space exploration.[54] In 2001 historian Alex Roland called the post-Sputnik controversy in the United States a turning point in the military's quest for 'more and better weapons, expansion of roles and missions, and mobilization of the civilian economy in the service of the state.'[55] At the center of the Sputnik controversy is the fact that the rockets that launched Sputnik and other early spacecraft had been originally developed for military purposes, although this fact has often been neglected in historical literature in favor of emphasizing the role of pioneers such as Tsiolkovsky, Oberth and Goddard.

McCurdy noted that these pioneers were little known to the public, but space promoters in the rocket societies of the early and mid-twentieth century in Germany, the Soviet Union, the United States and elsewhere translated their ideas into appealing visions for the public. American promoters of space travel tapped into popular American ideals such as those dealing with the value of exploration and opening new frontiers. Some of the new frontiers include fictional views of humans in space offered by novels, films and television series. And the idea of space colonies at Lagrangian points near earth and elsewhere has inspired promoters of space exploration through much of the twentieth century. In recent years, scholars and activists have questioned and criticized the impacts of exploration and colonization in history, and McCurdy and others have asserted that colonization of outer space will not match the hopes of its promoters.[56] Here it is argued that the military realities of missiles had more to do with the rise of space travel in the 1950s and 1960s than the utopian visions of humanity moving into space.

Examining the nuclear-influenced motivations of the leaders who directed the creation of the first generation of space launch vehicles will help historians better understand the development of missile programs and space programs during their early years. The 2009 publication of a major popular biography of the man most closely related to the development of Atlas, Titan and Thor, General Bernard Schriever, shows that the Air Force roots of the US space program are gaining more prominence in the literature on the early history of space exploration.[57] ICBMs such as Atlas were originally developed by military forces for military purposes, and so the evolution of nuclear weapons from fission to thermonuclear warheads must be considered as a technical precondition to the development of programs that used ICBMs as space launch vehicles. The US and Soviet governments decided in the spring of 1954 to proceed with ICBMs due to the military possibilities created by thermonuclear weapons. Without ICBMs becoming available in 1957 and 1958, space exploration in the twentieth century would have likely evolved at a much slower pace simply because the large rockets needed to make that exploration possible would not have been created so quickly just for the sake of science or exploration. The Soviet choice to use its first ICBM to launch a satellite in 1957 and the American choice not to that year helped create the competition that climaxed with footprints on the moon.

The links between military and civilian applications for strategic missiles are obvious. The United States, Soviet Union, United Kingdom, France, China and India developed their own nuclear weapons and missiles to carry those weapons during the time of the Cold War. They also developed rockets to launch spacecraft. It is also clear that the economic impacts of outer space, including sales from building spacecraft and offering launch services, have driven the work of these countries to build launch vehicles, especially since the end of the Cold War. These countries have also created space programs to symbolize their places as major powers in the world. The story of ICBMs, however, adds to the literature on nuclear arms, which has emphasized the weapons themselves at the expense of delivery systems. The creation of these missiles to deliver nuclear weapons had implications outside the military realm because they facilitated the creation of space programs. The primary factor leading to space launch vehicles was nuclear weapons, not love of space travel.

Notes

1. See Roger D. Launius, 'Competing Rationales for Spaceflight? History and the Search for Relevance,' in Steven J. Dick and Roger D. Launius, eds, *Critical Issues in the History of Spaceflight*, Washington, DC: NASA, 2006, 37–76. All Internet sources were last accessed on 15 July 2020.
2. Among their works are Willy Ley, *Rockets, Missiles, and Men in Space*, New York: Viking, 1968; and Wernher von Braun and Frederick I. Ordway, *The History of Rocketry and Space Travel*, New York: Thomas Y. Crowell, 1969. For an account of von Braun's life, see Michael J. Neufeld, *Von Braun: Dreamer of Space, Engineer of War*, New York: Alfred A. Knopf, 2007.
3. Michael J. Neufeld, 'The Three Heroes of Spaceflight: The Rise of the Tsiolkovskii-Goddard-Oberth Interpretation and Its Current Validity,' *Quest: The History of Spaceflight Quarterly* 19.4 (2012), 4–13; Frank H. Winter, *Prelude to the Space Age: The Rocket Societies, 1924–1940*, Washington, DC: Smithsonian Institution Press, 1983. For more on Tsiolkovsky, see Asif A. Siddiqi, *The Red Rockets' Glare: Spaceflight and the Soviet Imagination, 1857–1957*, Cambridge: Cambridge University Press, 2010. There are several books on Goddard, including David A. Clary, *Rocket Man: Robert H. Goddard and the Birth of the Space Age*, New York: Hyperion, 2003; and Milton Lehman, *This High Man: The Life of Robert H. Goddard*, New York: Farrar, Straus, 1963.
4. See Neufeld, 'Three Heroes of Spaceflight'; Christopher Gainor, *To a Distant Day: The Rocket Pioneers*, Lincoln: University of Nebraska Press, 2008; and Alexander C. T. Geppert, 'Space *Personae*: Cosmopolitan Networks of Peripheral Knowledge, 1927–1957,' *Journal of Modern European History* 6.2 (2008), 262–86. Potočnik was also known by his pseudonym, Hermann Noordung.
5. See Richard Rhodes, *The Making of the Atomic Bomb*, New York: Simon & Schuster, 1986; and Francis George Gosling, *The Manhattan Project: Making the Atomic Bomb*, Washington, DC: US Department of Energy, 2010.
6. Walter A. McDougall, *...The Heavens and the Earth: A Political History of the Space Age*, New York: Basic Books, 1985, 152; Yanek Mieczkowski, *Eisenhower's Sputnik Moment: The Race for Space and World Prestige*, Ithaca: Cornell University Press, 2013, 140.
7. McDougall, *...The Heavens and the Earth*, 6, 152–3; Roger D. Launius, *NASA: A History of the U.S. Civil Space Program*, Malabar: Krieger, 1994, 24–8.
8. See Gainor, *To a Distant Day*; McDougall, *...The Heavens and the Earth*; William E. Burrows, *This New Ocean: The Story of the First Space Age*, New York: Random House, 1998; and Launius, *NASA*, 65.
9. Asif A. Siddiqi, *Challenge to Apollo: The Soviet Union and the Space Race, 1945–1974*, Washington, DC: NASA, 2000, 210–12, 250, 590–6. See also Cathleen Lewis's contribution, Chapter 12 in this volume.
10. Siddiqi, *Red Rockets' Glare*, 365–8.
11. McDougall, *...The Heavens and the Earth*, 195–209. See also James E. David, *Spies and Shuttle: NASA's Secret Relationships with the DoD and CIA*, Gainesville: University Press of Florida, 2015.
12. See Harry R. Borowski, *A Hollow Threat: Strategic Air Power and Containment before Korea*, Westport: Greenwood Press, 1982; David Alan Rosenberg, 'U.S. Nuclear Stockpile, 1945 to 1980,' *Bulletin of the Atomic Scientists* 38.5 (May 1982), 25–30; Noel Francis Parrish, *Behind the Sheltering Bomb: Military*

Indecision from Alamogordo to Korea, New York: Arno Press, 1979; and Office of the Assistant to the Secretary of Defense (Atomic Energy), *History of the Custody and Deployment of Nuclear Weapons July 1945 Through September 1977*, Washington, DC: Department of Defense, 1978.

13. Michael J. Neufeld, *The Rocket and the Reich: Peenemünde and the Coming of the Ballistic Missile Era*, New York: Free Press, 1995, 272–5; Vannevar Bush, *Modern Arms and Free Men*, New York: Simon & Schuster, 1949, 83–4; and Major General Benjamin W. Chidlaw to Commanding General, AAF, 'AAF Guided Missiles Program,' 6 May 1947, attached to Mary R. Self, *History of the Development of Guided Missiles, 1946–1950*, Dayton: Historical Office of the Air Material Command, 1951.
14. The story of the development of thermonuclear weapons is told in Richard Rhodes, *Dark Sun: The Making of the Hydrogen Bomb*, New York: Simon & Schuster, 1995.
15. See Jacob Neufeld, *Ballistic Missiles in the United States Air Force 1945–1960*, Washington, DC: United States Air Force History Office, 1990 (which includes the text of the von Neumann committee report); Bruno W. Augenstein, *A Revised Development Program for Ballistic Missiles of Intercontinental Range*, Santa Monica: RAND Corporation, 1954; and Christopher Gainor, 'The Atlas and the Air Force: Reassessing the Beginnings of America's First Intercontinental Ballistic Missile,' *Technology and Culture* 54.2 (April 2013), 346–70.
16. Bernard Schriever, 'Testimony (20 June 1956),' in *Senate Subcommittee on the Air Force of the Committee on Armed Services, Hearings: Study of Airpower (2nd Session, 84th Congress)*, Washington DC: US Government Printing Office, 1956, 1156; idem, 'Testimony (21 February 1958),' in *House of Representatives, Committee on Armed Services: Investigation of National Defense Missiles (2nd Session, 85th Congress)*, Washington DC: US Government Printing Office, 1958, 4852; and idem, 'Testimony (4 February 1959),' in *House of Representatives, Subcommittee of the Committee on Government Operations: Organization and Management of Missile Programs (1st Session, 86th Congress)*, Washington, DC: US Government Printing Office, 1959, 10.
17. Ross Johnston, 'Initial Presentation of Atlas Project Made at the 40th Meeting of the Committee on Guided Missiles,' 21–22 May 1952, Records Group 330, Office of the Secretary of Defense, Box 396, 42nd meeting of GM committee, file '237/GM,' US National Archives, College Park, MD.
18. Steven J. Zaloga, *Target America: The Soviet Union and the Strategic Arms Race, 1945–1964*, Novato: Presidio Press, 1993, 134–41; Siddiqi, *Red Rockets' Glare*, 270–8.
19. For more on Sputnik, see Roger D. Launius, John M. Logsdon and Robert W. Smith, eds, *Reconsidering Sputnik: Forty Years since the Soviet Satellite*, Amsterdam: Harwood, 2000.
20. See Neil Sheehan, *A Fiery Peace in a Cold War: Bernard Schriever and the Ultimate Weapon*, New York: Random House, 2009, 404–9; Zaloga, *Target America*, 150–60, 213–17; and Edmund Beard, *Developing the ICBM: A Study in Bureaucratic Politics*, New York: Columbia University Press, 1976, 3–4. The missiles that set off this confrontation were Soviet medium-range missiles being deployed in Cuba, close to the United States.

21. Neufeld, *Ballistic Missiles*, 192, 233–8; Steven J. Zaloga, *The Kremlin's Nuclear Sword: The Rise and Fall of Russia's Strategic Nuclear Forces, 1945–2000*, Washington, DC: Smithsonian Institution Press, 2002, 58–68.
22. Frances FitzGerald, *Way Out There in the Blue: Reagan, Star Wars and the End of the Cold War*, New York: Simon & Schuster, 2000, 72–113; Alex Roland, *The Military-Industrial Complex*, Washington, DC: American Historical Association, 2001. See also Michael Neufeld's contribution, Chapter 2 in this volume.
23. Spencer R. Weart, *Nuclear Fear: A History of Images*, Cambridge, MA: Harvard University Press, 1988, 262–9.
24. Eugene Burdick and Harvey Wheeler, *Fail Safe*, New York: McGraw-Hill, 1962; *Fail-Safe*, directed by Sidney Lumet, USA 1964 (Columbia Pictures); *Dr. Strangelove or: How I Learned to Stop Worrying and Love the Bomb*, directed by Stanley Kubrick, UK/USA 1964 (Columbia Pictures).
25. Walter A. McDougall, 'Space Age Europe: Gaullism, Euro-Gaullism, and the American Dilemma,' *Technology and Culture* 26.2 (April 1985), 179–203.
26. John Krige and Arturo Russo, *A History of the European Space Agency, 1958–1987*, vol. 1: *The Story of ESRO and ELDO, 1958–1973*, Noordwijk: ESA, 2000, 81–119; eidem, *Europe in Space 1960–1973*, Noordwijk: ESA, 1994; John Krige, 'Building a Third Space Power: Western European Reactions to Sputnik at the Dawn of the Space Age,' in Launius, Logsdon and Smith, *Reconsidering Sputnik*, 289–307; Neil Whyte and Philip Gummett, 'The Military and Early United Kingdom Space Policy,' *Contemporary Record* 8.2 (Fall 1994), 343–69; and eidem, 'Far beyond the Bounds of Science: The Making of the United Kingdom's First Space Policy,' *Minerva* 35.2 (Summer 1997), 139–69.
27. Krige and Russo, *History of the European Space Agency*, 81–119; Johnson, *Secret of Apollo*, 154–78.
28. McDougall, 'Space Age Europe.' See also Krige and Russo, *History of the European Space Agency*, 89; eidem, *Europe in Space*; and Alain Souchier and Patrick Baudry, *Ariane*, Paris: Flammarion, 1986, 14–24.
29. Weart, *Nuclear Fear*, 273–80; FitzGerald, *Way Out There in the Blue*, 150–5. See also Alexander Geppert and Tilmann Siebeneichner's contribution, Chapter 1 in this volume.
30. FitzGerald, *Way Out There in the Blue*, 334–56. See in this context also Peter J. Westwick, 'From the Club of Rome to Star Wars: The Era of Limits, Space Colonization and the Origins of SDI,' in Alexander C. T. Geppert, ed., *Limiting Outer Space: Astroculture after Apollo*, London: Palgrave Macmillan, 2018, 283–302 (= *European Astroculture*, vol. 2).
31. Hans M. Kristensen, Robert S. Norris and Matthew G. McKinzie, *Chinese Nuclear Forces and U.S. Nuclear War Planning*, Washington, DC: Federation of American Scientists, 2006, 71–3.
32. N-I and N-II Japan Space Transportation Systems, available at https://www.globalsecurity.org/space/world/japan/ni.htm.
33. Manoranjan Rao, ed., *From Fishing Hamlet to Red Planet: India's Space Journey*, Noida, Uttar Pradesh: HarperCollins India, 2015, 36, 101, 121; 'Launchers: Overview,' Indian Space Research Organization, available at http://www.isro.gov.in/launchers; and Hans M. Kristensen and Robert S. Norris, 'Indian Nuclear Forces, 2015,' *Bulletin of the Atomic Scientists* 71.5 (1

September 2015), available at http://www.thebulletin.org/2015/september/indian-nuclear-forces-20158728.

34. See Stephen B. Johnson, 'The History and Historiography of National Security Space,' in Dick and Launius, *Critical Issues*, 481–548. On the distinction between militarization and weaponization, see Michael Neufeld's contribution, Chapter 2 in this volume; on military infrastructures, see Regina Peldszus's essay, Chapter 11.
35. Roger Handberg, 'Dual-Use as Unintended Policy Driver: The American Bubble,' in Steven J. Dick and Roger D. Launius, eds, *Societal Impact of Spaceflight*, Washington, DC: NASA, 2007, 353–68; Stephen B. Johnson, *The Secret of Apollo: Systems Management in American and European Space Programs*, Baltimore: Johns Hopkins University Press, 2002, 115–53; McDougall, *...The Heavens and the Earth*, 7–9; and Mieczkowski, *Eisenhower's Sputnik Moment*, 275.
36. Roger D. Launius and Dennis R. Jenkins, eds, *To Reach the High Frontier: A History of U.S. Launch Vehicles*, Lexington: University of Kentucky Press, 2002.
37. Beard, *Developing the ICBM*; Robert L. Perry, *The Ballistic Missile Decisions*, Santa Monica: RAND Corporation, 1967; and idem, 'The Atlas, Thor, Titan and Minuteman,' in Eugene M. Emme, ed., *The History of Rocket Technology*, Detroit: Wayne State University Press, 1964, 142–61. Perry wrote this essay while he was employed by the US Air Force.
38. Beard, *Developing the ICBM*, 4; Michael R. Beschloss, *The Crisis Years: Kennedy and Khrushchev 1960–1963*, New York: HarperCollins, 1991, 65–6.
39. Zaloga, *The Kremlin's Nuclear Sword*, 47–9; Mieczkowski, *Eisenhower's Sputnik Moment*, 20.
40. Beard, *Developing the ICBM*, 8, 218.
41. Perry, 'Atlas, Thor, Titan and Minuteman,' 142; Beard, *Developing the ICBM*, 8, 218.
42. Perry, *Ballistic Missile Decisions*, 5–11.
43. Robert A. Divine, *The Sputnik Challenge: Eisenhower's Response to the Soviet Satellite*, Oxford: Oxford University Press, 1993; Peter J. Roman, *Eisenhower and the Missile Gap*, Ithaca: Cornell University Press, 1995; Zaloga, *Target America*, 134–41; and Siddiqi, *Red Rockets' Glare*, 270–8.
44. John L. Chapman, *Atlas: The Story of a Missile*, New York: Harper, 1960.
45. Loyd S. Swenson Jr., James M. Grimwood and Charles C. Alexander, *This New Ocean: A History of Project Mercury*, Washington, DC: NASA, 1966, 23–6.
46. Beard, *Developing the ICBM*, 141–2.
47. Ibid., 124–8; McDougall, *...The Heavens and the Earth*, 97–107; Rip Bulkeley, *The Sputniks Crisis and Early United States Space Policy*, Bloomington: University of Indiana Press, 1991, 60–86; Carl H. Builder, *The Icarus Syndrome*, New Brunswick: Transaction, 1994, 155–78; and Thomas P. Hughes, *Rescuing Prometheus: Four Monumental Projects that Changed the Modern World*, New York: Pantheon, 1998, 76–9.
48. For examples of military, corporate and political support of US weapons programs, consult Roland, *Military-Industrial Complex*; but see also Matthew Evangelista, *Innovation and the Arms Race: How the United States and the Soviet Union Develop New Military Technologies*, Ithaca: Cornell University Press, 1988; and Michael H. Armacost, *The Politics of Weapons Innovation: The Thor-Jupiter*

Controversy, New York: Columbia University Press, 1969. For more on social forces in technological systems, see Thomas P. Hughes, 'The Evolution of Large Technological Systems,' in Wiebe Bijker, Thomas P. Hughes and Trevor Pinch, eds, *The Social Construction of Technological Systems*, Cambridge, MA: MIT Press, 1989, 51–82.

49. John Clayton Lonnquest, *The Face of Atlas: General Bernard Schriever and the Development of the Atlas Intercontinental Ballistic Missile, 1953–1960*, PhD thesis, Duke University, 1996, 65–99; Neufeld, *Ballistic Missiles*, 95–9; Robert Frank Futrell, *Ideas, Concepts, Doctrine: Basic Thinking in the United States Air Force 1907–1960*, vol. 1, Montgomery: Air University Press, 1989, 488–91; and Stephen B. Johnson, *The United States Air Force and the Culture of Innovation*, Washington, DC: Air Force, 2002, 60–3.
50. Walter A. McDougall, 'Introduction: Was Sputnik Really a Saltation?,' in Launius, Logsdon and Smith, *Reconsidering Sputnik*, xvi–xx.
51. See Mieczkowski, *Eisenhower's Sputnik Moment*, 2, 20–2; Kim McQuaid, 'Sputnik Reconsidered: Image and Reality in the Early Space Age,' *Canadian Review of American Studies* 37.3 (Winter 2007), 371–401; Alexander C. T. Geppert, 'Anfang – oder Ende des planetarischen Zeitalters? Der Sputnikschock als Realitätseffekt, 1945–1957,' in Igor J. Polianski and Matthias Schwartz, eds, *Die Spur des Sputnik: Kulturhistorische Expeditionen ins kosmische Zeitalter*, Frankfurt am Main: Campus, 2009, 74–94; and Roger D. Launius, 'An Unintended Consequence of the IGY: Eisenhower, Sputnik, the Founding of NASA,' *Acta Astronautica* 67.1/2 (July 2010), 254–63.
52. Ira Chernus, *Apocalypse Management: Eisenhower and the Discourse of National Security*, Stanford: Stanford University Press, 2008, 177; Mieczkowski, *Eisenhower's Sputnik Moment*, 2, 6. Other examples of fresh views of Sputnik can be found in Launius, Logsdon and Smith, *Reconsidering Sputnik*; Columba Peoples, '*Sputnik* and "Skill Thinking" Revisited: Technological Determinism in American Responses to the Soviet Missile Threat,' *Cold War History* 8.1 (February 2008), 55–75; Alice Gorman and Beth O'Leary, 'An Ideological Vacuum: The Cold War in Outer Space,' in John Schofield and Wayne D. Cocroft, eds, *A Fearsome Heritage: Diverse Legacies of the Cold War*, Tucson: Left Coast Press, 2007, 73–92; and Asif A. Siddiqi, 'Sputnik 50 Years Later: New Evidence of Its Origins,' *Acta Astronautica* 63.1–4 (July–August 2008), 529–39.
53. Roger D. Launius, 'The Historical Dimension of Space Exploration: Reflections and Possibilities,' *Space Policy* 16.1 (February 2000), 23–8. The term 'Huntsville School' was coined by historian Rip Bulkeley in idem, *Sputniks Crisis*, 204–8. Once settled in the United States, the von Braun team was based at Huntsville, Alabama.
54. Howard E. McCurdy, *Space and the American Imagination*, Washington, DC: Smithsonian Institution Press, 1997, 53–82.
55. Roland, *Military-Industrial Complex*, 8.
56. McCurdy, *Space and the American Imagination*, 139–61. For an explicitly European perspective, see the contributions in the first volume of this *European Astroculture* trilogy, Alexander C. T. Geppert, ed., *Imagining Outer Space: European Astroculture in the Twentieth Century*, Basingstoke: Palgrave Macmillan, 2012 (2nd edn, London: Palgrave Macmillan, 2018) (= *European Astroculture*, vol. 1). Siddiqi, *Red Rockets' Glare*, 74–133 explores Soviet visions of space utopias.
57. Sheehan, *Fiery Peace in a Cold War*.

CHAPTER 4

West European Integration and the Militarization of Outer Space, 1945–70

Michael Sheehan

This chapter seeks to analyze the process of West European space cooperation in the early 1960s from an interdisciplinary perspective. In taking this approach, it aims at explaining the interaction of underpinning ideas in politics, technology policy and literature, and explores the impact of astroculture in European technological cooperation.[1] In the late 1950s a number of features could be identified in West European culture and politics. The appalling death and destruction experienced in the two world wars had inspired a generation of European politicians to initiate the move towards European economic integration, particularly with the creation of the European Economic Community (EEC) in March 1957. The Community represented an aspiration to encourage future political integration.[2] At the same time, a dramatic new phase of the Space Age had begun with the launch of Sputnik 1 by the Soviet Union in October 1957, which encouraged European scientific and political leaders to extend the idea of European international cooperation to the realm of space science and technology.

The early West European space effort was thus influenced by the same aspirations to build a durable peace in Europe that had led to the creation of the EEC. However, the attempt to emulate the process of integration using non-military tools in the space realm was made more complex and ambiguous by a variety of factors that pulled the European NATO countries towards the military use of space. The NATO states continued to shelter under the

Michael Sheehan (✉)
Swansea University, Swansea, Wales, UK
e-mail: m.sheehan@swansea.ac.uk

Alexander C. T. Geppert et al. (eds), *Militarizing Outer Space*
European Astroculture, vol. 3
https://doi.org/10.1057/978-1-349-95851-1_4

military shield provided by the United States, and as the Cold War continued, the United States would increasingly rely on the strategic intelligence and force multiplication advantages provided by its military satellites. In addition, the United Kingdom and France sought to develop their own military space capabilities, while for those states that did not, the realities of dual-use technology meant that the militarization of space by West European states would become a reality as the technology matured. In the background to these political and technological developments there was a significant outpouring of European science fiction in the 1950s and 1960s, and this too encompassed both military space themes and dystopian nuclear nightmares, as well as alternative narratives suggesting a future marked by peace and the non-military use of space. These phenomena can be treated separately, but they were part of a common political and historical context to which politicians, scientists and producers of science fiction were all responding.[3]

Because space policy was a blank slate in 1957, its development by European states was influenced not just by astropolitics, but by what Alexander Geppert has christened astroculture. Geppert defines astroculture as 'a heterogeneous array of images and artifacts, media and practices that all aim to ascribe meaning to outer space while stirring both the individual and the collective imagination.'[4] Defined thus, astroculture focuses on the cultural dimensions of the relationship between West Europeans and outer space, and Geppert emphasizes that it is to be seen as 'an explicitly culture-related counterpart to such better known and firmly established notions as "astrophysics" [and] "astropolitics".'[5]

In this sense, there was a pre-existing astroculture already exerting an influence when the launch of Sputnik in 1957 transformed space exploration from the imagination to practical reality. To understand the form that European space policy took in the 1960s it is necessary to examine the interrelationships between national security policy in relation to the Cold War, the political pressures promoting efforts at peaceful West European integration and Franco-German reconciliation, and the influence of the broader astroculture, exemplified particularly for the purposes of this chapter by contemporary European science fiction.[6]

As these dynamics – strategy, astroculture and integration – began to converge in the early 1960s, the individuals influenced by them were also interacting with what John Krige has called the emerging European Space System. Krige defines it as

> that complex of institutions, artifacts, national and international networks, production facilities and commercial activities, which have been built up over the past three decades through a collaborative European presence in space. Its outputs include satellites, sounding rockets and launchers, scientific results, satellite photographs, and telecommunications linkages, scientists and engineers with their embodied technical and managerial expertise, and an ideology which links the conquest of space with industrial progress.[7]

Some of the components of this system would clearly overlap with those of astroculture, though culture itself is not included in Krige's description. The 'space system' is important because key European states were responding to several, sometimes contradictory, pressures in developing their policy on space. One factor was the understanding that space technology was emerging as one of the new criteria for great power status, and for states such as Britain and France, the desire to possess this technology in order to maintain their perceived great power position was an important driver of policy.[8] Krige notes that the European Space System was not closed, but was rather 'a dynamic and unstable system, subject to crosscurrents which originated outside it but which could profoundly affect its path.'[9] Here he is referring to both domestic/regional political pressures and technological developments, particularly in the United States. But it is also true that European astroculture was another 'externality' influencing developments, and one of the most significant elements of that externality was science fiction.

Western Europe's relationship with outer space in the late 1950s and early 1960s was an example of intertextuality, in which any one text, in this case the political imperative towards West European integration, 'is necessarily read in relationship to others, and [...] a range of textual knowledges is brought to bear upon it.'[10] The different 'texts' interacting for European space policy were those noted already – the integrationist in pursuit of stable intra-European peace, the technocratic impulse to embrace cutting edge technology in pursuit of economic development and great power status, the binary East-West confrontation and the resultant pressures to exploit launcher and satellite technology for military purposes in response to the Soviet threat and the explorations of alternative possible futures found in contemporary science fiction.

These interacting narratives meant that conceptualizations of outer space were contested, and background cultural influences became significant in influencing individual world views. Individuals and groups acted in terms of beliefs, values, theories and understandings of the 'reality' they perceived. In this sense, what individuals understood by 'outer space' was a cultural construction, and that construction was at issue in the late 1950s and early 1960s as was demonstrated by the disputes between sections of the scientific and political communities over the appropriate structure for pursuing European space collaboration.

The way in which such a consensus on understandings of reality is constructed is never an entirely innocent exercise, as 'theory is always *for* someone and *for* some purpose.'[11] By firmly establishing a specific perception of outer space, a dominant narrative helps to shape a particular reality. It allows outer space to be understood in a particular way, as a particular kind of realm in which certain types of activity are possible or even expected, while others are frowned upon, specifically forbidden, or simply never considered. It was just such an attempt to embed a particular understanding of outer space, as a field of purely peaceful and scientific activity, that lay behind the ultimately unsuccessful efforts of the Italian scientist Edoardo Amaldi (1908–89; Figure 4.1) to define the European space enterprise in ways that would eliminate the possibility of military use.[12] More than any other leading figure in the creation of the cooperative European space program, Amaldi sought to prevent it having any military dimension whatsoever. He was an

Figure 4.1 Edoardo Amaldi (1908–89) believed that scientists needed to be politically active on peace issues and that they should be energetic in informing public opinion on the dangers posed by nuclear weaponry.
Source: Fototeca Gilardi. Courtesy of akg-images.

active member of the Pugwash disarmament organization, founded the International School on Disarmament and Research on Conflicts in 1966, and spoke frequently on the need for international cooperation, subscribing to the contemporary pacifist slogan, 'one world or none.'[13]

Thus this chapter explores the common elements and tropes seen in European astroculture of the late 1950s and early 1960s, specifically as reflected in science fiction, the political movement towards West European unity and peace, and the development of a common space program. The military pressures produced by the Cold War were interacting with two fundamental tropes emerging in the late 1950s which were the related ideas of a 'united Europe' and of Europe as a historically novel 'zone of peace.' There was, in Maarten Van Alstein's phrase, 'a triadic link' between the core ideas of the memory of the two world wars, the process of intra-European integration, and the construction of a durable peace.[14] These were reflected in foreign policy strategies, in science fiction and in the emerging European space program, but they were in competition with the counternarrative of military modernization in response to a perceived external threat.

I European astroculture and the political context of ELDO and ESRO

Outer space was a uniquely distinctive realm at the outset of the Cold War. Among the cultural forces helping to shape perceptions of space policy, science fiction had the potential to be highly significant.[15] Science fiction is typically thought of as an art form addressing fictional constructions rather than objective reality.[16] However, the relationship is much more complex. As Jutta Weldes has argued, while in theory science fiction describes alternatives to reality, in practice the dividing line between science fiction and world politics is actually far from clear. And, as H. Bruce Franklin notes, when fantasies such as science fiction 'shape the thinking of inventors and leaders and common people, they become a material force.'[17] There is evidence of the shaping effect of science fiction on the imagining of political and technological possibilities in the processes that led to the formation of the European Space Research Organisation (ESRO) and the European Launcher Development Organisation (ELDO) in the 1960s.

The world views of decision-makers were necessarily shaped by the culture in which they operated. Culture provided the psychological background which allowed individuals to make sense of their external environment and their place within it, and therefore in relation to the background influence of science fiction, popular culture was one of the influences which constructed international politics for state officials, since understandings were drawn from societies' total cultural resources.[18] The significance of science fiction in regard to the specific realm of space was that it was a crucial part of the way in which external reality was mediated by culture so as to influence the ways populations understood policies, and the ways in which governments devised and explained those policies. This was particularly important in the second half of the twentieth century because, as Geppert notes, space 'was key to the self-image of public, governmental and technical elites, and to modernist narratives of progress.'[19]

Science fiction could thus help to make a particular possibility seem plausible. Before something could be designed and brought into existence, it first had to be imagined, and this was as true of political communities as it was of technology. Because science fiction appeared to indicate where the world might be heading, it could also help shape that future by encouraging people to imagine it as either utopia or dystopia. Since meanings in society were always capable of being contested, science fiction could help endorse and thereby reproduce power structures or policies, or it could challenge and contest them, or suggest the possibility of new ones. This can be seen in European science fiction during the Cold War period. Novels such as Gérard Klein's *Les Seigneurs de la guerre* (1970) or Carlos Saiz Cidoncha's space opera *La caída del imperio galáctico* (1978) made future space warfare seem inevitable, even desirable.[20] But science fiction could also become a vehicle for critiquing or subverting the national military ethos as it applied to space. European science fiction produced examples of key anti-war texts, such as Nevil Shute's *On the Beach* (1957), René Barjavel's bestseller *La Nuit des temps* (1968) and Michael Moorcock's *A Cure for Cancer* (1971), all of which warned of the catastrophic dangers involved in future high technology war.[21] Even works that had militaristic space opera themes could include an anti-war subtext, for example, Peter Randa's *Survive* (1960), where visitors from earth encounter the ruins of a highly advanced civilization annihilated by war.[22] Such warnings were a major topic in the European science fiction of this period.

The post-apocalyptic theme was also evident in the classic French films *La Jetée* (1962) and *Le Dernier combat* (1983), and novels such as Stefan Wul's *Niourk* (1957) and Pierre Boulle's *La Planète des singes* (1963).[23] In Germany, the apocalypse of the Second World War was influential and post-apocalyptic fiction prominent, as well as dystopian works such as Walter Jens's *Nein! Die Welt der Angeklagten* (1950), a *1984*-like piece of grim futurology.[24] The apocalyptic theme is well represented by Hans Wörner's *Wir fanden Menschen* (1948) and Hellmuth Lange's *Blumen wachsen im Himmel* (1948), amongst others.[25] The prevalence of dystopian science fiction in France and Germany projected forward in time the recent memories of the Second World War, the threat posed by nuclear weapons and the shattered European civilization that was painfully emerging from the debris.

In addition to apocalyptic warnings about the grim future facing Europe if it could not transcend its warlike past, there were a number of accompanying themes, particularly in French science fiction, which would also play out in subsequent European efforts to develop political-economic unity and to construct a collaborative space program. One such theme was uneasiness about the implications of the military dominating space research and exploration. Science-fiction writers suggested the possibilities of regional and global unity enabling the ending of war and of the military being marginalized. In Jean-Gaston Vandel's *Les Chevaliers de l'espace* (1952), when war breaks out between America, Asia and a united Europe, the mysterious 'space knights' organization intervenes causing all the combatant major weapons to malfunction (Figure 4.2).[26] The Knights are not interested in dominance or political power, only in ensuring the maintenance of international peace.[27] In B. R. Bruss's *Rideau magnetic* (1956), the experience of fighting united against a Martian invasion causes the worlds powers to form a planetary government and usher in an era of global peace.[28] The same behavior pattern occurs in Jimmy Guieu's *L'Agonie du verre* (1955) and Henri-Richard Bessière's *Route de néant* (1956).[29]

A further theme in these novels and films was the idea of Europe as a formerly dominant continent now reduced to political impotence.[30] This was a significant subtext of the political efforts to unite Europe after 1957, which was driven not simply by a desire to avoid future intra-European war, but also by a wish to collectively regain global influence that the former individual European great powers had lost. Thus the peaceful but ineffectual Europe depicted by Bessière in *Les Conquerants des l'univers* (1951), and also by Guieu in *L'Agonie du verre*, would have struck an immediate chord with their European readers, whose sympathy for efforts at international cooperation reflected this same perception.[31]

French science fiction in particular frequently presented a Europe that was united and at peace. In Vandel's *Les Chevaliers de l'espace* Europe is one of three global powers. In Bessière's *Les Conquerants des l'univers* earth is slowly recovering from an earlier nuclear war between the superpowers, where the efforts of the united European bloc had been unable to secure the peace. While by no means all authors reflected this theme of European unity, much of the rest of the writing still supported its logic by describing a political unity that had embraced the entire planet Earth. In Bruss's *Rideau magnetic*, the pressures produced by an attempted Martian invasion of earth have led to the creation of a 'Planetary Union' and world peace. Bessière describes a world in which global unity has been achieved, borders erased and global peace attained in his 1956 novel, *Route du néant*. A pattern in his works is the characterization of alien societies as having achieved global unity after long wars, a unity that alone makes peace and civilization possible. This is seen with the Martians in *Les Conquerants des l'univers*, the Venusians in *Retour de météore* (1951) and many of his other novels.[32] By consistently arguing that only political unity could guarantee peace and socio-economic progress and that such unity was characteristic of future and more advanced societies, the authors were encouraging their readers to view this possible future for Europe and the world in a positive light. Publishing houses including Fleuve Noir

Figure 4.2 Front cover of Jean Libert's (1913–95) and Gaston Vandenpanhuyse's (1913–81) *Les Chevaliers de l'espace* (1952), published under their shared pen name Jean-Gaston Vandel. The Knights are a powerful military order, with their own spacecraft, who impose a universal peace on earth's warring nations. In this illustration the astronaut's space suit resembles a medieval suit of armor, emphasizing the military theme of the book.

Source: Jean-Gaston Vandel, *Les Chevaliers de l'espace*, Paris: Fleuve Noir, 1952, cover image. Courtesy of illustrator René Brantonne's family.

published 'pulp' fiction, which reached a comparatively large audience.[33] Works such as these presented an imagined future Europe both united and at peace. French science fiction was a highly political genre and the political context was a major inspiration for authors as is seen in speculations about a future nuclear holocaust. This is in line with the emergent process of West European integration that was taking place from 1951 onwards as a direct response to the catastrophic death and destruction Europe had experienced in the two world wars.[34]

The tropes of European unity and peace therefore represent an intertext between politics, space policy and science fiction. An important element of all three is the recognition that there is nothing inevitable about the process. It has to be the result of conscious, goal-directed activity, without which an alternative and more warlike course of history is likely to unfold. As French foreign minister Robert Schuman (1886–1963) put it in his historic 1950 declaration, 'a united Europe will not be achieved all at once, nor in a single framework; it will be formed by concrete measures which first of all create a solidarity in fact.'[35] Since it was important to encourage integration in as many fields as possible, there was a clear logic in including space research and development as one of the integrating areas of policy. The intertextual image bank such as those from European science fiction in Weldes's terminology 'pre-oriented' readers, encouraging them to construct meanings in particular ways in order to bring about a more favorable historical outcome and to see space cooperation as one of Schuman's multiple frameworks for building European unity. While it is not possible to directly measure the impact of such speculative fiction on the politically active segment of national populations, Tom Moylan argues that the kind of 'world-building' characteristic of science fiction has a subversive potential: 'Imaginatively and cognitively engaging with such works can bring willing readers back to their own worlds with new or clearer perceptions, possibly helping them to raise their consciousness about what is right and wrong in that world, and even to think about what is to be done, especially in concert with others, to change it for the better.'[36]

One further noteworthy theme from science-fiction works of the period is the idea of the leading role played by scientists, as distinct from government or military leaders. This trope appeared with great frequency in European science fiction of the period. Frequently in this literature it is scientists rather than governments which take the key interplanetary initiatives, often using technology that they have developed independently of the state. In Jimmy Guieu's *L'Invasion de la terre* (1952), a group of earth scientists travel to the planet Glamora seeking help for earth, which has been attacked by Mars.[37] In William Earl John's series of teen novels, Professor Brane develops his own spacecraft to take the protagonists on voyages through the solar system.[38] Strikingly, when the international movement to create a West European space program began in the late 1950s, it was just such a group of scientists, rather than political leaders, who took the initiative and prompted governments to create ESRO. Several of the same scientists, including Pierre Auger (1899–1993) and Edoardo Amaldi, had led a similar initiative in the early 1950s to create the European Centre for Nuclear Research (CERN) as a collaborative European nuclear research organization.[39]

II ESRO: the European approach to space and peace

The pioneers of European space cooperation including Amaldi shared the thinking of many scientists and politicians in postwar Western Europe, with powerful themes of aspiration to European unity. This was partly because of a desire to avoid another cataclysmic war and partly because of a strong perception that the Second World War had left Europe weak and in a clear position of relative political, military and economic inferiority to the superpowers. For the proponents of European unity in the early 1950s, such as the French political economist Jean Monnet (1888–1979), it was 'precisely because the countries of Western Europe play no part in the great decisions of the world that we face instability.'[40] The European concern with relative decline and fear of being left behind manifested itself in politics, space policy and, as has already been shown, in science fiction.

A perception of the comparative weakness represented by individual European states attempting to compete with the superpowers was most clearly seen in the subsequent formation of ESRO and ELDO, but it was not limited to these endeavors. It could also be seen elsewhere in the European Space System, for example, the aerospace industry, which was also strongly aware of these issues and was beginning to contemplate the formation of European-scale entities. In October 1960 the British Member of Parliament David Price (1924–2014), a lobbyist for the aerospace industry and a member of the Council of Europe Parliamentary Assembly, authored a report for that body on *European Cooperation in Space Research and Technology*. Price argued that Europe could only compete successfully through cooperation, and that relying on American technology would mean that Europe would not develop the latest technology itself. Price's report stressed the potential for using former military technologies such as the Blue Streak ballistic missile for peaceful purposes. Price's advice ran counter to the views of Amaldi, who wished to avoid any reliance on technology developed for military purposes. The British Interplanetary Society (BIS) presented a very similar set of proposals to the British government in March 1960.[41] This specialist advice would be one of the factors that encouraged the government to explore the possibility of developing Blue Streak as a satellite launcher after its cancellation as a military project was announced to Parliament in April 1960.[42] By the summer the British government had begun to sound out France on the possibility of cooperating on a European satellite launcher based on Blue Streak.[43]

This political significance of astroculture can be further seen in terms of the European Symposium organized by the BIS in February 1961. Attended by over 200 British and European delegates, it also included representatives of the aerospace industry. Significantly, speakers from companies such as Rolls Royce and Hawker-Siddeley argued against the pursuit of a 'narrow and selfish national interest' and in favor of a European space program. The idea that a collaborative European program was within the realms of possibility was becoming more widely held. President De Gaulle (1890–1970) of France had encouraged such thinking by arguing in 1961 that Europe was an 'attractive candidate' for the role of 'the third space power' after the United States and the Soviet Union. France encouraged European collaboration because it saw it as a natural extension of its national space program and believed it necessary to build European autonomy in space technology.[44] One of the outcomes of the desire for greater European cooperation was the formation, in September 1961, of Eurospace, a supranational lobbying organization. This grouped together a number of leading European companies involved in the related fields of aircraft, missiles and associated electronics, again demonstrating the impossibility of sealing space research off from the military realm given the common technological origins of both. It embraced a specific objective of developing the space sector in Western Europe. Initially comprising 47 member companies, within three years it had reached 1,000.[45]

The original European space effort was strongly influenced by an international community of scientists who helped to shape the European avoidance of collaborative military space that became characteristic of subsequent integration. The space scientists were operating as an epistemic community, 'a network of professionals with recognized expertise and competence in a particular domain and an authoritative claim to policy relevant knowledge within that domain or issue area.'[46] They exercised an unusual degree of influence on the political and technological movement towards European integration in this period, reflecting one of the sub-themes of European science fiction, of individual or groups of scientists guiding policy at the national and international level.

The initial enthusiasm for a joint European space program came from a collective of scientists involved with CERN, which had been established by twelve West European states at Meyrin on the Franco-Swiss border near Geneva in 1954. Amaldi acted as General Secretary during its provisional phase.[47] CERN represented a successful example of the ability of European states to pool their resources in order to do 'big science,' and in the late 1950s Amaldi initiated a vigorous correspondence between European scientists about the possibility of similar international cooperation in the space field.[48] These included leading national scientific figures, such as Sir Harrie Massey (1908–83), head of the British space research committee, and Pierre Auger of France, who was Director of UNESCO's Department of Science and in the process of creating France's national space agency, the

Centre National d'Etudes Spatiales (CNES). Auger was also an early leader of CERN. The origin of the group within CERN was important because it encouraged those seeking to create a European space organization to see the scientist-led CERN, rather than the government-led EURATOM, as the appropriate model for a future European space organization. A providential platform for discussing this possibility emerged in 1958 when the International Council of Scientific Unions created a Committee on Space Research (COSPAR). At this meeting, Massey proposed the idea of using the British Blue Streak ballistic missile as the basis of a European satellite launcher. The Blue Streak had become a powerful symbol of international scientific status and British national prestige in the late 1950s, alongside other technological achievements such as the Vulcan bomber aircraft and the Jodrell Bank radio telescope. These were visually iconic and became important cultural symbols (Figure 4.3).

The British offer of Blue Streak as the core of a new European satellite launcher proved to be unacceptable to the other participants. This was a technology that had been developed in order to carry nuclear warheads. The

Figure 4.3 British Mark II radio telescope at Jodrell Bank, University of Manchester, as viewed from control room, ca. 1950. It became an example of astrocultural architecture, both singular and dramatic in its own right, and emblematic of the status Britain sought as a key player in the Space Age. Alongside its significance as a pioneering technology, Jodrell Bank became a visible national icon on a par with existing symbols like Stonehenge and Tower Bridge.
Source: Courtesy of Universal History Archive/UIG/Getty Images.

United Kingdom had decided that its surface-basing mode meant that it was vulnerable to attack and would quickly become obsolete, and so had decided to cancel the military program, but were keen to get some return on the investment already made in the Blue Streak technology. The British wanted an arrangement where the future European rocket would essentially be the existing British technology, but developed with input and funding from the other partners. However, the representatives of several other states, such as Italy and Denmark, maintained that their governments would expect British technology to be entirely subsumed within a broader European program.[49]

Amaldi insisted that any European space program should have 'a real European character' and a 'peaceful character.'[50] His attitude reflected the perceived lessons of Europe's recent history as well as the idealism of European and wider international cooperation. It was a vehicle with which to embrace the outsider, particularly in relation to the bitter historical Franco-German conflicts which had seen three wars between the two countries between 1870 and 1945, a vision similarly reflected in both politics and popular culture. It was felt that the ability of European states, particularly France and Germany, to cooperate in the peaceful utilization of a technology that in the form of the V-2 'Vengeance weapon' had been originally developed by Nazi Germany for military purposes, would project a powerful positive symbolism.[51]

In calling for the creation of an international organization of European countries to enable Europe to participate in space exploration, Amaldi was very clear about the kind of objectives and mandate he felt the organization should have. The space scientists in general wanted a European equivalent of NASA, but were unenthusiastic about the development of a European launcher, because for them the science carried out in space was the crucial objective. What rocket carried the experiments was a secondary consideration. They were aware that the United States had already offered its Scout launcher for European satellites, and this system would be considerably less expensive than developing the Blue Streak missile into a European launcher. However, both the British and French felt that Scout would be too small to carry the future large satellites the Europeans would want to launch.[52]

Like the key European economic and political integrationists such as Monnet and science-fiction writers such as Bessière and Vandel, Amaldi felt that European organizations should be committed to the peaceful, that is, non-violent exploration of outer space. The proposed European Space Research Organization, he argued, 'should have no other purpose than research and should, therefore, be independent of any kind of military organization and free from Any Official Secret Act [*sic*].'[53] This would enable it to maintain what Amaldi referred to as its 'moral authority' and also enable a wide cross section of European states to take part, including the neutral states outside NATO such as Sweden and Switzerland. Switzerland in fact took the lead in hosting the key international meeting of government representatives that created the working groups who drafted the founding documents for the proposed organization. From the outset, therefore, the leading role played by the neutral states helped shape the future direction of European space cooperation. Because the military origins of the British Blue Streak program ran counter to this specific preference at that same meeting, the United Kingdom and France proposed that the development of launchers should be the responsibility of a separate organization.[54]

Not all the scientists opposed military involvement, however, any more than all the science-fiction writers proposed pacific unions. One such was Amaldi's fellow countryman Luigi Broglio (1911–2001), director of the Institute of Aeronautical Engineering at Sapienza University of Rome and Italy's leading rocket designer. Broglio felt that military interest in space was so logical and inevitable that there was little realistic prospect of creating an organization without a military dimension. Amaldi, however, believed that military involvement would hold back scientific and technological progress and run counter to the efforts to establish Europe as a zone of peace. Amaldi's position was supported by many of the European scientists because they had fairly recent experience of having to work under tight military restrictions during the Second World War.[55]

Amaldi's insistence on a purely non-military European space program was significant because when European space cooperation was first being discussed in the late 1950s, there were moves within NATO to shape the future development of European space collaboration. In June 1957 NATO set up a Task Force on Scientific and Technical Cooperation, and a Science Committee, which recommended a NATO space program, while the NATO Secretary-General's science advisor urged the creation of a European NASA to work in partnership with its American equivalent.[56] Although Amaldi opposed NATO involvement in a European space organization, his attitude was not shared by some of the other leading scientists involved in discussions on European space cooperation. In addition to Broglio's lack of hostility, Amaldi's friend and collaborator Isidor Isaac Rabi (1898–1988) argued that NATO could be a useful ally and proposed approaching it in the same way that he had arranged for the United Nations Educational, Scientific and Cultural Organization (UNESCO) to support CERN when it was created.[57] Similarly, the European defense organization, the Western European Union (WEU), was supportive of a military space program. In November 1964 the WEU established the Committee on Space Questions. Its initial report argued that Western Europe needed to develop its own military space capabilities in order to maintain its position vis-à-vis the United States and the Soviet Union.[58] It was against this background of NATO and WEU interest in European space activities that European scientists sought to promote space cooperation for purely 'peaceful' purposes.

The efforts to create ESRO came to fruition in 1964 when its Convention entered into force.[59] This founding document stipulated in Article II that the organization's purpose was 'to provide for, and to promote, collaboration among European states, exclusively for peaceful purposes.' The creation of ESRO was a remarkable development and a significant contribution to European integration, and was understood in this light by many of its creators and staff. It was seen as being, among other things, a vehicle for the political unification of Europe via functional integration. Amaldi had claimed that 'the establishment of an independent European space organization would be one of the mosaic stones which would lead to a European unity.'[60]

Amaldi was ultimately disappointed that the European space collaboration had not taken the form he had hoped for, and after the formation of ELDO in 1964 he subsequently ceased being a leader on the European stage and focused on developing the Italian national program. Amaldi had wanted the space organization to be modeled on CERN and led by scientists, who would develop a rocket specifically for satellite launches, with no involvement by the military. Eventually, the ELDO launcher decided upon drew on existing British and French ballistic missile technology. The dual-use nature of such technology made this inevitable, but Amaldi could not reconcile himself to this reality. However, while his utopian vision was never fully realized, his views were influential and played a significant part in shaping the subsequent perception of the military involvement in European space cooperation, even within the ELDO organization whose creation he had opposed. ESRO, ELDO and the later European Space Agency (ESA) all contained a ban on direct military activity in their Conventions, a legacy of the influence of Amaldi and like-minded scientists in their formation. The European Space System as it developed, however, would embrace technologies with military as well as civilian potentials, and the 'purely military' European program was more an unjustified fear than reality.

III ELDO and the military implications of dual-use technology

By the end of 1960 the British and French governments were approaching agreement on the form the new European launcher should take and had accepted the need for two separate European space organizations.[61] While ESRO represented a bridge between the states of Western Europe, the development of the necessary launcher posed an obvious difficulty for the neutral states. While little thought had been given to the ultimate purposes to which the launcher would be put, it was clear from the outset that its development would draw on existing British and French military technology. In the longer term, such an asset would also be capable of orbiting military satellites. Developing such technology in cooperation with NATO states would clearly compromise the political status of neutral countries. Thus the creation of ELDO allowed the neutrals to become members of ESRO, while remaining outside ELDO and the rocket development program.[62]

ELDO, the European Launcher Development Organisation, would be a very different body from ESRO, and whereas the former had been formed with the intent of excluding military activities, ELDO was formed on the basis of the potential represented by pre-existing military capabilities. In this sense it did not represent the same dovetailing of themes from science fiction, culture and politics as did ESRO. While as another European space operation it clearly met the aspiration seen in politics and science fiction for a more united Europe growing together through functional cooperation, it was less convincing in terms of the second key trope, the aspiration to unify Western Europe peacefully using integration in non-military activities. It is noticeable that many of the actors who energetically promoted the formation of ESRO were unenthusiastic or even hostile towards ELDO, and advised their own governments against becoming participants. Edoardo Amaldi was the leading example of this phenomenon, but many others within government and the epistemic scientific community shared his reservations, though often for very different reasons. Amaldi led a vigorous campaign of opposition to Italian participation in ELDO and lobbied other European scientists in the hope of encouraging other countries to oppose the organization.[63]

In the beginning of the 1960s, the Space Race between the superpowers was a competition for prestige and technological and political leadership that appeared almost as a surrogate for war. The military and civilian dimensions overlapped, so that the European neutral states such as Switzerland and Sweden that had been leading players in the development of ESRO felt unable to participate in ELDO because it could be perceived as taking sides in the military-strategic confrontation between the NATO and Warsaw Pact countries.[64] In addition, the link between military rationales and technology and the rockets to be developed by ELDO meant that the enterprise was viewed in a very different light by both governments and the scientific community. Because missile technology was so clearly connected to military aspirations, the member governments of ELDO were unable to agree strong organizational powers to the body, which hampered its development.

Several of the problems associated with ELDO derived from its part in the broader project of European integration. Genuine European autonomy in space policy seemed to require a European launcher, but also suggested that it should be developed and built by several West European countries, so as to make it a truly European project and spread the technological benefits between participating states. This logic worked against the British suggestion of simply taking the existing Blue Streak system and modifying it, perhaps using another British rocket, Black Knight, as the second stage. Black Knight had been developed for a military purpose, as a nuclear warhead re-entry test-vehicle. The technology was drawn on when Britain developed the Black Arrow satellite launcher in the 1960s.[65] Such a combined vehicle, called Black Prince, had been evaluated by the British government and only rejected on cost grounds.[66] The desire for a truly European launcher led to the decision to develop a multi-stage rocket with the different stages built by different countries. This was a further deterrent for some countries such as Denmark and Switzerland who did not feel that their industries were in a position to benefit from such a project in the early 1960s.[67] Moreover, the purely European dimension of the project was compromised by the membership of Australia, necessary for the use of the Woomera launch site. ELDO's narrow membership base was problematic in terms of European unity. One of the great virtues of ESRO, in contrast, was that it was seen as bridging existing divides between states in Western Europe, for example, those between the NATO and neutral states, and between the rival memberships of the EEC and the European Free Trade Association.[68]

The ELDO Convention of 1962 committed the signatories to the development and operation of a space launcher and ancillary equipment. While ELDO clearly elicited reservations from neutral states, who feared a military dimension, the ELDO Convention, like that of ESRO, stipulated that the organization should concern itself only with peaceful applications of the launchers and equipment, so that even ELDO was influenced by the political and astrocultural aspirations towards a Europe whose terrestrial and space policies reflected a vocation for peace. The Convention's preamble noted the signatories' desire 'to harmonize their policies in space matters with a view to common action for peaceful purposes,' while Article II of the document states specifically that 'the Organization shall concern itself only with peaceful applications of such launchers and equipment.'[69] There was, however, no elaboration of what kinds of activities were clearly 'peaceful' and which were not.

Had ELDO successfully developed a launcher, this would have been a difficult restriction to adhere to, though much would have turned on how the term 'peaceful' was interpreted. In inheriting ELDO's mandate in 1973, ESA subsequently chose to understand it as 'non-weapon' so that force enhancement satellites for navigation, early warning, military communication and so on, could be launched by the Ariane rockets. ELDO would have almost certainly adopted a similar approach. Its members were relaxed about the use of dual-use technology and drawing on military systems for the European launcher. This was essential given the dual-use technology that it represented. Thus in modifying the existing Blue Streak missile in order to make it the first stage of the proposed three-stage Europa satellite launcher, the British replaced the warhead section of the military version with a transition bay which would allow the connection of the French second stage.[70] This was true also in terms of the European attitude towards the parallel national military and dual-use space projects being pursued by France and the United Kingdom. In an address to the British House of Commons in February 1966,

Ambassador Renzo Carrobio di Carrobio (1905–94), Secretary-General of ELDO from 1964 to 1972, praised the successes of the French Diamant satellite launcher and looked forward to the success of the British Black Arrow.[71]

Both Britain and France sought to use the proposed cooperative European rocket to sustain or enhance their existing military missile development programs. Ironically, the first rocket launched by ELDO was actually a military Black Knight, which carried electronic equipment to test the safety of the Woomera test range.[72] One of the British rationales for offering Blue Streak for the European venture was the belief that Britain would then 'retain current first-hand experience of the design and construction of large rockets, and would be free to develop them for military purposes' if later desired.[73] Similarly, the need for secure communications with the components of its nuclear deterrent was a key reason for the subsequent British development of military communications satellites. In addition, the Royal Air Force had been considering the potential of reconnaissance satellites since 1955.[74] France was similarly seeking to protect its military investment. France insisted that it should provide the second stage of the proposed satellite launcher, rather than using the British Black Knight, since 'they would want it to correspond to a type for which the French military authorities had already made provision in their plans.'[75] The French support for ELDO was driven by President De Gaulle's wish to develop missile technology for the French nuclear deterrent as rapidly as possible, in order to help rebuild French 'sovereignty and self-respect' in response to the disasters of the Second World War and post-war decolonization process. France had begun developing long-range missiles in 1960, drawing on the lessons of the German V-2 military program. Its Véronique rocket would be the basis of the proposed French Coralie second stage of the European rocket. By 1959 France had already initiated development of a new series of solid-fuel long-range military missiles, the 'precious stone' series, begun with Agate and ending with Diamant in 1973 (Figure 4.4).[76]

Figure 4.4 French postage stamp Fusée Diamant, showing the Diamant rocket which orbited the first French satellite, the A-1, from Hammaguìr, Algeria, in 1965. Engraver Claude Durrens (1921–2002) designed this stamp in 1966.
Source: Courtesy of Janine Durrens.

IV Venus and Mars: the ambiguity of European space cooperation

In European culture, Venus and Mars are not simply planets in the solar system, through their association with the Roman goddess of love and the god of war, they are metaphors for peace and war, and pacific and warlike policies. Metaphors play a crucial role in both politics and fiction, in both realms they are tools to shape the perceptions of others.[77] The emergence of West European space organizations in the 1960s can be seen as the result of the intertextual consensus on a set of interrelated themes – of Europe's terrible history of division and war, of the potential represented by a Europe united and at peace, and of the use of functional international cooperation and gradual integration to get from one situation to the other. As Stacia Zabusky put it in her ethnological study of ESA staff, 'the space organization is a child of the post-1945 impulse towards West European integration,' in which 'integration takes on a moral imperative – it will be the savior of Europe, delivering the continent from war, from poverty and from backwardness. In this ideology, cooperation appears as the embodiment of peace, symbolizing the transcendence of the incessant warfare that has characterized European international relations for centuries.'[78] This sense of a historic purpose was a crucial determinant of the European space project of the early 1960s, with many states sharing a sense that 'collaboration in space was just one dimension of a general historic tendency towards European "unity" and that failure here would inevitably have repercussions in other sectors, scientific, technological, economic and political.'[79]

The drive for European integration in space technology reflected not only the political lessons drawn from the disastrous European experience of the two world wars, but also the framing effect of European astroculture. The symbolism represented by the development of space technology and space exploration lent itself to meeting the goal of reviving European political leadership in the Cold War world, while the normalization of the idea of European, and indeed global political unity, in the science fiction of the period helped to create a sense of the possible and an imagined contrast to Europe's warlike past. But just as science fiction offered examples of warlike as well as peaceful possible futures, so too was European space cooperation an ambiguous enterprise. The reality was a complex set of motivations in which some of the cooperating European states such as Britain and France were partly motivated by the desire to use European cooperation as a vehicle for developing military space technologies, while others sought to build industrial capacity in the aerospace field which would inevitably generate dual-use capabilities in the future. While there were those in the scientific community who sought a purely non-military space program for Europe, for the governments that created ESRO and ELDO this could never be the core consideration. As the latter part of the Cold War would show, European space technology was inevitably dual use and would be exploited for both civilian and military purposes.

Notes

1. Alexander C. T. Geppert, ed., *Imagining Outer Space: European Astroculture in the Twentieth Century*, Basingstoke: Palgrave Macmillan, 2012 (2nd edn, London: Palgrave Macmillan, 2018) (= *European Astroculture*, vol. 1). All Internet sources were last accessed on 15 July 2020.
2. Treaty of Rome establishing the EEC. The Preamble noted that the signatory states were 'determined to lay the foundations of an ever-closer union among the peoples of Europe.' The treaty text can be found at http://www.cfr.org/eu/treaty-establishing-european-economic-community-rome/p18964.
3. See also Christopher Gainor's contribution, Chapter 3 in this volume.
4. Alexander C. T. Geppert, 'European Astrofuturism, Cosmic Provincialism: Historicizing the Space Age,' in idem, ed., *Imagining Outer Space*, 3–24, here 6.
5. Ibid., 6–7. The term 'Western Europe' here is understood in its geostrategic rather than geographical sense, that is, those democratic states in Europe outside the control of the Soviet Union. Some were geopolitically aligned with the United States within NATO, some were non-aligned neutrals.
6. Science fiction, particularly books and films, will be used here to trace the overlaps with the political and military debates shaping European space policy. However, science fiction embraced more than just these media, for example, comics and art, and astroculture itself was a much wider and more multi-faceted phenomenon than science fiction alone.
7. John Krige, 'The European Space System,' in idem and Arturo Russo, eds, *Reflections on Europe in Space*, Noordwijk: ESA, 1994, 1–11, here 1.
8. Walter A. McDougall, *...The Heavens and the Earth: A Political History of the Space Age*, New York: Basic Books, 1985, 424–6.
9. Krige, 'The European Space System,' 2.
10. Jutta Weldes, 'Globalisation Is Science Fiction,' *Millennium: Journal of International Studies* 30.3 (December 2001), 647–67, here 649.
11. Robert W. Cox, 'Social Forces, States and World Orders: Beyond International Relations Theory,' *Millennium: Journal of International Studies* 10.2 (June 1981), 126–55, here 128 (emphasis in original).
12. Niklas Reinke, *Geschichte der deutschen Raumfahrtpolitik: Konzepte, Einflußfaktoren und Interdependenzen 1923–2002*, Munich: Oldenbourg, 2004, 52.
13. Lodovica Clavarino, '"Many Countries Will Have the Bomb: There Will Be Hell": Edoardo Amaldi and the Italian Physicists Committed to Disarmament, Arms Control and Détente,' in Elisabetta Bini and Igor Londero, eds, *Nuclear Italy: An International History of Italian Nuclear Policies During the Cold War*, Trieste: Edizioni Università di Trieste, 2017, 245–57, here 251.
14. Maarten Van Alstein, 'From Ypres to Brussels: Europe, Peace, and the Commemoration of WWI,' available at http://www.eu.boell.org/sites/default/files/uploads/2014/06/from_ypres_to_brussels.pdf.
15. Pascal Ory refers to science fiction as being 'probably the pearl of the culture of the Cold War'; see idem, 'The Introduction of Science Fiction into France,' in Brian Rigby and Nicholas Hewitt, eds, *France and the Mass Media*, Basingstoke: Macmillan, 1991, 98–110, here 107; and Pascal Ory, 'Peut-on parler d'une culture français de Guerre Froid,' *Bulletin de la Société d'Histoire Moderne* 37.1 (January 1988), 24–9.

16. Darko Suvin, 'The State of the Art in Science Fiction Theory: Determining and Delimiting the Genre,' *Science Fiction Studies* 6.1 (March 1979), 32–45, here 37.
17. Jutta Weldes, 'Popular Culture, Science Fiction and World Politics: Exploring International Relations,' in idem, ed., *To Seek Out New Worlds: Science Fiction and World Politics*, Basingstoke: Palgrave Macmillan, 2003, 1–30, here 2; H. Bruce Franklin, *War Stars: The Superweapon and the American Imagination* [1988], 2nd edn, Amherst: University of Massachusetts Press, 2008, 5.
18. Weldes, 'Popular Culture, Science Fiction and World Politics,' 7.
19. Alexander C. T. Geppert, 'Rethinking the Space Age: Astroculture and Technoscience,' *History and Technology* 28.3 (September 2012), 219–23, here 219.
20. Gérard Klein, *Les Seigneurs de la guerre*, Paris: Robert Laffont, 1970; Carlos Saiz Cidoncha, *La caída del imperio galáctico*, Bilbao: Albia, 1978.
21. Nevil Shute, *On the Beach*, London: Heinemann, 1957; René Barjavel, *La Nuit des temps*, Paris: Presses de la Cité, 1968 (Eng. *The Ice People*, London: William Morrow, 1971); and Michael Moorcock, *A Cure for Cancer*, London: Allison & Busby, 1971.
22. Peter Randa, *Survive*, Paris: Fleuve Noir, 1960. Peter Randa was a pseudonym for André Duquesne.
23. *La Jetée*, directed by Chris Marker, FR 1962 (Argos Films/Radio-Télévision Français); *Le Dernier combat*, directed by Luc Besson, FR 1983 (Les Films du Loup); Stefan Wul, *Niourk*, Paris: Fleuve Noir, 1957; and Pierre Boulle, *La Planète des singes*, Paris: Editions René Julliard, 1963.
24. Walter Jens, *Nein! Die Welt der Angeklagten*, Hamburg: Rowohlt, 1950.
25. Hans Wörner, *Wir fanden Menschen*, Braunschweig: Limbach, 1948; Hellmuth Lange, *Blumen wachsen im Himmel*, Berlin: Minerva, 1948.
26. Jean-Gaston Vandel was the collaborative pseudonym for two Belgian authors, Jean Libert and Gaston Vandenpanhuyse. Several of their science-fiction novels were published in other language editions including German, Dutch, Spanish, Italian and Portuguese. Unusually for the anticipation novelists, one work, *Cosmic Attack* (1953), was also published in English by Evergreen Books under the title *Enemy beyond Pluto*.
27. Jean-Gaston Vandel, *Les Chevaliers de l'espace*, Paris: Fleuve Noir, 1952.
28. B. R. Bruss, *Rideau magnetic*, Paris: Fleuve Noir, 1956. B. R. Bruss was the pseudonym of Rene Bonnefoy (1895–1980). Bonnefoy used other pen name including Roger Blondel, George Brass, Marcel Castillan and Roger Fairelle. He never wrote under his own name because he had served as a Minister of Information in the wartime Vichy government and was sentenced to death *in absentia* in 1945. He re-emerged with a new identity after the war. Despite his Vichy background Bruss's postwar science-fiction writings had a strongly anti-militaristic tone.
29. Jimmy Guieu, *L'Agonie du verre*, Paris: Fleuve Noir, 1955; Henri-Richard Bessière, *Route du néant*, Paris: Fleuve Noir, 1956. Jimmy Guieu was the pseudonym of Henri-René Guieu (1926–2000). Guieu had an astonishing literary output, producing more than a hundred science-fiction novels between 1952 and 1991. He also found time to publish 17 detective novels writing as Jimmy Quint. Henri-Richard Bessière was the pseudonym for a collaboration between Henri Bessière and François Richard. Bessière (1923–2011) insisted in later life that he had written all the books, with Richard merely providing some editorial advice.

Bessière, like Guieu, was a prolific author. He wrote 98 science-fiction novels as well as several spy thrillers (using the pseudonym F. H. Ribes). His first book *Conquerors of the Universe* was written when Bessière was only 18 years old.

30. The definitive studies of the Fleuve Noir novels, on which the following section draws, are the work of Bradford Lyau; see idem, 'Technocratic Anxiety in France: The Fleuve Noir "Anticipation" Novels, 1951–60,' *Science Fiction Studies* 16.3 (November 1989), 277–97; and idem, *The Anticipation Science Fiction Novelists of 1950s France: Stepchildren of Voltaire*, Jefferson: McFarland, 2010.
31. Richard Bessière, *Les Conquerants des l'univers*, Paris: Fleuve Noir, 1951.
32. Lyau, 'Technocratic Anxiety,' 279–80.
33. Ory, 'Introduction of Science Fiction,' 99; Arthur B. Evans, 'Science Fiction,' in Pierre L. Horn, ed., *Handbook of French Popular Culture*, New York: Greenwood Publishing, 1991, 229–51, here 235.
34. The initial step of this process began with the signing in 1951 of the treaty creating the European Coal and Steel Community. See John Gillingham, *Coal, Steel, and the Rebirth of Europe, 1945–1955: The Germans and French from Ruhr Conflict to European Community*, Cambridge: Cambridge University Press, 1991; Klaus Schwabe, ed., *Die Anfänge des Schuman-Plans, 1950/51*, Baden-Baden: Nomos, 1988; Edmund Dell, *The Schuman Plan and the British Abdication of Leadership in Europe*, Oxford: Clarendon Press, 1995; and Douglas A. Brinkley and Clifford Hackett, eds, *Jean Monnet and the Path to European Unity*, Basingstoke: Palgrave Macmillan, 1991.
35. Schuman Declaration, 9 May 1950; available at https://europa.eu/european-union/about-eu/symbols/europe-day/schuman-declaration_en.
36. Tom Moylan, *Scraps of the Untainted Sky: Science Fiction, Utopia, Dystopia*, Boulder: Westview Press, 2000, xvii.
37. Jimmy Guieu, *L'Invasion de la terre*, Paris: Fleuve Noir, 1952.
38. This was a series of eleven novels, beginning with *Kings of Space*, London: Hodder & Stoughton, 1954, and ending with *The Man Who Vanished into Space*, London: Hodder & Stoughton, 1963.
39. Armin Hermann, John Krige, Ulrike Mersits and Dominique Pestre, *History of CERN*, vol. 1: *Launching the European Organisation for Nuclear Research*, Amsterdam: North-Holland, 1987.
40. Jean Monnet, *Memoirs: The Architect and Master Builder of the European Economic Community*, New York: Doubleday, 1979, 298; originally published as idem, *Mémoires*, Paris: Fayard, 1976.
41. The proposals are outlined in the Society's magazine *Spaceflight* 3.1 (January 1961), 5–8.
42. John Krige, *The Launch of ELDO*, Noordwijk: ESA, 1993, 6.
43. Michelangelo de Maria, *The History of ELDO*, part 1: *1961–1964*, Noordwijk: ESA, 1993, 3.
44. 'European 10-years Space Program Proposed,' *Aviation Week* (3 July 1961), 30–1, here 30; John Krige, 'Building a Third Space Power,' in Roger D. Launius, John M. Logsdon and Robert W. Smith, eds, *Reconsidering Sputnik: Forty Years since the Soviet Satellite*, Amsterdam: Harwood, 2000, 289–308, here 304; and Alasdair McLean and Michael Sheehan, 'A Hare Turned Tortoise: 40 Years of UK Space Policy,' *Quest: The History of Spaceflight Quarterly* 6.4 (Winter 1998), 15–24, here 17.

45. John Krige, 'Politicians, Experts, and Industrialists at the Launch of ELDO: Some Pitfalls and How to Avoid Them,' in idem and Russo, *Reflections on Europe in Space*, 13–25, here 21; Maria, *History of ELDO*, 16.
46. Peter M. Haas, 'Epistemic Communities and International Policy Coordination,' *International Organisation* 46.1 (Winter 1992), 1–35, here 3.
47. Alasdair McLean, *Western European Military Space Policy*, Aldershot: Dartmouth, 1992, 62.
48. Dominique Pestre's work represents the most comprehensive study of the history of CERN in this period. See, for example, idem, 'Aux origines du CERN: Politiques scientifiques et relations internationales 1949–1951,' *Gesnerus* 41.3–4 (1984), 279–89; idem, 'La naissance du CERN, le comment et le pourquoi,' *Relations internationales* 46 (Summer 1986), 209–26; and idem and John Krige, 'Some Thoughts on the Early History of CERN,' in John Krige and Luca Guzzetti, eds, *History of European Scientific and Technological Cooperation*, Luxembourg: European Commission, 1997, 37–57.
49. Walter A. McDougall, 'Space-Age Europe: Gaullism, Euro-Gaullism, and the American Dilemma,' *Technology and Culture* 26.2 (April 1985), 179–203. In this context see also Doug Millard, 'A Grounding in Space: Were the 1970s a Period of Transition in Britain's Exploration of Outer Space?,' in Alexander C. T. Geppert, ed., *Limiting Outer Space: Astroculture after Apollo*, London: Palgrave Macmillan, 2018, 79–99 (= *European Astroculture*, vol. 2).
50. John Krige, *Europe into Space: The Auger Years (1959–1967)*, Noordwijk: ESA, 1993, 5.
51. Michael J. Neufeld, *The Rocket and the Reich: Peenemünde and the Coming of the Ballistic Missile Era*, New York: Free Press, 1995; see also his contribution to the present volume, Chapter 2.
52. Kazuto Suzuki, *Policy Logics and Institutions of European Space Collaboration*, Burlington: Ashgate, 2003, 46.
53. Edoardo Amaldi, 'Introduction to the Discussion on Space Research in Europe,' 30 April 1959, Historical Archives of the European Union/European Space Agency, COPERS 0001, 4.
54. Krige, *Europe into Space*, 12.
55. Michelangelo de Maria, *Europe in Space: Edoardo Amaldi and the Inception of ESRO*, Noordwijk: ESA, 1992, 6–7.
56. Peter Fischer, *The Origins of the Federal Republic of Germany's Space Policy 1959–1965: European and National Dimensions*, Noordwijk: ESA, 1992, 4.
57. Maria, *Europe in Space*, 8–12.
58. Alasdair McLean and Fraser Lovie, *Europe's Final Frontier: The Search for Security Through Space*, Commack: Nova Science, 1997, 43–5.
59. Convention for the Establishment of a European Space Research Organisation: Opened for Signature Paris, 14 June 1962, Entered into Force 20 March 1964; full text available at http://ops-alaska.com/IOSL/V3P1/1962.convention_EN.pdf.
60. Amaldi, quoted in Blandina Baranes, 'European Identity: Attempt of a Definition,' in idem and Christophe Venet, eds, *European Identity Through Space: Space Activities and Programmes as a Tool to Reinvigorate the European Identity*, Vienna: Springer, 2013, 10–24, here 18. See also Norman Longdon and Duc Guyenne, eds, *Twenty Years of European Cooperation in Space: An ESA Report*, Noordwijk: ESA, 1984.

61. Krige, *Europe into Space*, 11.
62. The ESRO members were Belgium, Denmark, West Germany, France, Italy, the Netherlands and the United Kingdom, all NATO states. In addition the non-NATO states of Spain, Sweden and Switzerland joined. ELDO membership was limited to Australia (which provided the Woomera launch site), Belgium, West Germany, France, Italy, the Netherlands and the United Kingdom.
63. Maria, *History of ELDO*, 4.
64. Peter Creola, *Switzerland in Space: A Brief History*, Noordwijk: ESA, 2003, 3.
65. H. G. R. Robinson, 'Suggested Developments of Black Knight,' *Journal of the British Interplanetary Society* 43.7 (July 1990), 317–18, here 317.
66. *ELDO Report, 1960–1965, to the Council of Europe*, Paris: ELDO, 1965, 7; Geoffrey K. C. Pardoe, *The Challenge of Space*, London: Chatto & Windus, 1964, 121.
67. Preben Gudmandsen, *ESRO/ELDO and Denmark: Participation by Research and Industry*, Noordwijk: ESA, 2003, 3–5; Creola, *Switzerland in Space*, 4. Switzerland did not join, and Denmark opted for 'Observer' status.
68. Michael Sheehan, *The International Politics of Space*, London: Routledge, 2007, 76.
69. Convention for the Establishment of a European Organisation for the Development and Construction of Space Vehicle Launchers: Opened for Signature, London, 29 March 1962, Entered into Force 29 February 1964. The full text is available in A. H. Robertson, *European Institutions: Cooperation, Integration, Unification*, 3rd edn, London: Stevens, 1973, 438–53; https://ops-alaska.com/IOSL/V3P1/1962_ELDOConvention_EN.pdf.
70. Kenneth W. Gatland, 'Woomera and the ELDO Launcher,' *Aircraft* 42.5 (May 1962), 16–17, here 16.
71. 'European Space Policy and ELDO,' *Flight International* (10 February 1966), 237–8, here 237.
72. Charles N. Hill, *A Vertical Empire: The History of the UK Rocket and Space Programme, 1950–1971*, London: Imperial College Press, 2001, 7.
73. 'Space Research: Blue Streak. Report by Officials,' The National Archives, File FO371/149657 (April 1960), 11; quoted in John Krige, *The Launch of ELDO*, Noordwijk: ESA, 1993, 7.
74. Keith Hayward, *British Military Space Programs*, London: Royal United Services Institute, 1996, 5; Hill, *Vertical Empire*, 119.
75. Telegram from French Foreign Office to the British Foreign Office, 24 November 1960; quoted in Krige, *Launch of ELDO*, 13.
76. See John Krige, 'Embedding the National in the Global: US-French Relationships in Space Science and Rocketry in the 1960s,' in Naomi Oreskes and John Krige, eds, *Science and Technology in the Global Cold War*, Cambridge, MA: MIT Press, 2014, 227–50, here 228; Bruno Gire and Jacques Schibler, 'The French National Space Program, 1950–1975,' *Journal of the British Interplanetary Society* 40.2 (February 1987), 51–66, here 60.
77. George Lakoff and Mark Johnson, *Metaphors We Live By*, Chicago: University of Chicago Press, 1980; Elena Semino, *Metaphor in Discourse*, Cambridge: Cambridge University Press, 2008.
78. Stacia E. Zabusky, *Launching Europe: An Ethnography of European Cooperation in Space*, Princeton: Princeton University Press, 1995, 5.
79. Krige, 'The European Space System,' 10.

PART II

Waging Future Wars

CHAPTER 5

In Space, Violence Rules: Clashes and Conquests in Science-Fiction Cinema

Natalija Majsova

Twentieth-century cinematography offers fruitful grounds for space-themed science fiction of various sub-genres and formats, from metaphorical shorts and animated films to allegorical political sagas, serials, introspective investigations of the human psyche and offbeat humoresques. In retrospect, this transnational cinematic landscape is not only an exciting aspect of global astroculture, but also a compelling vantage point for research in cinematic constructions of popular imaginaries of outer space and space exploration. Many recent studies on Western astroculture highlight that the conception of outer space thematized in policies and cultural products – for both entertainment and education – is usually a reflection, or extrapolation, of earthly concerns.[1] On the other hand, it is a common contemporary expectation to challenge familiar stereotypes, preconceptions and expectations of the future in space. Indeed, why imagine the future to be akin to the present if the present is not 'the best of all possible worlds'? Why imagine outer space as an extrapolation of earth or strive to turn it into such an extrapolation? Why transpose terrestrial conflicts into outer space, transforming it into a game-like battlefield?

As an unexplored, unconquered horizon, a vast and virtually unlivable environment, outer space may serve as a good prop to spice up a box office blockbuster or as a didactic aide, allowing the director to convey, in allegorical terms, a certain political message. From this perspective, it is unsurprising that numerous twentieth-century Western science-fiction films set in

Natalija Majsova (✉)
Université Catholique de Louvain, Ottignies-Louvain-la-Neuve, Belgium
e-mail: nmajsova@gmail.com

Alexander C. T. Geppert et al. (eds), *Militarizing Outer Space*
European Astroculture, vol. 3
https://doi.org/10.1057/978-1-349-95851-1_5

outer space demonstrate an awareness of connections that space exploration had with the development of military technologies and with the context of contemporary military and political conflicts, namely the two world wars, the Cold War and the disintegration of the Eastern bloc. However, this chapter aims to demonstrate that science-fiction cinema portrayed outer space and the militaristic dimension of space programs in terms of a set of thematic preoccupations.

Political and legal experts usually define space militarization as policies and acts that involve placing weapons and technologies that may serve aggressive ends into space beyond the atmosphere of planet Earth.[2] From the perspective of popular culture analysis, this definition appears rather narrow. This is because military or violent conflicts in space as created within the science-fiction genre are only partly conditioned by the actual presence of weapons and different kinds of technologies in outer space. The space of fiction usually focuses less on creating precise analogies to political and technological realities and more on conveying allegorical meanings. Therefore, we can speak of the militarization of space in cinema as early as 1902 when the first images of the dream of spaceflight were introduced to cinematic screens and decades before the construction of the first rockets, let alone spaceflight. This chapter argues that a look at popular culture allows us to broaden the definition of space militarization. Science-fiction films of the twentieth century demonstrate how the presence of violence and military weapons in space is conditioned by sets of relations: political and international relations as well as those between genders, races and individuals.

This chapter refers to a wide array of space-themed fiction films from the archives of transnational film heritage in order to explore at least four different evident categories of antagonisms that underpin the producers' approaches to activities in space: ideological, geopolitical, psychological and biopolitical antagonisms. These metaphorical battlefields are predictably in line with major global political preoccupations of the twentieth century, that is, the Russian revolution and the communist threat, the Cold War divide, the rise of minority rights movements and the global expansion of capitalism. Furthermore, they produce specific versions of outer space as constellations of space and time or 'timespaces,' to use a term from film studies. While violence plays an important role in all these constructions, it does so with reference to the specific dimensions of astroculture: its varying ideological backdrops and geopolitical references, its interplay with the human imagination and the underlying presumptions about values and morals, and, finally, its attitude towards the very basic definition of life.

Rather than aiming at providing a systematic overview of space-fiction cinematography, this chapter foregrounds a number of selected films in order to approach, illustrate and develop the overarching conceptual issue: the interplay of representations of current socio-political reality and imagination in cinematic portrayals of militarized outer space. Relying on relatively well-known films, I argue that while science-fiction cinema of the previous century represented current political preoccupations, it also contributed to space militarization by normalizing the notion of an omnipresence of danger and violence in space or coming from space. Over the course of the century, the image of outer space as a potentially dangerous and therefore violent setting, a battlefield gradually turned into a common presumption and a widely used trope in popular culture. It is on this very level that a lot of the films discussed below offer a strong critique of space programs, either with reference to their political connotations or by highlighting the basic social structures that created the conditions for the first spaceflights. In this sense, some of the films discussed are not representations but alternative visions of these societies and their respective space efforts. While providing a geographically decentralized perspective on the heritage of space cinema and granting limited attention to the most commonly discussed classics including *Destination Moon* (1950), *The Day the Earth Stood Still* (1951) and the *Alien* film series (1979–97), the chapter remains within the confines of European (including both so-called Western and Eastern European) and, to an extent, US film production.[3]

I Setting the scene: the 'non-place' fantasy

Stevphen Shukaitis has argued that contemporary approaches to space programs rely on a representational conceptualization of outer space, employing it as an answer to terrestrial political and economic crises and circumstances.[4] Shukaitis discusses this apprehension of space as a '(non)place.' Space as a non-place functions both in Michel de Certeau's terms, that is, as a place constructed on rules established by ideological apparatuses such as legal institutions, schools and political actors, and as discussed by Marc Augé, as a space of transition regulated by texts and signs, like an airport or a shopping mall.[5] In contrast with these formulations, Shukaitis stresses that outer space could be conceptualized as a Deleuzian 'imaginal machine,' that is, a milieu that fosters the emergence of new sets of rules, ideas and worlds. According to Shukaitis, outer space could be such a Deleuzian space of radical imagination: it could be a space for novel concepts, ideas and social structures, but in fact currently is not one. Rather, it is akin to a floating signifier, which is systematically filled with humanity's projections that reflect contemporaneous socio-political circumstances, such as romantic utopianism, which favors the search for new worlds, or the Cold War, which induces paranoia about the possible omnipresence of the ideological enemy. In this regard, the fact that it often – in theory, practice and popular cultural representations – becomes filled with weapons and military mechanisms of various kinds comes as no surprise.

In the domain of cinema, outer space fulfils another function. Aside from being a grateful terrain for playing out terrestrial conflicts, it serves as the arena for the ultimate spectacle. This has been pointed out by Susan Sontag in her classic 1965 discussion of American science-fiction cinema:

> Science fiction films are one of the purest forms of spectacle. [...] Things, objects, machinery play a major role in these films. A greater range of ethical values is embodied in the décor of these films than in the people. Things, rather than the helpless humans, are the locus of values because we experience them, rather than people, as the sources of power. According to science fiction films, man is naked without his artifacts. *They* stand for different values, they are potent, they are what gets destroyed, and they are the indispensable tools for the repulse of the alien invaders or the repair of the damaged environment.[6]

Sontag's argument foregrounds an important point. Despite the typically heavy reliance on narratives, which are often adaptations of science-fiction literature, the cinematic medium does much more than merely 'adapt' narrative-based scripts. Being an audio-visual production, science-fiction cinema provides a certain 'immediacy' of experience, which arguably cannot be reproduced in the experience of reading. Cinema does not represent experience; rather, it constructs it. Hence, in science-fiction films, particularly in the more introspective productions, artifacts such as machines, monuments or buildings sometimes function as elongations of thought. A monumental building, such as the Empire State Building in *The Day the Earth Stood Still*, may hence center the world around itself, turning it into a specific place which points to the center and to the periphery of the world, to the good and the evil actors, to which decisions are acceptable and which are not. Humans are no longer the only carriers of subjectivity or the only possible loci of the viewer's identification. Anthropomorphic subjectivity is only one of many possible variants, alongside others, such as inorganic cultural artifacts. In these cases, the militarization of outer space can no longer be attributed to certain actors, for example, to states or armies. The violent disposition marked by threats and armor is generated by the texture of the cinematic world, which is constructed out of cinematic factors such as the mise-en-scène, the script, the characters and extra-cinematic references, pertaining to contemporary and historical issues. This is particularly evident in European productions, where space movies such as *Kin-Dza-Dza!* (1986) and *Na srebrnym globie* (1988) were seen as an opportunity for experimentation with genre conventions and subversion of established social conventions.[7]

Certainly, the argument on the imaginal capacities of space-themed science-fiction cinema is limited by the realities of film production. NASA played an active role in the Hollywood film industry, in accordance with the 1958 National Aeronautics and Space Act that obliged the Administration to keep the public as informed as possible about its activities. In the 1960s the agency formed a special Entertainment Industry Liaison in charge of collaboration with the popular culture industry. Until the 1990s, when direct participation between NASA and film production began, the former was predominantly concerned with the film industry's accuracy in its portrayal of the development of space technologies and depiction of key milestones of scientific progress. For example, it played the role of a consultant, overseeing the representation of space travel in the space travel films of the 1960s and 1970s, like *Marooned* (1969), and space shuttles in the James Bond-movies *Diamonds Are Forever* (1971) and *Moonraker* (1979). The space station in *2001: A Space Odyssey* was created by former NASA associate Harry Lange (1930–2008). Due to Lange's previous work at NASA alongside chief rocket designer Wernher von Braun (1912–77), the drawings had to receive security clearance before they could be used on set. Collaboration with the entertainment industry also helped NASA maintain its positive public image even in the light of space tragedies such as the Space Shuttle *Challenger* catastrophe in 1986.[8] While Western European space programs did not focus on systematic collaboration with the film industry, an indirect relation between the two can be noted in the Soviet Union. The movie industry was subordinated to the State Committee of Cinematography which authorized production of all films.[9] Goskino generally supported space travel films with unambiguous narratives and treated science fiction as a potentially educational genre, supporting ideologically plausible productions that celebrated the Soviet space program and its future, and were aimed at younger audiences, especially in the 1970s.[10]

A large proportion of space-themed science-fiction cinema was also subordinated to representational functions, namely to representing the ambitions and achievements of space programs and thereby to conveying subliminal ideology-infused messages, such as value hierarchies, national stereotypes or visions of the future. As such, twentieth-century science-fiction cinema typically constructed outer space in terms of an inside/outside binary. In doing so, it used space-related tropes (space exploration and/or alien invasion from outer space) in order to construct this binary along ideological, geopolitical, gender, psychological and biopolitical axes. The ways in which these axes were constructed and worked will be explored in the selected films discussed below, their chronological succession reflecting certain political preoccupations and historical developments. Tracing this evolution of cinematic portrayals of outer space from an arena of ideological conflict into a vast, unknown terrain that calls for a new model of governance on planet Earth and beyond will allow us to explore the symbolic, rather than literal aspects of space militarization in the twentieth century.

II Dispelling the discreet charm of the bourgeoisie: the pre-canonical years

The history of space science-fiction cinema began with Georges Méliès's (1861–1938) innovative short film *Voyage dans la lune* (1902), and what could be seen as its 1898 prequel, *La Lune à un mètre*, a short reverie about an astronomer's dream about 'looking the moon straight in the eye.'[11] Méliès's two *fin-de-siècle* depictions of man's first ventures into outer space were products of, and comments on, the advent of modernity, accompanied by the emergence of various workers' rights movements and searches for alternatives to the existent economic system. Moreover, the lengthier and more elaborate *Voyage dans la lune* also appeared as a prototype for different sub-genres of contemporary science-fiction cinema.[12] Both shorts located the ambitions to reach the moon at the intersection of the then contemporary 'state-of-the-art' in science with the imagination. Numerous critics have noted the parodic appearance of the astronomers in the two films. Their visions and apparent lunacy are coupled with imperialist ambitions of the state. Not only does no one have a problem with launching a cannonball-shaped capsule into outer space, aiming at the anthropomorphic moon and hitting the man in the moon in his eye (Figure 5.1), but Professor Barbenfouillis also receives a monument erected in his honor upon his successful return to earth with a captive selenite.

The short silent film was met with great enthusiasm by French audiences and was equally successful in the United States, where pirated versions by Lubin, Selig and Edison ensured its great distribution. By 1904 it had also been widely screened in Germany, Canada and Italy.[13] The film provided

Figure 5.1 A still from *Voyage dans la lune*, showing the bullet-like rocket from earth spectacularly hitting the man in the moon in the eye, alluding to the violence inherent to human pursuits of knowledge and new horizons.

Source: *Voyage dans la lune*, directed by Georges Méliès, FR 1902. Courtesy of Star Film Company.

a satirical critique of political order and pointed to the limits and dangers inherent to blind faith in scientific progress. Humanity only ventures to the moon to find out what it looks like to be atop a foreign planet. The earthlings are completely ignorant of all kinds of non-anthropomorphic extraterrestrial life that it encounters along the way, like the man in the moon or the selenites, which appear to explode when physically attacked. Humanity, represented in full only by mature bourgeois males, flies off into space absolutely certain of its intellectual and physical supremacy. It nearly destroys a foreign culture and returns to earth rejoicing in the triumphs based on the astronomers' rational calculations as well as human courage and supremacy over the selenites. The narrative, commenting on the colonialist and imperialist political context of the late nineteenth and early twentieth centuries, is suspiciously familiar to later science-fiction films, based on the premise that science and military power are key to humanity's victory over all possible extraterrestrial threats. The explicitly satirical tone adopted by *Voyage dans la lune*, however, would be much less familiar to later science-fiction films. Critics have suggested that it may be seen as an anti-imperialist parody, reminiscent of later films of very different science-fiction sub-genres, such as space musicals or late 1980s Eastern European films, for instance, the afore-mentioned *Kin-Dza-Dza!*[14]

Furthermore, *Voyage dans la lune* was produced long before the codification of the now-classical cinematic gestures, such as close-ups, juxtapositions, the tracking shot and the shot-reverse-shot sequence. It does not operate with a conventional linear narrative, but rather treats time and space as flexible parameters, which allowed showing the voyage to the moon in two sequences from two different perspectives.[15] Méliès created a visually rich spectacle which allowed the spectator to enjoy fantastic, innovative shots of imagined situations, places and times, transposing the tradition established in the literary genre of science fiction to a new medium.[16] Equally important, Méliès's film predicted that space exploration would be the result of the development of military technologies. Therefore, even this forerunner of the science-fiction spectacle linked the very idea of space exploration to military pursuits. Aggressive ventures into space were depicted as a consequence of a colonialist attitude exhibited by the protagonists of the film. They arrive on the lunar surface unarmed, but are happy to use their umbrellas against the selenites once they realize this to be an efficient means of destroying them.

Early space science-fiction films also include the first space opera prototype, the Danish silent film *Himmelskibet* (1918); literary adaptations, such as H. G. Wells's *The First Men in the Moon* (1919); the revolutionary allegory *Aelita* (1924); as well as the scientifically rigorous first space melodrama *Frau im Mond* (1929).[17] With regard to their formal features, aesthetics and narratives, these films were much less ambiguous than *Voyage dans la lune*. They signaled a solidification of a transnational visual culture of outer space within the Western imaginary, demonstrating that, although outer space might be portrayed in an expressionist (*Frau im Mond*, *Aelita*) or a more realist-futurist manner (*The First Men in the Moon*), it is already constructed as the ultimate battlefield, featuring the struggles of a prevalent conception of society against extraterrestrials, imperialists, capitalists or others. Notably, for contemporary analysts these films demonstrated a heightened awareness of the need of socio-political critique. While the critical attitude remained relatively covert in Western European productions, early, pre-Second World War Soviet films featured a much more pronounced degree of socio-political engagement. Of course, an ample amount of anti-capitalist and pro-communist propaganda in the Soviet Union in the 1920s and early 1930s is hardly surprising; much more so are cases where such propaganda was openly mocked.

The year 1924 was marked by the release of two films based on Tolstoy's 1923 novel *Aelita*, which imagined a communist revolution on Mars: the homonymic *Aelita*, directed by Yakov Protazanov, and *Mezhplanetnaya revoliutsia*, which directly parodied Protazanov's adaptation.[18] In the latter silent animated short, Red Army soldier comrade Kominternov follows a bourgeois mission to Mars in order to prevent capitalist rule over the red planet, initiating a communist revolution instead. Interestingly enough, both parties in the class struggle are portrayed with the same amount of mockery. The bourgeois resemble bulldogs which regard outer space as their new 'final frontier,' full of resources and therefore ready for their lucrative ventures, whereas the ideologically-drugged communists appear one-dimensional and emotionally void. The ways in which the directors – young avant-garde Soviet animators – played with narrative structures, explicit and implicit social critique as well as outworldly fantasies, made this experimental animation function in a very similar way to *Voyage dans la lune*. However, in contrast to Méliès's work, *Mezhplanetnaya revoliutsia* was never finished and did not make it to film theaters, leaving *Aelita*, a slightly less politically problematic adaptation of Tolstoy's novel, the only mark of early Soviet cinematic astroculture.[19]

The combination of swift technological progress (coupled with enthusiasm) in the domain of rocketry and the absence of a strictly codified cinematic language allowed for an unprecedented degree of experimentation on screen. Nevertheless, space in these early science-fiction films, which were produced long before the first tangible results of space programs, consistently – and very much in line with contemporary colonialist/socialist scenarios – appeared as a somewhat generic, unexplored terrain that needed to be tamed by military means. The decades that followed led to significant changes in the imagery of outer space, which offered conflict, warfare and violence within very specific timespaces.[20] This can be explained by the gradual establishment of film genres with distinct features, audiences and marketing strategies as well as the appearance of distinct film policies, particularly in Eastern Europe, where film policy was governed by state communist parties.

III The classical age of the space science-fiction blockbuster: the search for the ultimate other

While pre-Second World War space-themed science-fiction cinema focused on the ways outer space futures could broaden humanity's horizons beyond the constraints of contemporary political regimes, postwar science fiction took a different turn. As the day of the launch of the first satellite and human spaceflights drew closer, Hollywood released spectacular, albeit somewhat dystopian feature films including *When Worlds Collide* (1951), *The Day the Earth Stood Still* (1951), *Invaders from Mars* (1953) and *War of the Worlds* (1953), known today as some of the most iconic science-fiction films of the 1950s.[21] Although at first glance these genre films utilized the very same recipe perfected by their prewar predecessors – a pinch of threat and the atmosphere of an impending disaster – tone and message were starkly different. 1950s science fiction emphasized the possible existence of a serious external threat to a familiar society and justified the need for further internal social consolidation. The spectacle portrayed by many 1950s space-themed science-fiction films was ultimately the feat of order, collaboration and transparency over an unidentified or a vaguely identified threat. As noted by Vivian Sobchack, the trope of the city, an ordered and consolidated urban space, plays an important role in science-fiction films such as *The Day the Earth Stood Still*, that is, films that aim at reconciling human and alien (or unknown) power.[22] Miranda Banks has argued that '[c]ountless science fiction films use easily recognizable foreign landmarks in montage to show the international scale of the events taking place within the film. These single spaces speak not only to the residents' experience but rather show the solidarity of all nations: What happens to one will happen to all.'[23]

Alongside this internationalist undercurrent, the universe of Hollywood 1950s science-fiction cinema heavily relied on the dichotomy of inside versus outside. Contemporary analysts agreed that the increased rate of science-fiction film production of the 1950s was in direct relationship to the increasing public concern about communism and the fear of nuclear disaster.[24] For instance, the 1950 production *Destination Moon*, praised for its technological accuracy and box office popularity, states that private investments in the space program are necessary for the United States to "win" the moon race against the Soviet Union, turning outer space into an extrapolation of a political conflict (the early Cold War) on earth.[25] In the film, fictional General Thayer states, quoting Wernher von Braun, 'The first country that can use the moon for the launching of missiles will control the earth. That, gentlemen, is the most important military fact of this century.'[26]

The Day the Earth Stood Still is an equally illustrative case.[27] The alien Klaatu arrives on earth to warn earthlings that their escalating belligerence and inability to coexist peacefully will not be tolerated by the interstellar community. In contrast, Klaatu's own futuristic alien community protects itself from armed conflict by subjugating itself to robot guardians. Utopia comes at a price: the technologically advanced alien race buys peace at the expense of confining itself to a panopticon. The film suggests that this is the only alternative to incessant wars which eventually lead to the elimination of the society in question. In this respect, *The Day the Earth Stood Still* offers a complex analysis of the problem of technological progress, coupling it with the context of the eternal and unchanging problem of social antagonisms and political conflicts. According to director Robert Wise (1914–2005), the film offered elements of biblical allegory, with Klaatu functioning as a Christ-like figure reacting both to the recent Second World War and the Cold War (Figure 5.2).

A similarly nuanced perspective on the future of modernity, characterized by technological advances, was provided by the 1960 East German production *Der schweigende Stern.*[28] Distributed as *First Spaceship on Venus* in the United States and set in 1985, it focuses on the role of the international community of scientists in discovering evidence of a Venusian presence on earth

Figure 5.2 A still from *The Day the Earth Stood Still*, depicting Klaatu and Gort descending from the alien spaceship. The image conveys a feeling of uneasiness; the alien and his robot guardian appear to be eerily alike. Gort's pose mimics Klaatu's, while Klaatu's expression is so reserved that it resembles Gort's motionless, featureless face. Gort is the externalization of Klaatu's conscience, and the pair embodies the somber triumph of rationalism over affect.

Source: *The Day the Earth Stood Still*, directed by Robert Wise, USA 1951. Courtesy of Twentieth Century Fox.

and sending a spaceship to Venus, only to realize that the Venusians, once a powerful race that had harbored a plan to eradicate humanity and invade earth, have gone. The race destroyed itself with its own atomic weapons, its ruins serving as a warning to the earthlings. In contrast to productions such as *The Day the Earth Stood Still*, this film, an adaptation of Stanisław Łem's (1921–2006) novel, had a more pronounced internationalist note. Not only does the space crew consist of scientists of various nationalities, but their journey through the ruins of Venusian civilization serves as a visual allegory to post-Second World War earth. *Der schweigende Stern* used the genre of science fiction and the trope of the Venusians to create an otherworldly allegory of the perils inherent in humanity's thirst for physical, symbolic and technological domination.

Other space-themed science-fiction films from this period were less far-reaching in this regard, portraying outer space as a simple extrapolation of the Cold War antagonism. *The War of the Worlds* (1953) used a Martian invasion as a cue for a visually excessive thriller, wherein humanity, embodied by the inhabitants of a town in Arizona, is attacked by a violent alien aggressor, only to seek salvation in love and prayer. The threat is removed by nature itself: The Martians succumb to microbes and bacteria present in the earth's atmosphere. *Invaders from Mars* (1953) went a step further to more clearly couple the Martian threat with the Soviet one. Martian invaders, like a McCarthyist 'red scare,' indoctrinate humans and take over their brains.[29] Incidentally, Soviet space cinema's preoccupation with the Cold War divide was notably different. Films such as *Nebo zovyot* (*Battle beyond the Sun*, 1959), featuring the imagined future of the Soviet space program (that is, a flight to Mars), envisaged eventual collaboration between the United States and the Soviet Union, albeit clearly under the aegis of the latter.[30]

In contrast to these Soviet and American films, Western European productions of this period rarely depict the militarization of outer space as a result of current political constellations or use the metaphor of an alien threat to demonize a contemporary political other. Rather, outer space was a fantastic arena for mixing genres, intertextual references and, ultimately, for producing 'notoriously bad' films.[31] Such productions were evidently inspired by genres such as horror and satire, and often focused on how space exploration might foreground and question traditional dichotomies, sometimes with unconventional results. In the UK low-budget production *Devil Girl from Mars* (1954), alien space commander Nyah ('Devil Girl') from Mars, dressed in black vinyl, is on a mission to abduct the earth's male population and exploit it for the revitalization of her own, male-deprived planet.[32] While the film's narrative is humorous in its absurdity, it also defends a fairly conservative point: Earth's peaceful future depends on sustaining an equilibrium between what appear to be established, fixed gender roles. Proactive, physically attractive and dominant Nyah is a threat to this equilibrium, and it is the male earthlings' duty to disempower her.

Another low-budget 1956 British production, *Fire Maidens from Outer Space*, foregrounded the question of gender roles in space, but with a plot twist.[33] Here, a crew of stereotypically macho, heavily smoking astronauts from earth lands on the thirteenth moon of Jupiter to explore signs of life that have recently been discovered there. They encounter a dying civilization, consisting of 16 women and their middle-aged symbolic father Prasus. The people are being terrorized by a beastly-headed monster. This dying population is led by Duessa, who also initially wishes to retain the astronauts in captivity, using them as mates. However, they prove to be very useful, killing the monster who has murdered Prasus and several women in his latest attack. The astronauts leave the hostile moon with one of the women and promise to send spaceships full of men to the forsaken celestial body. Once again, male physical strength and virility proved to be crucial for the future of civilization, both terrestrial and alien. Moreover, both films portrayed the possibly alien inhabited pockets in outer space as dangerous places that must be approached bearing firearms. Even in these humorous films of the first half of the 1950s, the visions of space exploration and extraterrestrial encounters depict space programs as the result of technological progress, brought about by a social order organized and ruled by males. In this situation, space exploration is limited to being a colonialist pursuit. Technological progress is not aligned with a pluralist, open attitude toward novel forms of life or social order. Both the aliens and the expeditions from earth are, by default, armed with lethal weapons intended for the destruction of anything considered to be alien.

Forbidden Planet was released in 1956, a year after both the United States and the Soviet Union had declared their intentions to launch artificial satellites in the near future, catapulting them into the beginning of the Space Race.[34] Apart from combining many of the themes and tropes highlighted by the films already mentioned, *Forbidden Planet* set new precedents that later films in this genre would successfully continue to exploit, along with a fully electronic score composed by Louis and Bebe Barron. Much like *Destination Moon*, *Forbidden Planet* glorified the technological advances of the human race, rating them higher than those of alien life forms. The film, set in the twenty-third century, features a human crew traveling in a spaceship faster-than-light, accompanied by Robby the Robot, the first-ever opinionated, anthropomorphic robot to appear in a space science-fiction film. The expedition is set out to check on an earlier mission to Altair IV. They find out that the entire expedition on Altair IV succumbed to Monsters of the Id, produced by scientist Edward Morbius's subconscious – actual monsters released into reality by Morbius who has access to an ancient machine.[35] The monsters of Morbius's Id are only defeated after Morbius acknowledges his complicity in the elimination of the first expedition to Altair. The price he has to pay for his power is his life.

Forbidden Planet set a number of trajectories for the development of space-themed science-fiction cinema for several decades. First, the narrative turns outer space into a mixture of perils and challenges 'from the outside' and of 'perils from within,' which are most often incarnations of an individual's unconscious and fall into Sigmund Freud's category of the uncanny. The film suggests that space exploration could have the capacity to exceed our expectations and may be the beginning of developments that humans cannot foresee. According to the societal allegory presented in the film, advances in science, which are not followed by changes in the psyche, can result in the demise of an entire civilization.

Second, the film's treatment of women is frequently echoed in films that followed. The science-fiction universe of the twentieth century is emphatically phallocentric, even when films such as the aforementioned *Devil Girl from Mars* or the later *Barbarella* (1968) focus on female protagonists.[36] In the specific case of *Forbidden Planet*, the entire crew is male, and they allow themselves to seduce Altaira, Morbius's teenage daughter, whose only possible choice is to follow the scientist-astronauts back to earth. Women were typically depicted as either passive and submissive, motherly or alien. Outer space was reserved for men who had to combat all alien elements that came their way, regardless of the latter's sex or gender.

Thus, Western outer space-themed films of the 1950s demonstrated a clear resonance with the general societal preoccupations of the period, functioning as a critical allegory, at the same time wrapped into the form of entertainment, often described as pulp by critics at the time. Outer space was linked to the idea of an approaching apocalypse. However, apart from simply demonstrating the danger of the possibility of ideological conflicts turning into tangible armed confrontations on earth and beyond, films like *Forbidden Planet* also satirically highlighted the role of social norms and rules in the development of such grave and destructive disputes. *Forbidden Planet* and *Devil Girl from Mars* hinted at resolving political and ideological conflicts on earth ultimately not providing true peace of mind, and suggested that more antagonisms, such as a gender conflict, were to be found within democratic societies.

IV Post-1957 science-fiction cinema: the danger from within

If the 1950s and early 1960s have at times been characterized as the golden age of space-themed cinematography, the post-Apollo period was marked by a different atmosphere, for instance, in the films of Italian director Antonio Margheriti (1930–2002) such as *Space Men* (1960), known as *Assignment: Outer Space* on the US market.[37] Operating within a narrative framework similar to *Forbidden Planet*, *Space Men* featured a much more nuanced set of characters, suggesting that man's perception of outer space does not merely depend on outer space as a dark, empty and potentially perilous place, or on its ability to bring out our deepest fears, but also on the particular qualities of the characters who end up there. The astronauts aboard the space station are no longer the perfect heroes of previous films. Neither are they immune to interpersonal quarrels, amorous delusions or impulsive decisions, nor do they automatically sacrifice their personal dignity and well-being for the betterment of humanity. Rather than constituting unambiguously 'good' or 'bad' characters, they are emotional, imperfect human beings who find it difficult to distinguish between their personal desires and the greater good of humankind.[38]

The trajectory of the problem posed to space exploration by the ambiguous and irrational nature of humanity itself resurfaced in numerous space-themed science-fiction films of the 1960s and 1970s. Interestingly enough, there was an evident decline in Western space film production after the Apollo moon landings. A turn to inner space, materialized in the horror genre and, a little later, to cyberspace, was notable in both Hollywood and Western European science-fiction cinema. Outer space-themed films did not disappear, but they were more frequently used as a scenic prop for either political satire or explorations of the human psyche, or both. In the 1950s the main fascination exhibited by cinematography in relation to outer space had been the very *possibility* of spaceflight and its implications, that is, its philosophical significance, as well as its attractiveness for traditional genres, such as horror, comedy and science fiction. However, films released after the launch of Sputnik 1 and the first spaceflights were typically less preoccupied with the first theme and more focused on visualizing outer space in as horrifying, humorous or politically allegorical terms as possible.

For many film enthusiasts, the late 1960s and the 1970s produced some of the most profound cinematic explorations of the universality and significance of spaceflight that eventually acquired cult status among science-fiction fans. *2001: A Space Odyssey* (1968), *Solaris* (1972) and *The Man Who Fell to Earth* (1976) are perhaps the first titles that spring to mind.[39] These films presented a stark diversification of the genre of space fiction. Rather than focusing on outer space as an uncharted territory that contains perils and needs to be rendered safe, they concentrated on the people from or in space. In these three films, space served as a device that allowed the protagonists to explore their psyche in previously unforeseen circumstances. The cinematic proclamation that the real danger of space lay in the opportunity to encounter oneself echoed the imperative of the United Nations' Outer Space Treaty from 1967. While the treaty called for 'peaceful' use of outer space, *2001: A Space Odyssey*, *Solaris* and *The Man Who Fell to Earth* implied that militarization, that is, trying to save humanity by placing weapons or even surveillance technology into space, would not provide any far-reaching solutions.

Moreover, these three films foregrounded a distinct feature of post-spaceflight science-fiction cinema: a familiarity with space, uncommon for earlier works. The diversity of science-fiction sub-genres hardly increased in the years following the launch of Sputnik. On the other hand, films within individual existing sub-genres evolved in terms of their relative narrative, visual and cinematic complexity. Building upon the legacy of *The Day the Earth Stood Still* as well as *Der schweigende Stern*, Kubrick's *2001: A Space Odyssey* plunged straight into the question of the birth of a new kind of

Space Age subjectivity. It was embodied by a child of technology, the computer HAL 9000, on the one hand transcultural and ahistorical in its rational logic, yet carrying the weight of the European Enlightenment on the other. *Solaris* built on an altogether different legacy, the one of Eastern Europe's rich science-fiction literature tradition, coupled with Tarkovsky's aspirations to transform cinema into a means of 'sculpting in time,' that is, carving out timeless timespaces of universal significance.[40] Viewed from the perspective of space-related violence, both *Solaris* and *2001: A Space Odyssey* highlighted the same idea: that violence, here epitomized by human-created sentinel computer HAL 9000, was inherent to thought, language and culture, and may therefore not be prevented in human space exploration.

Both *Solaris* and *2001: A Space Odyssey* distinguished between violence and militarization. This distinction was a symptom of post-Sputnik science-fiction space-related cinematography. Productions such as *The Man Who Fell to Earth* also emphasized an increasing militarization of social life as not the greatest peril posed by the technological progress that brought about and set the limits and horizons of space exploration. *The Man Who Fell to Earth* explored the fate of intellectually superior humanoid alien Thomas Jerome Newton (played by David Bowie; Figure 5.3) who crashes on earth and is trying to find a way to send water to his own exhausted planet, which is on the brink of environmental collapse. However, underneath this relatively simple narrative the film also addressed the paradoxes of the core values directing terrestrial human lives. Newton feels torn between his mission and his commitment to his home planet, and his current life on earth, where he is kept in captivity in a luxurious apartment by Dr. Nathan Bryce, who wishes to exploit Newton's alien powers. Newton's story, visualized as an exploration of earth

Figure 5.3 A still from *The Man Who Fell to Earth*, demonstrating that the humanoid alien was portrayed as an inverted human being: human in form but different in terms of color, perhaps as a reflection of a difference in values.
Source: *The Man Who Fell to Earth*, directed by Nicolas Roeg, UK 1976. Courtesy of British Lion Films.

through the intersection of two perspectives, an alien and a human one, juxtaposes fleeting emotions induced by momentary experiences, such as sexual desire, alcohol, drugs and exciting events like travels and parties, to structural existential problems. This has a hierarchizing effect: individual psychology is exposed as inadequate in the face of problems such as global drought. As long as human psychology is subjugated to the imperative of momentary enjoyment, it is powerless against greater challenges, which, paradoxically, will eventually be brought about as a consequence of humanity's negligent attitude to its social organization and to its environment. The world that Newton crashes into to seek a cure for his own planet is depicted as full of temptations, which are much more accessible to the wealthy, but essentially do not entail any kind of revelations and do not motivate their consumers for action. Rather, they numb individuals and retain the social status quo, at the expense of the well-being of planet Earth.

While early post-Second World War cinematography tended to align humanity's expansion into outer space with militarization – that is, expansion into outer space accompanied by heavy armament, usually portrayed as necessary to protect humanity from external threats – late 1960s and 1970s space-themed science-fiction cinema witnessed a turn 'inward.' In this turn, the human being as the subject of space exploration appeared to be the source of violence and the precursor for militarization. Space exploration was portrayed as a violent enterprise because it had been enabled by a human culture that grew out of violent military conflicts and subordination or suppression of foreign species and element.[41]

Contemplative reflections on the nature of the human psyche, the essential coordinates of social and cultural relations on earth and their implications for the Space Age, provided by *2001: A Space Odyssey*, *Solaris*, *The Man Who Fell to Earth* and others, stood in juxtaposition to an altogether different approach to outer space science-fiction cinema in the 1960s and 1970s.[42] In contrast to emphasizing violence, inherent to the human condition, less philosophically sophisticated productions focused on spectacular special effects and action-driven plots continued the logic set out by earlier productions such as *Invaders from Mars.* In comparison to the 1950s, the transnational spectrum of such productions grew significantly broader. 1970s productions treated outer space and alien worlds as a means for elevating otherwise relatively simple plots, usually adaptations of conventional genre narratives (horror, thriller, melodrama, historical drama) to 'galactic' proportions, thus providing room for excessive visual effects.[43] The militarized image of space, usually presented as a silent threat and a reason for numerous safety precautions, therefore related to a commonplace awareness that international, let alone interplanetary, politics is conducted in order to prevent military conflict. War in space remained, to echo Prussian General Carl von Clausewitz's (1780–1831) famous line, the 'continuation of politics by other means.' According to such productions, the threat of war in space was ever greater due to the potential presence of alien races which are difficult to reason with on earthly terms and due to the cataclysmic effect produced by the harsh conditions of outer space as such.

V The end of the Cold War and the advent of space biopolitics

Ridley Scott's *Alien* franchise (1979–) is an example of a cinematic production maneuvering between these two tendencies, that is, intricate philosophical contemplation on the human condition in outer space on the one hand, and the relatively simple, formally uncomplicated narrative of a visual spectacle on the other. Interestingly enough, writer Dan O'Bannon (1946–2009) developed the idea for the *Alien* series based on *Dark Star* (1974), another science-fiction production he had co-written with the film's director John Carpenter (1948–).[44] *Dark Star*, a somber comedy about an overdue space mission to destroy 'unstable planets' which might jeopardize humanity's colonization of the universe, involves an episode with an alien life form aboard the ship. This theme was molded into a fully-fledged plot in *Alien*, where the entire narrative focuses on an alien life form threatening a human space mission.

There is another parallel between these two projects. The quirky, visually straightforward, humorous *Dark Star*, which in its finale depicts a philosophical conversation between an astronaut and an advanced, thinking bomb called thermostellar Bomb #20, shares an important focus with the thrilling and lavish *Alien* franchise. Both productions treat life itself as an abstract concept, taking into consideration the possibility of its perfect alien-ness and thereby expanding the need for governance to the very core of existence. Ever since the mid-1960s, space-themed science-fiction films lost their preoccupation with spaceflight programs and the possible implications of space exploration. Instead, they mainly used space as a milieu for testing various special effects on the one hand and discussing philosophical questions, such as the status of life and the role of humanity in the universe on the other. Rather than replaying various possible threats that might emerge from an expansion of human politics into outer space, philosophically-concerned films shifted their focus to the status of life in outer space on a more abstract level.

As aliens and technology grew less defined in terms of form, outer space cinema did not turn any less violent or militarized. Rather, it tilted its focus to biopolitics, in a Foucauldian sense, a politics which 'deals with the population, with the population as a political problem, as a problem that is at once scientific and political, as a biological problem and as power's problem' – or, rather, biopolitics in the realms of outer space.[45] Here, the subject of governance is the notion of life at its most abstract; life in its most trivial incarnations; life defined as all the processes that define existence, including nutrition, reproduction, excretion and production. If we expand the reach of biopolitics to outer space, the concept of life should also be extrapolated to involve all artificial intelligence and alien life forms.

In *Dark Star*, Bomb #20 is granted autonomy in its decision-making. It is therefore reasoned and negotiated with as if it were human or at least rational. The conversation between the astronaut and the bomb affects the bomb's later operations and the consequences of its reasoning are felt by the entire crew. Eventually the bomb explodes, after deciding that its demise might be the only evidence of its own existence. While Bomb #20 appears not nearly as autonomous as, for example, the computer HAL from *2001* and does not possess the elaborate personality of *Forbidden Planet*'s Robby the Robot, it is more than just a sophisticated instrument. The Bomb, whose decision-making processes affect life and death, relies on simple deductive reasoning. Furthermore, it is depicted as a life form and life force that needs to be governed, rather than an aide whose actions need to be directed. The *Alien* franchise extrapolates this issue of a foreign life force requiring governance rather than simple instructions or management of an external threat. Here, the alien life form, in contrast to, for example, depictions in *Invaders from Mars*, is no longer a direct analogy of a contemporary political antagonist. Rather, the physicality of the alien presence appears in startling contrast to all kinds of rationality and organization governing life on earth. The alien life form is abstract and dangerous in its infinite potential for destruction that is not curbed by any kind of social norms. From two different perspectives, the two films highlighted the topics of body politics and biopolitics in relation to space exploration.

Dark Star depicted how the disintegration of a social organization and, with it, of normal daily routines, gave way to the emergence of alternative value hierarchies. The *Alien* franchise, on the other hand, foregrounded the dangers of agency, represented by a life form not subjugated to our value system. Such sporadic philosophical interpretations of the political implications of the Space Age were aligned with a general shift in approaches to the nexus of space exploration, science-fiction cinema and space militarization. The shift is particularly discernable in certain productions of the 1980s. This was a period marked by the end of the Cold War, accompanied by a notable halt in the Soviet space program, as well as by a general global decreased popular interest in space exploration, related to an increasing environmental consciousness since the 1970s. At the same time, a growing popular uneasiness emerged in the light of then US President Ronald Reagan's new national defense policy, based on plans for a Strategic Defense Initiative – a missile defense system intended to protect the United States from attacks by ballistic nuclear weapons.[46]

This time brought some important space-related developments on the cinematic front, arguably rounding off the spectrum of sub-genres of space-themed science-fiction cinema. Aside from thrillers such as the *Alien* franchise and entertainment-oriented space operas such as *Star Wars*, a rapprochement between Hollywood, Western-European and so-called Eastern bloc cinematography, facilitated by looser censorship and more favorable film policies in the latter, contributed to some productions that transgressed the boundaries of national genre conventions in an unprecedented way.[47] Apart from the release of the first and only Soviet satirical response to the Soviet space program, *Kin-Dza-Dza!*, the period was characterized by the first cinematic version of the Soviet novel by Arkady (1925–91) and Boris Strugatsky (1933–2012), *Trudno byt' bogom* (*Es ist nicht leicht, ein Gott zu sein/Hard to Be a God* [1964]), and by an unconventional film interpretation of Jerzy Żuławski's (1874–1915) book *Na srebrnym globie* (1988).[48] The two films were innovative contributions to the archive of space science-fiction cinema and may be seen as the last important developments of the genre in the twentieth century. Both films approached the theme of the militarization of space exploration using the conventional form of the science-fiction epos as the basis for their plots, at the same time subverting this basis to highlight the body-political aspect of such epics.

Es ist nicht leicht, ein Gott zu sein (1989) was an epic Soviet-German feature film based on *Trudno byt' bogom*. The original version of the novel, an allegory of terrestrial totalitarian regimes, was heavily censored due to its ideological inadequacy and published in 1964. It quickly acquired cult status among Soviet science-fiction fans. The release of the film was a symptom of the fall of the Iron Curtain and of the politics of Glasnost – a democratization of the public sphere advocated by the USSR Communist Party Secretary-General Mikhail Gorbachev (1931–). At the same time, it was a symptom of the drawbacks of these very political proclamations. The producers did not respect the Strugatsky brothers' wish for the film director to be Aleksey German and eventually stopped collaborating with Peter Fleischmann, who had been entrusted with the job.[49]

Es ist nicht leicht, ein Gott zu sein, a linearly constructed cinematic tale, tells the story of an inhabited alien planet, where the humanoid population is going through its medieval stage, characterized by an authoritarian social order, feudalism, patriarchy and a set of peculiar beliefs spread by local doctors and scientists. The planet is monitored by a spaceship from earth, and several crew members are infiltrated into the local society, conducting an odd experiment: observing the durability and the transformations of social bonds and political structures in order to determine historical laws. The plot is secondary to the highlighted trait of all-pervasive violence that marks the society and social dynamics on said alien planet. Society is governed by violence exerted by those in possession of power. The ruling classes are completely oblivious of their subordinates' humanity, instrumentalizing them entirely, regardless of their age and gender. *Es ist nicht leicht, ein Gott zu sein* reserves rational judgment and rationalist compassion for the space mission orbiting the planet and obliterates emotions altogether. Technologically advanced earth is committed to peace and rationality, whereas Arkanar is being torn apart by crude medieval body politics. Governance over life is induced by violence and sealed by a feudal social order.

An altogether different perspective on space colonialist sagas is offered by *Na srebrnym globie* (*On the Silver Globe*), directed by Andrzej Żuławski (1940–2016), known for his direct aesthetics, daring technique and irreverence toward social norms. The film was an adaptation of the first part of the 1900 *The Lunar Trilogy* written by the director's grand-uncle. The film was ultimately released in 1988 after the end of communist rule in Poland and a decade's halt in production due to political blockages.[50] The movie, a combination of original footage – shot before the production had been interrupted in the 1970s – and Żulawski's commentary filling in the gaps, features the story of an alien civilization. The latter comes into being after a group of astronauts, equipped with video recorders, has landed on a distant planet with no way to return home. They build a village by the seashore and eventually create a mythology consisting of tales of earth. The last remaining crew member, Jerzy, also known as the Old Man, watches this civilization grow and battle the indigenous alien Szerns (Figure 5.4). Before his death, he sends a video recording of these developments to earth. The recording is intercepted by Marek, an astronaut who has come to the planet a long time ago. He had first been received as the Messiah, about to save the civilization

Figure 5.4 A still from *Na srebrnym globie* (1988) shows a close-up of an alien human belonging to the civilization that grew out of the crew of astronauts from earth which got stranded on a faraway planet.
Source: *Na srebrnym globie* (*On the Silver Globe*), directed by Andrzej Żuławski, PL 1988. Courtesy of B. A. Produktion/Garance.

from the Szerns, but was later re-interpreted as an impostor, an outcast from earth and sent away to make up for his sins. The narrative alternates between the past and present, between earth and the distant alien planet, and between perspectives provided by different characters. Żuławski's daring decision to provide sequences of point-of-view shots, referring to different characters, enhanced the impression of dealing with a foreign, alien world. Hijacking the spectator's gaze, convincing the audience that all reality is subjective, the director went even further to direct the gaze towards various explicit displays of corporeal, physical violence, inherent to the society on the foreign planet.

As contemporary cinematic interlocutors, *Alien* and *Na srebrnym globie* highlighted a thematic development in late twentieth-century science-fiction cinematography. While 1980s spacescapes continued to be marked by perceived threats of encounters with extraterrestrial life, the angle that this threat was approached from differed greatly from earlier productions. The menace that characterized extraterrestrial life in 1950s and 1960s space science-fiction movies was no longer just a simple fear of being eliminated by the alien. Rather, *Alien* and *Na srebrnym globie* showed this threat stemming from an alien model of life, which could not be incorporated into terrestrial models of governance and administration due to its very alien-ness and was difficult to eliminate due to its physical strength. In the cases discussed earlier, the militarization of outer space was necessary to either portray ideological, geopolitical, gender-political and psychological antagonisms or, in response, to protect very specific and very tangible values, such as an ideology, peace on the planet and a certain constellation of gender roles. The 1980s science-fiction epic, however, related to a different issue. The arena of militarization was now life itself: life on earth related to different biopolitics than destructive alien forces. Furthermore, these alien forces did not only target certain aspects of life on earth, but rather wished to impose a wholly different conception of biopolitics, incompatible to the one on earth. Therefore, they were untamable, impossible to govern or subjugate and should be destroyed. In this sense, 1980s science-fiction cinematic productions moved away from the traditional Cold War views on human enhancement and astronauts as modern-day cyborg heroes and toward the trans- and posthumanist debates on the future of the human condition and the human species, which played an important role within the then current humanities and social sciences.[51]

VI Space militarization: from ideological critique to biopolitical spectacle

This chapter has discerned four thematic trajectories that marked the astro-cinematic landscape of twentieth-century science fiction, pointing to distinct modes of the militarization of outer space within the cinematic medium. While these modes were related to certain socio-political developments pertaining to the history of the Space Age, such as the Space Race in the 1960s, it should also be noted that the main stylistic and thematic trajectories outlined did not follow one another in chronological succession. Nor were they necessarily subordinated to the policies and politics that accompanied the Space Age and the Space Race. Rather, science-fiction film gradually inscribed into the setting of outer space a varied spectrum of meanings, providing a nuanced critical stance on space exploration and its links to militarization.

The big feats of the space programs of the 1950s and 1960s, such as the first human spaceflight and the Apollo moon landings, were represented and extrapolated particularly in big budget films, supported by major film studios in the West and government film agencies and departments of culture in the Eastern bloc. These films, such as *The Day the Earth Stood Still*, *Destination Moon*, *Invaders from Mars*, as well as *Nebo zovyot* and *Der schweigende Stern*, are now considered iconic by film experts and enthusiasts alike.[52] At the same time, apart from reflecting the history of the Space Age, space science-fiction cinema was also greatly influenced by broader preoccupations that marked the evolution of the science-fiction genre on the one hand, and pertinent intellectual debates on the future of humanity on the other. For instance, while 1950s American science-fiction cinema was indeed dominated by the fear of a nuclear Armageddon and communism, this thematic preoccupation was not dealt with homogeneously on the level of cinematic aesthetics. Films that addressed the same topics (for example, *Plan 9 from Outer Space* and *Invaders from Mars*) often did so through different genres, coexisting but not equaling one another in terms of popularity and frequency of appearance.

Granting the fantastic, satirical and epic trajectories equal analytical attention allows us to uncover the deeper synchronic and temporal transformations that took place within the genre of space science-fiction cinema over the past century. From this perspective, it becomes clear that in the domain of fiction, the militarization of outer space was not merely a byproduct of political antagonisms of the Cold War. The idea of using outer space for aggressive purposes to achieve political supremacy was screened as early as in 1902. Pre-Second World War science-fiction cinema, such as *Voyage dans la lune*, typically considered spaceflight in terms of a fantastic, yet instructive allegory and constructed military conflict around questions related to ideology, such as class antagonisms within a certain society or state. The ruling classes were the agents of militarization, never questioning their right to go to space carrying weapons, or to attack the inhabitants of other planets. At the same time, they did not consider using outer space as a medium for military pursuits aimed at their antagonists on earth. Post-Second World War science-fiction cinema highlighted a different milieu of armed conflict. The agents of space militarization in these films were professional politicians, who saw it as a means of solving their problems on earth. Hollywood films in particular often related conflict in outer space (or involving visitors from outer space) to external, yet earthly threats, such as an armed conflict with the Soviet Union.

Furthermore, invaders from outer space or aliens found in outer space were used as mere metaphors for "outsiders" on earth. This created the idealistic impression of the existence of a united society on earth, which needed to combat outside threats.

Therefore, 1950s and 1960s cinematography tended to portray alien-ness by employing the concept of 'the other' in terms of species, gender or political convictions. However, the 1960s also gave rise to a different focus in space science-fiction films. Rather than concentrating on conflict with an external enemy, films began to foreground humanity's inner struggles and conflicts, demonstrating that militarization aimed at stopping external threats could not save it from its own inner imperfections. In these movies, armed confrontations and violent ventures in space were depicted as the consequence of the drives and antagonisms of the human psyche. *2001: A Space Odyssey* highlighted violence as inherent to human culture, including its version of rational thought as basis for technological progress. Films exploring the conflictual nature of the human psyche and its implications for interactions with outer space flourished in the 1970s, the thematic undercurrent running through various genres, from spectacular thrillers (such as Margheriti's *Space Men*) to reflective dramas (such as *The Man Who Fell to Earth*).

The 1970s also involved an internationalization of space-fiction cinema, related to political détente, which also meant greater access to Western films in the Eastern bloc.[53] Eventually, this contributed to, if not encouraged, the production in socialist governed countries of more commercial films including melodramas, quasi-historical sagas, humorous satires about space programs and space detective films. Moreover, this period foregrounds another turn in the way militarization and violence were treated in space-related science-fiction cinema on a global level. In extrapolating this turn towards the human psyche, science-fiction films turned to the question of how to govern bodies and lives in the Space Age. The emphasis on body politics (for example, in *Es ist nicht leicht, ein Gott zu sein*) and biopolitics (for instance, in the *Alien* franchise and *Na srebrnym globie*), which paralleled the debates on the future of the human condition and the 1980s popularization of discussions on cyborgs, trans- and posthumanism, as well as reflections on the falling totalitarian regimes of the Eastern bloc, can be considered the final twist in the double-bind of the human and the alien. Space militarization in science-fiction cinema no longer referred to placing weapons in space for the sake of protecting humanity from various threats. Rather, it followed the scientifically and politically interesting idea that every life form participated in governance structures. Therefore, in order to protect humanity, humans needed to learn how to govern other life forms as well.

As arguably the most important astrocultural medium of the twentieth century, space films provided – and still provide – nuanced insights into the history of the Space Age and its militaristic aspects, highlighting violence at the structural core of social order. In their own differing ways, all of the films explored in this chapter emphasized the link between space exploration and space militarization as not natural or spontaneous, but reflecting underlying social antagonisms and fears. In doing so, they uncovered outer space as much more than a non-place, either the kind governed by textual and pictorial rules, as examined by Augé, or the kind left unanalyzed due to a prevalent ideology, as defined by De Certeau. Despite various national and regional constraints placed on cinematic productions including censorship in the Eastern bloc and direct producer and indirect stakeholder expectations in the West, the films testified to the constructive and reflective dimensions of popular culture. Regarded as a thematically grounded corpus of cinematic works, these productions did not just reproduce and represent the conflicting, militaristic aspect of space programs and their particular milestones. They also constructed metaphorical extrapolations of the state-of-the-art of spaceflight, pointing to the consequences of possible encounters with the human psyche and with the ultimate otherness within various historical contexts, from the imperialist *fin-de-siècle* to the rise of environmentalist movements in the 1970s. Apart from referring to relevant political events, they harnessed contemporary philosophical and intellectual debates in order to point to the potentially problematic, covertly violent presumptions guiding the development of space technologies.

Notes

1. See, for example, Alexander C. T. Geppert, ed., *Imagining Outer Space: European Astroculture in the Twentieth Century*, Basingstoke: Palgrave Macmillan, 2012 (2nd edn, London: Palgrave Macmillan, 2018) (= *European Astroculture*, vol. 1); Eva Maurer et al., eds, *Soviet Space Culture: Cosmic Enthusiasm in Socialist Societies*, Basingstoke: Palgrave Macmillan, 2011; and Jutta Weldes, ed., *To Seek out New Worlds: Science Fiction and World Politics*, Basingstoke: Palgrave Macmillan, 2003.
2. Arjen Vermeer, 'The Laws of War in Outer Space: Some Legal Implications for Jus ad Bellum and Jus in Bello of the Militarisation and Weaponisation of Outer Space,' in Bob Brecher, ed., *The New Order of War*, Amsterdam: Rodopi, 2010, 69–87, here 70.
3. *Destination Moon*, directed by Irving Pichel, USA 1950 (George Pal Productions); *The Day the Earth Stood Still*, directed by Robert Wise, USA 1951 (Twentieth Century Fox). For the purposes of this text, I will primarily refer to the first two parts of the franchise, namely *Alien* (1979) and *Aliens* (1986); see *Alien*, directed by Ridley Scott, UK/USA 1979 (Brandywine/Twentieth Century Fox) and *Aliens*, directed by James Cameron, USA 1986 (Brandywine). This choice has been made in order to foreground and argue for certain continuities within this cinematic landscape, as well as to point to its internal heterogeneity and intertextual references that reach across the boundaries of national cinematographies.
4. Stevphen Shukaitis, 'Space is the (Non)Place: Martians, Marxists, and the Outer Space of the Radical Imagination,' *Sociological Review* 57.s1 (2009), 98–113, here 101.

5. Michel de Certeau, *L'Invention du quotidien*, vol. 1: *Arts de faire*, Paris: UGE, 1980, 208; Marc Augé, *Non-lieux: Introduction à une anthropologie de la surmodernité*, Paris: Seuil, 1992, 100.
6. Susan Sontag, 'The Imagination of Disaster,' *Commentary* 40.4 (October 1965), 42–8, here 45 (emphasis in original).
7. *Kin-Dza-Dza!*, directed by Georgiy Daneliya, USSR 1986 (Mosfilm); *Na srebrnym globie* (*On the Silver Globe*), directed by Andrzej Żuławski, PL 1988 (Zespół Filmowy 'Kadr').
8. See David A. Kirby, *Lab Coats in Hollywood: Science, Scientists, and Cinema*, Cambridge, MA: MIT Press, 2010, 52–3; and idem, 'Final Frontiers? Envisioning Utopia in the Era of Limits,' in Alexander C. T. Geppert, ed., *Limiting Outer Space: Astroculture after Apollo*, London: Palgrave Macmillan, 2018, 305–17 (= *European Astroculture*, vol. 2).
9. The Committee was known as Gosfilm until 1963 and as Goskino after 1963.
10. Anna Lawton, *Kinoglasnost: Soviet Cinema in Our Time*, Cambridge: Cambridge University Press, 1992, 106.
11. *Voyage dans la lune*, directed by Georges Méliès, FR 1902 (Star Film Company); *La Lune à un mètre*, directed by idem, FR 1898 (Star Film Company).
12. See Geoff King and Tanya Krzywinska, *Science Fiction Cinema: From Outerspace to Cyberspace: Short Cuts*, New York: Wallflower Press, 2000, 23, 67.
13. Richard Abel, '*A Trip to the Moon* as an American Phenomenon,' in Matthew Solomon, ed., *Fantastic Voyages of the Cinematic Imagination: Georges Méliès's* Trip to the Moon, Albany: SUNY Press, 2011, 129–42, here 133.
14. For the political undercurrents of Méliès's short film, see Philippe Mather, 'A Brief Typology of French Science Fiction Film,' in idem and Sylvain Rheault, eds, *Rediscovering French Science-Fiction in Literature, Film and Comics: From Cyrano to Barbarella*, Newcastle-upon-Tyne: Cambridge Scholars, 2015, 143–60, here 143.
15. David Sandner, 'Shooting for the Moon: Méliès, Verne, Wells, and the Imperial Satire,' *Extrapolation* 39.1 (April 1998), 5–25, here 5.
16. For examples of such analyses, see the contributions to Solomon, *Fantastic Voyages*.
17. *Himmelskibet* (*A Trip to Mars*), directed by Holger-Madsen, DK 1918 (Nordisk Film); *The First Men in the Moon*, directed by Bruce Gordon and J. L. V. Leigh, UK 1919 (Gaumont British); *Aelita*, directed by Yakov Protazanov, USSR 1924 (Mezhrabpom-Rus); *Frau im Mond*, directed by Fritz Lang, DE 1929 (Ufa). On *Himmelskibet*, see Thore Bjørnvig, 'The Holy Grail of Outer Space: Pluralism, Druidry, and the Religion of Cinema in *The Sky Ship*,' *Astrobiology* 12.10 (October 2012), 998–1014; on *Frau im Mond*, see Alexander C. T. Geppert and Tilmann Siebeneichner, '*Lieux de l'Avenir*: Zur Lokalgeschichte des Weltraumdenkens,' *Technikgeschichte* 84.4 (2017), 285–304, here 285–8.
18. *Aelita*; *Mezhplanetnaya revoliutsia* (*Interplanetary Revolution*), directed by Zenon Komissarenko, Yury Merkulov and Nikolay Khodataev, USSR 1924 (Mezhrabprom-Rus).
19. For a detailed discussion, see Peter G. Christensen, 'Women as Princesses or Comrades: Ambivalence in Yakov Protazanov's *Aelita* (1924),' *New Zealand Slavonic Journal* 26.1 (November 2000), 107–22.

20. For an overview of Eastern European postwar film industries and policies, see Mira Liehm and Antonín J. Liehm, *The Most Important Art: Soviet and Eastern European Film after 1945*, Berkeley: University of California Press, 1980.
21. *When Worlds Collide*, directed by Rudolph Maté, USA 1951 (Paramount Pictures); *The Day the Earth Stood Still*; *Invaders from Mars*, directed by William Cameron Menzies, USA 1953 (Edward L. Alperson Productions); *War of the Worlds*, directed by Byron Haskin, USA 1953 (Paramount Pictures).
22. Vivian Sobchack, *Screening Space: The American Science Fiction Film* [1987], 2nd edn, New Brunswick: Rutgers University Press, 1997, 63.
23. Miranda J. Banks, 'Monumental Fictions: National Monument as a Science Fiction Space,' *Journal of Popular Film and Television* 30.3 (Fall 2002), 136–45, here 138.
24. Robert E. Hunter, 'Expecting the Unexpected: Nuclear Terrorism in 1950s Hollywood Films,' in Rosemary B. Mariner and G. Kurt Piehler, eds, *The Atomic Bomb and American Society: New Perspectives*, Knoxville: University of Tennessee Press, 2009, 211–40, here 230.
25. *Destination Moon*, directed by Irving Pichel, USA 1950 (George Pal Productions). David A. Kirby, 'The Future is Now: Diegetic Prototypes and the Role of Popular Films in Generating Real-World Technological Development,' *Social Studies of Science* 40.1 (January 2010), 41–70, here 47.
26. Wernher von Braun, 'Crossing the Last Frontier,' *Collier's* (22 March 1952), 24–9, 72–3, here 25: 'Within the next 10 or 15 years, the earth will have a new companion in the skies, a man-made satellite that could be either the greatest force for peace ever devised, or one the most terrible weapons of war-depending on who makes and controls it. Inhabited by humans, and visible from the ground as a fast-moving star, it will sweep around the earth at an incredible rate of speed in that dark void beyond the atmosphere which is known as "space".'
27. *The Day the Earth Stood Still*; Blair Davis, 'Singing Sci-Fi Cowboys: Gene Autry and Genre Amalgamation in *The Phantom Empire* (1935),' *Historical Journal of Film, Radio and Television* 33.4 (November 2013), 552–75.
28. *Der schweigende Stern* (*First Spaceship on Venus*), directed by Kurt Maetzig, DDR/PL 1960 (VEB DEFA-Studio für Spielfilme/DEFA Gruppe Roter Kreis/Film Polski/Iluzjon). For more on the film, see Steven J. Dick, 'Space, Time, and Aliens: The Role of Imagination in Outer Space,' in Geppert, *Imagining Outer Space*, 27–44, here 27.
29. Soon enough, such obvious allegories of the Cold War political divide became fruitful grounds for overtly satirical productions such as *Plan 9 from Outer Space*, directed by Ed Wood, USA 1959 (Reynolds Pictures).
30. *Nebo zovyot* (*Battle beyond the Sun*), directed by Mikhail Karyukov and Aleksandr Kozyr, USSR 1959 (A. P. Dovzenko Filmstudio/Mosfilm).
31. Landon Brooks, 'Doubling Down on Double Vision,' *Science Fiction Studies* 44.3 (November 2017), 592–8, here 593.
32. *Devil Girl from Mars*, directed by David MacDonald, UK 1954 (Danziger Productions Ltd.).
33. *Fire Maidens from Outer Space*, directed by Cy Roth, UK 1956 (Criterion Films). See Steve Chibnall, '4 Alien Women,' in Ian Hunter, ed., *British Science Fiction Cinema*, London: Routledge, 1999, 57–74, here 57.

34. *Forbidden Planet*, directed by Fred M. Wilcox, USA 1956 (Metro-Goldwyn-Mayer).
35. The film itself refers to a 'subconscious' rather than an 'unconscious.'
36. *Barbarella*, directed by Roger Vadim, FR/IT 1968 (Marianne Productions).
37. David G. Hartwell, *Age of Wonders: Exploring the World of Science Fiction*, London: Palgrave Macmillan, 2017, 181. *Space Men*, directed by Antonio Margheriti, IT 1960 (Titanus/Ultra Film). Margheriti was the most prominent Italian science-fiction director; see Dennis Fischer, ed., *Science Fiction Film Directors, 1895–1998*, Jefferson: McFarland, 2011, 422–6.
38. This trait is much more characteristic of European cinematography than it is of US or Soviet science-fiction films. Arguably, one of the reasons for this discrepancy may be the traditionally smaller emphasis placed on the military dimension of the Space Age in Europe.
39. *2001: A Space Odyssey*, directed by Stanley Kubrick, USA 1968 (Metro-Goldwyn-Mayer); *Solaris*, directed by Andrey Tarkovsky, USSR 1972 (Creative Unit of Writers & Cinema Workers/Mosfilm/Unit Four); *The Man Who Fell to Earth*, directed by Nicolas Roeg, UK 1976 (British Lion Films). For more on *2001*, see Robert Poole, 'The Myth of Progress: *2001 – A Space Odyssey*,' in Geppert, *Limiting Outer Space*, 103–29.
40. Andrey Tarkovsky, *Sculpting in Time: Reflections on the Cinema*, London: Bodley Head, 1986.
41. For more on the post-Apollo inward turn, see Alexander C. T. Geppert, 'The Post-Apollo Paradox: Envisioning Limits during the Planetized 1970s,' in idem, *Limiting Outer Space*, 3–26, here 5–9.
42. See, for example, the Soviet-Polish production *Test Pilota Pirxa* (*Pilot Pirx's Inquest*), directed by Marek Piestrak, PL/USSR 1979 (Przedsiebiorstwo Realizacji Filmów 'Zespóły Filmowe'/Filmistuudio 'Tallinnfilm'/Dovzhenko Film Studios).
43. *Moskva – Kassiopeya* (*Moscow – Cassiopeia*) and *Otroki vo vselennoy* (*Teens in the Universe*), both directed by Richard Viktorov, USSR 1945 and 1975 (Gorky Studio); *Bol'shoye kosmicheskoye puteshestviye* (*The Big Space Travel*), directed by Valentin Selivanov, USSR 1975 (Gorky Studio); *Terrore nello spazio* (*Planet of the Vampires*), directed by Mario Bava, IT/ESP/USA 1965 (Italian International Film/Castilla Cooperative Cinematográfica/American International Pictures); *Space Battleship Yamato*, directed by Leiji Matsumoto, JP 1977 (Office Academy).
44. *Dark Star*, directed by John Carpenter, USA 1974 (Jack H. Harris Enterprises).
45. Michel Foucault, 'Lecture 11, 17 March 1976,' in idem, *Society Must Be Defended: Lectures at the Collège de France 1975–1976*, eds. Mauro Bertani and Alessandro Fontana, New York: Picard, 2003, 239–64, here 243; originally published as *Il faut défendre la société: Cours au Collège de France, 1976*, Paris: Gallimard, 1997.
46. Steven M. Sanders, 'An Introduction to the Philosophy of Science Fiction Film,' in idem, ed., *The Philosophy of Science Fiction Film*, Lexington: University Press of Kentucky, 2007, 1–17, here 17. For more on SDI, see the introduction and Michael Neufeld's contribution, Chapters 1 and 2, respectively.
47. *Star Wars* is an American epic space opera created by George Lucas. The franchise began in 1977 with the release of the film *Star Wars* (later subtitled *Episode IV: A New Hope* in 1981), which became a worldwide pop culture phenomenon. It was followed by the sequels *The Empire Strikes Back* (1980) and *Return of the*

Jedi (1983). These three films constitute the original *Star Wars* trilogy. A prequel trilogy was released between 1999 and 2005; see *Star Wars*, directed by George Lucas, USA 1977–2005 (Lucasfilm/Twentieth Century Fox). More spin-offs have been produced since the franchise was bought by Disney in 2012.

48. *Es ist nicht leicht, ein Gott zu sein* (*Hard to Be a God*), directed by Peter Fleischmann, BRD/FR/USSR/CH 1989 (B. A. Produktion/Garance/Hallelujah Films/Mediactuel/Sovinfilm/Studio Dowschenko Kiew/Zweites Deutsches Fernsehen); *Na srebrnym globie* (*On the Silver Globe*). For more on Soviet perestroika science fiction, see Natalija Majsova, 'Articulating Dissonance between Man and the Cosmos: Soviet Scientific Fantasy in the 1980s and Its Legacy,' in Birgit Beumers and Eugenie Zvonkine, eds, *Ruptures and Continuity in Soviet/Russian Cinema: Styles, Characters and Genres before and after the Collapse of the USSR*, London: Routledge, 2017, 184–99.
49. Elena V. Boroda, 'Ot blagodetelya k progressoru: modifikatsiza obraza sverkh-cheloveka v otechestvennoi fantastike XX veka,' *Filologiya i chelovek* 3.4 (December 2008), 21–7, here 21.
50. Anna Misiak, 'The Polish Film Industry under Communist Control: Conceptions and Misconceptions of Censorship,' *Illuminance* 24.4 (2012), 61–83, here 81.
51. On the astronaut as the Cold War cyborg, see Patrick Kilian's contribution, Chapter 8 in this volume. For more on the cyborg debate of the 1980s, see Linda Howell, 'The Cyborg Manifesto, Revisited: Issues and Methods for Technocultural Feminism,' in Richard Dellamora, ed., *Postmodern Apocalypse: Theory and Cultural Practice at the End*, Philadelphia: University of Pennsylvania Press, 1995, 199–218.
52. Eva Näripea, 'Work in Outer Space: Notes on Eastern European Science Fiction Cinema,' in Ewa Mazierska, ed., *Work in Cinema: Labour and the Human Condition*, Basingstoke: Palgrave Macmillan, 2013, 209–26, here 211–12; Robert A. Jones, 'They Came in Peace for All Mankind: Popular Culture as a Reflection of Public Attitudes to Space,' *Space Policy* 20.1 (February 2004), 45–8, here 46.
53. See also Kirby, 'Final Frontiers.'

CHAPTER 6

C. S. Lewis and the Moral Threat of Space Exploration, 1938–64

Oliver Dunnett

Clive Staples Lewis (1898–1963) is seen today as a celebrated author of fantasy fiction, his *Narnia* novels (published 1950–56) recognized as some of the most popular and well-loved works of children's literature, which still attract a wide readership in the twenty-first century. Lewis also had a successful academic career, teaching medieval and Renaissance literature as a Fellow of Magdalen College, Oxford, from 1925 to 1954, and thereafter as the holder of a newly created chair at Magdalene College, Cambridge (Figure 6.1).[1] Lewis's life, documented through his own extensive letter-writing, alongside numerous biographical and critical works, can be read as a chronicle of the twentieth century, witnessing events including the First World War through to the advent of spaceflight in the late 1950s and early 1960s. However, Lewis is said to have felt out of place in contemporary times, regarding himself as 'a dinosaur still surviving from the lost age when [...] the material and the spiritual interacted to create a vital culture.'[2] Indeed, Lewis's primary scholarly achievement was arguably the reconstruction of a harmonious and all-encompassing 'medieval model' of the cosmos that incorporated a spiritual understanding of 'the heavens' into the theology, science and history of the Middle Ages.[3] In his academic work Lewis saw value in promoting an understanding of older ways of thinking about the universe, which he read as being lost or 'discarded' in the face of the overwhelming march of modern science and culture.

Oliver Dunnett (✉)
Queen's University Belfast, Belfast, Northern Ireland, UK
e-mail: O.Dunnett@qub.ac.uk

Alexander C. T. Geppert et al. (eds), *Militarizing Outer Space*
European Astroculture, vol. 3
https://doi.org/10.1057/978-1-349-95851-1_6

Figure 6.1 C. S. Lewis at his desk in his home, The Kilns, Oxford, August 1960.
Source: Courtesy of Marion E. Wade Center, Wheaton College.

This is not to say that Lewis's mindset was entirely focused on peace, harmony and a romantic vision of the past. Lewis saw active military service as an officer in the First World War, and throughout his life thereafter was afflicted by a shrapnel wound suffered at the Battle of Arras in 1917.[4] Although, as an Irishman, he was not drafted in to compulsory service in the British Army, Lewis signed up voluntarily out of a sense of duty, and later in his career he strongly defended the notion of a just war in an address to the Oxford Pacifist Society. Although Lewis never sentimentalized or glorified war, he saw it as 'a fact of human existence that we must accept if we are to be rational.'[5] Furthermore, throughout his writing he is said to have favored a certain notion of chivalric valor in battle, and in a 1946 letter he referred to 'the *good* element in the martial spirit, the discipline and freedom from anxiety' that can be found in fighting a just war.[6] It is this combination of factors

– his renown as an author of speculative fiction, his expertise as a scholar of past cultures, and his experience of key events of the twentieth century – that puts C. S. Lewis in a unique position in terms of understanding the developing relationship between humankind and the cosmos in the twentieth century, a relationship that can be defined, from Lewis's perspective, by what he saw as a 'great division' in the history of Western culture.

In his inaugural address at Cambridge in 1954 Lewis explained that, whereas past scholars had tended to draw a 'great division' in Western culture between the medieval era and the Renaissance, for Lewis, if there should be any such division, it should be placed at a more recent point, 'somewhere towards the end of the seventeenth century, with the general acceptance of Copernicanism, the dominance of Descartes, and (in England) the foundation of the Royal Society.'[7] For Lewis, this would help to signify a number of fundamental shifts in society, including changes in politics and art, the transition from a 'Christian to post-Christian' society, and, Lewis's 'trump-card,' the 'birth of the machines.' The results of these changes in systems of thought would, according to Lewis, be felt in a number of ways throughout the twentieth century and ought to be a subject of concern. Ominously, Lewis suggested that, 'when Watt makes his engine, when Darwin starts monkeying with the ancestry of Man, and Freud with his soul, and the economists with all that is his, then indeed the lion will have got out of its cage,' highlighting a whole raft of social and cultural developments that would contribute to defining the modern age.[8]

One of the most revealing ways in which this implied schism between the medieval 'lost age' of Lewis's scholarly imagination and the incoming modernity of life in the twentieth century can be explored is by examining the ways in which outer space, or 'the heavens,' has been conceptualized by Lewis in these alternative contexts. This seems to be a particularly relevant part of Lewis's œuvre, given his clear affinity for the overarching medieval model of the cosmos which he spent so much of his scholarly life uncovering. Especially in *The Discarded Image* (1964) Lewis often referred to a distinction between medieval and modern understandings of the cosmos. This should be understood in the context of his 'great division' thesis and was perhaps best defined when he stated that, whereas 'the "space" of modern astronomy [that is, post-Copernicus] may rouse terror, or bewilderment or vague reverie; the spheres of the old present us with an object in which the mind can rest, overwhelming in its greatness but satisfying in its harmony.'[9] This distinction is key to understanding Lewis's moral and ethical vision of the cosmos, and therefore his outlook on violence and warfare in the cosmic realm, and herein will be referred to as the tension between the 'medieval model' and the 'modern understanding' of the cosmos.

The concept of 'moral geographies' can be usefully employed as a conceptual framework for further understanding the cosmos as a geographical space to which moral and ethical values can be assigned. As such, geographer David M.

Smith has understood moral geographies as the ways in which ethics and morality 'are embedded within specific sets of social and physical relationships manifest in geographical space.'[10] Contemporaneously, a number of scholars in cultural, historical and political geography have started to conceptualize outer space as a realm of geographical enquiry, both in terms of the grounded geographies of space-related activity and the imaginative geographies of outer space, including other planets.[11] Particularly relevant is the work of Denis Cosgrove, who understood that modern configurations of the earth's surface that constituted geographical hegemony in the modern era had side-lined earlier mappings of 'the heavens' and associated parallel understandings of earth and cosmos.[12] Drawing influence from contemporary geographical enquiry in thinking about Lewis's works, both in terms of moral geographies and geographies of outer space, it is possible to identify a critical current running through Lewis's writings that conceptualizes the cosmos according to moral geographies, implicates space exploration as a morally questionable act and offers a warning against the militarization of outer space.

In exploring Lewis's stance on space exploration, this chapter adopts Lewis as a historical exemplar of a critical engagement with the concept of human space exploration and the associated militarization of space. In doing this, aspects of Lewis's 'cosmic trilogy' of science-fiction novels (published 1938–45) are examined, whilst other written works including his academic texts, poetry and published correspondence are also consulted in an intertextual approach. This approach is crucially important, in that Lewis's whole mindset was configured around a premodern understanding of ethics, morality and science, through which a comprehensive 'exposition of the past' was deployed as 'a critique of the present.'[13] Therefore it is not only through Lewis's science-fiction novels, but also through his scholarly and other writings, that his stance towards space exploration can be fully elucidated. As well as considering these different registers of writing, it is also important to understand Lewis's fictional and scholarly works as woven together and inter-reliant, as it is this merging between myth and reality that lies at the center of Lewis's work. Indeed, Lewis scholar Michael Ward has applied such an analysis to the *Narnia* series, demonstrating how a medieval understanding of the cosmos was wholly integrated into Lewis's most famous novels.[14] In order to fully contextualize Lewis's cosmic imagination, other related texts shall also be examined, particularly the works of British writers David Lindsay (1876–1945), Olaf Stapledon (1886–1950) and J. B. S. Haldane (1892–1964), who were all known influences on the cosmic trilogy and other related texts of Lewis.

Lewis's significance in this analysis lies with his status as an influential figure in twentieth-century culture, including his popularity as both an imaginative author of fantasy and science-fiction novels, and a non-fiction author on religious, ethical and historical topics.[15] This status has meant that there have been a substantial number of critical engagements with Lewis's writings, particularly in the past twenty years or so, including accounts of the cosmic trilogy.[16] Whereas such works have examined these novels with a focus on understanding their roots in eighteenth-century literature, their adoption of Arcadian fantasies of the rural and exploring their Christian sense of morality, this chapter will focus specifically on Lewis's understanding of interplanetary space and space exploration. It examines the cosmic trilogy alongside Lewis's poetry, correspondence and scholarly work, arguing that together these texts amount to a warning by Lewis of the threat of space exploration and the associated militarization of outer space. In doing so, the chapter will also address pertinent issues in understandings of astroculture, such as the interplay between the 'real' and the 'imagined' in cultures of outer space, critiques of space science that can be found in imaginative representations, as well as helping to decenter histories of outer space away from typical understandings of the American-Soviet Space Race binary to foreground British space culture, which played its own significant role in understanding outer space in the twentieth century.[17]

I The medieval and the modern in science fiction and fantasy

Before the Second World War, Lewis and his friend J. R. R. Tolkien (1892–1973) thought that in the world of anglophone imaginative literature there was a distinct lack of the kind of stories that they were interested in. To remedy this situation they decided that each should write an 'excursionary thriller,' Lewis's being a space-story and Tolkien's a time-story.[18] Whilst Tolkien's side of the bargain eventually became part of the *History of Middle-Earth* anthology, Lewis's space-story was to be his first major work of fiction, *Out of the Silent Planet*, published in 1938. Along with its two sequels, *Perelandra* (1943) and *That Hideous Strength* (1945), the cosmic trilogy became one of the most enduring successes of London-based publisher The Bodley Head, despite the challenging economic circumstances under which they were initially published.[19] The novels can partly be categorized as space adventure stories, but also overlap into "terrestrial" science fiction, fantasy and mythology. Whilst they have remained regularly in print since their publication, they have largely been overshadowed by Lewis's later *Narnia* novels in terms of their popularity. Although stories set in outer space are usually identified as science fiction, Lewis and Tolkien were proponents of fantasy as a narrative form. More than simply something that 'couldn't happen in the real world,' as it was later characterized by Arthur C. Clarke (1917–2008), Tolkien and Lewis drew their conception of fantasy from the literary theory of the English Romantic poet Samuel Taylor Coleridge (1772–1834), who saw literature as a lamp rather than a mirror.[20] That is, literature created 'Secondary Worlds which illuminate, rather than reflect, reality,' drawing on certain pre-existing conventions and being rational by its own standards.[21] Fantasy literature can therefore be understood as offering relevant commentary on the life worlds of their authors, even though the 'secondary worlds' they create are typically less familiar than those of more realistic fictional prose.

Although Lewis and Tolkien bemoaned the absence of this kind of speculative fiction in the interwar period, Lewis later revealed that one of the principal positive influences on his cosmic trilogy was David Lindsay's dystopian scientific romance *Voyage to Arcturus* (1920), which he acclaimed as 'entirely on the imaginative and not at all on the scientific wing' of science fiction. Lewis appreciated the novel for its creative achievements, primarily because it took science fiction beyond the simple extension of future technology to interplanetary settings, and considered more cerebral themes such as evolutionary philosophy and the possibilities of extraterrestrial life.[22] 'From Lindsay,' Lewis later recalled, 'I first learned what other planets in fiction are really good for: for *spiritual* adventures. Only they can satisfy the craving which sends our imaginations off the Earth.'[23] In this way Lewis hinted at the real value of speculative fiction, which he saw as an opportunity to "smuggle" spiritual and philosophical messages into the more familiar narrative structures of adventure and science fiction.

The best-known stories of interplanetary adventure in the early twentieth century were undoubtedly the novels of H. G. Wells (1866–1946), Jules Verne (1828–95) and Edgar Rice Burroughs (1875–1950). Although Lewis admired the works of these 'fathers of science fiction,' he was completely against their moral and ethical tone, which seemed to promote humanity's mastery over time and space.[24] Indeed, he once described Wells's novels as '*first* class pure fantasy [...] and *third* class didacticism,' and further explained that 'I like the whole interplanetary idea as a *mythology* and simply wanted to conquer for my own (Christian) point of view what had always hitherto been used by the opposite side.' Here the 'opposite side' refers to the prevailing culture of science-fiction narratives in what has been referred to as the 'Wellsian decade' of science fiction in 1930s Britain.[25] For Lewis, this meant stories that promoted science and technology, but were lacking in the spiritual or mythical elements that he thought were so important. This was significant in Lewis's view because of its cultural effect, promoting not science per se but what he referred to as 'scientism – a certain outlook on the world which is casually connected with the popularisation of the sciences.'[26]

Lewis was becoming aware of this kind of scientific thinking and its presence in speculative fiction. 'What immediately spurred me to write [the cosmic trilogy] was Olaf Stapledon's *Last and First Men*, and an essay in J. B. S. Haldane's *Possible Worlds*, both of which seemed to take the idea of [space] travel seriously and to have [a] desperately immoral outlook,' he stated in a 1938 letter to Lancelyn Green.[27] In this way Lewis's moral geography of the imagined cosmos was beginning to take shape, as he conflated the 'immoral outlook' of Haldane and Stapledon's stories with the notion of space travel, which by 1938 had become the goal of rocketry groups in parts of Europe and the United States.[28] If the moral tone of Lewis's imaginative engagement with the cosmos is to be fully appreciated in this historical context of emerging cultures of spaceflight technology and science-fiction writing, some further understanding is needed of these two key texts, their bio-militaristic overtones and their connection to wider scientific cultures.

The relevant essay by J. B. S. Haldane was entitled 'The Last Judgement,' which Lewis later characterized as 'brilliant, though [...] depraved.' The collection in which it appeared, *Possible Worlds*, brought together articles originally written by Haldane for a range of popular publications, covering topics including 'Blood Transfusion,' 'Eugenics and Social Reform' and 'Scientific Research for Amateurs,' which helped make him known in the United Kingdom as one of the great popularizers of science.[29] Historian of science Mark Adams has suggested that Haldane's 'popular' essays allowed him to articulate 'the fullest record of his thinking' and were 'the place where his originality and special genius is most clearly manifest.'[30] Furthermore, whilst Haldane was convinced that 'the average man should attempt to realise what is happening to-day [*sic*] in the laboratories,' he also believed that 'in scientific work the imagination must work in harness,' thereby opening up the possibility for speculative accounts of the development of science and its application in society. Haldane was contributing to a new wave of popular science writing in early twentieth-century Britain, whereby scientists 'were happy to use popular writing as a way of trying to influence the public's attitude.'[31] As a representative of this wider trend, Haldane was seen by Lewis as a focus for his arguments against the prevailing 'scientism' that he saw as so problematic.

It is in 'The Last Judgement,' the final essay in *Possible Worlds*, that Haldane turned his attention to space exploration, adopting the 'history of the future' narrative style that had earlier been popularized by H. G. Wells. Indeed, Wells's novel *The Time Machine* (1895) was a known influence on Haldane, with its long-range projections of the future of humankind, including its evolution to altered biological states.[32] In 'The Last Judgement' Haldane described a future earth threatened by the fall of the moon, and narrated the rise of a new human species that has taken refuge on the planet Venus. In extrapolating technologies such as rocketry as a means of space travel, Haldane promoted a 'hard science' approach to science fiction, in contrast to Lewis, whose cosmic trilogy Haldane was later to criticize scathingly for its lack of scientific authenticity.[33] In his short story Haldane further described how, through the propagation of biochemically conditioned bacteria on Venus by the human settlers, 'the previous life on that planet was destroyed,' whilst the ongoing colonization necessitated the 'deliberate evolution' of the human species to suit a rarefied atmosphere and warmer climate.[34] Here Haldane applied his knowledge of contemporary biological research, which was at the time moving into new areas such as genetics and eugenics in the context of developments in evolutionary theory.[35] Haldane's colonizers continue to thrive on Venus, their physiology evolving to such an extent that the final remaining refugees from the dying earth were 'incapable of fertile unions' with their genetic cousins on Venus, and 'were therefore used for experimental purposes,' in a flourish that betrays Haldane's clinical view of the role of ethics in science.[36] The speculative potential of science fiction, fusing imaginative and scientific thought, along with a prevailing culture of popular science in the prewar period, allowed Haldane to explore the possibilities of posthumanism in outer space, a theme that would be directly criticized by Lewis in his own space adventure fiction.

The other piece of writing mentioned by Lewis in his letter in 1938 will be more familiar to science-fiction readers and is known to have been guided by Haldane's striking essay.[37] Olaf Stapledon's first major novel, *Last and First Men*, is said to have caused something of a sensation among writers and critics upon its publication in 1930, before being nearly forgotten in the postwar period due to publishing restrictions and the waning of eugenic theory in mainstream culture.[38] Its writing style is impersonal and detached as a result of the galloping pace of the narrative, and the story begins on earth in contemporary times with the 'First Men.' The narrative outlines a global apocalypse which leaves only thirty-five survivors stationed at the North Pole, out of which develop the 'Second Men,' and the pattern continues, witnessing the development of new human species across the solar system, until finally the reader arrives at the 'Eighteenth Men' two billion years in the future, who inhabit the planet Neptune.

Also adopting the 'future history' style, *Last and First Men* follows the Hegelian dialectic in narrating the consecutive rise and fall of human civilizations. Indeed, Stapledon made clear that in Hegel's idealist theory of history he saw the possibility of a civilization being 'thrown into logical conflict with itself,' which would 'give birth to a new form of culture in which the conflict is resolved in a new synthesis [...] and so on,' thereby positing conflict as a key driver of human history.[39] Stapledon evidently saw the potential of applying this thesis to interplanetary space in a 'future history' scenario, and fused this idea with developments in evolutionary theory that were gaining popularity at the time. Of particular interest in Stapledon's novel are his 'Fifth Men,' who have been genetically 'perfected through and through' in a triumph of eugenic science.[40] Having emerged from an existential conflict with their predecessors the 'Fourth Men,' they wipe out the main indigenous species on Venus in 'a vast slaughter' in order to establish an earth-like environment, much like Haldane's colonizers.[41] As representatives of Hegel's 'logical conflict,' Stapledon's imagined future 'Men' come to bear upon Lewis's view that, as a result of technological and scientific developments in modern society (the 'lion being let out of the cage'), unpredictable and disturbing consequences would result for humanity.[42] Indeed, as the narrative progresses, the form of Stapledon's 'Men' becomes yet more strange, and, ultimately, as the exploding sun envelops the 'Eighteenth Men' at the end of the novel, they dispatch a virus to other parts of the galaxy in their attempt to maintain a derivative of human life in the universe. It is clear that what drives human evolution across the solar system in *Last and First Men* is a combination of eugenics and total war, highlighting a prevailing bio-militarism in Stapledon's œuvre.[43]

Haldane and Stapledon clearly shared an anticipation of interplanetary travel, and evidence to support Lewis's charge that these writers wanted to take spaceflight seriously can be found in Stapledon's 1948 address to the British Interplanetary Society.[44] Arthur Clarke, who had issued the invitation on behalf of the BIS, flattered Stapledon by claiming that 'your writings are in no small degree responsible for the outlook of many of the younger technicians in this society; they have also profoundly influenced the substantial number of "non-technical" members who are interested in astronautics [...] because of its implications to humanity.'[45] The last thing Lewis would have wanted is Stapledon's ideas having a tangible influence on spaceflight research, and in this case it is possible to get a sense of the overlap between fictional speculation and scientific progress in spaceflight, with Stapledon's novel fostering just the kind of spreading 'scientism' that Lewis was wary of.

In his talk, Stapledon envisaged a utopian future in which what he called 'the highly specialized venture of interplanetary travel' would take precedence.[46] In parallel to *Last and First Men*, Stapledon speculated that '[g]iven sufficient biological knowledge and eugenical technique, it might be possible to breed new human types of men to people the planets.'[47] Stapledon again imaginatively extrapolated contemporary advances in the biological sciences to the anticipated environmental conditions of interplanetary space, conflating his imaginative and scientific thinking. Furthermore, Stapledon identified interplanetary conflict as a somewhat likely scenario if other planets were found to be inhabited, even suggesting that such a 'war of the worlds' might 'cause at least a temporary unification of mankind in the face of common danger, much as Russia and the West united against Hitler.'[48] Here, and throughout his talk, Stapledon maintained that the main goal of space exploration should be to make the most of humankind, in both a temporal evolutionary sense as well as in an expansionary spatial sense. In doing this, he was articulating Lewis's worst nightmare: a perfect storm of imperialistic space conquest and liberal eugenic meddling that was foregrounded in *Last and First Men*, Lewis's exemplar of a morally dubious science-fiction narrative.

In seeking to install a more ethically and morally justifiable vision of humankind's engagement with cosmic spaces, Lewis turned away from the prevalent trends of science fiction in the early twentieth century, their overlap with discourses of popular science and, in particular, the morally questionable association of space exploration with eugenics and war that was propounded by the likes of Haldane and Stapledon. As such, instead of looking to the future for inspiration, Lewis looked to the past and the perceived harmony of the medieval model of the cosmos. Yet his response to these issues did not simply constitute a utopian narrative based on the medieval model, but positioned modern science and space exploration as a menacing presence right from the opening pages of the cosmic trilogy. It is in Lewis's response to this threat that his articulation of a morally just engagement with the cosmos can be most clearly understood, and this can be traced by examining three aspects of this counternarrative: the nature of heavenly bodies; the conceptualization of human bodies in space; and the act of space exploration.

II Heavenly bodies: outer space as sacred space

A prominent feature in Lewis's reconstructed medieval model of the cosmos was the division of 'the heavens' into seven concentric spheres, each of which came under the influence of one of the heavenly bodies: the Moon, Mercury, Venus, the Sun, Mars, Jupiter and Saturn (Figure 6.2). Lewis's wider affinity for this cosmic heptarchy is notable, in the first instance as the subject of a posthumously published poem entitled 'The Planets,' and also embedded in his most famous fictional works, the *Narnia* series.[49] Indeed, as well as enrolling the symbolic potential of the seven heavenly bodies through these

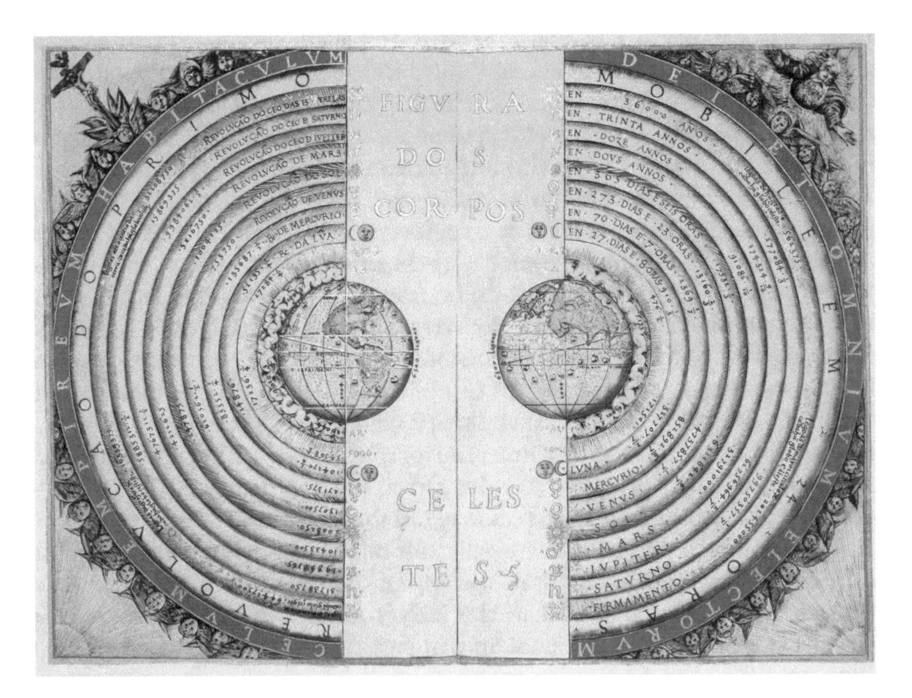

Figure 6.2 The Ptolemaic geocentric model of the universe according to the Portuguese cosmographer and cartographer Bartolomeu Velho, 1568, illustrating the seven heavenly bodies.
Source: Wikimedia.

creative registers, Lewis also communicated this 'model' to scholarly audiences in his academic work. But in what ways did Lewis's understanding of premodern cosmic mythology relate to his science-fictional narratives, his critique of the modern understanding of the cosmos and the anticipation of space exploration? In order to trace this line of enquiry the key texts are the two interplanetary adventures in his cosmic trilogy, *Out of the Silent Planet* (1938) and *Perelandra* (1943). Here the protagonist Ransom, a middle-aged scholar of an unnamed Cambridge college, experiences space travel, life on other planets and learns to appreciate the presence of a sublime cosmic order. It is in these parts of the trilogy's narrative that we can establish Lewis's alternative to the modern understanding of outer space that he outlined in his other writings.

In the opening pages of *Out of the Silent Planet*, Ransom is kidnapped from a walking holiday somewhere in the English midlands and taken onboard a rocket ship against his will by the devious scientist Weston and his collaborator Devine. In Ransom's reactions to being in outer space, Lewis's thoughts about the nature of the cosmos can start to be explored. In particular it appears that Ransom's initial response of sickness, disorientation and 'general lassitude' is symptomatic of the extraction of earthly sin that is affected by an exposure to the cosmic radiance of outer space.[50] This imaginative effect of spaceflight has been noted in the writings of Arthur Clarke, in relation to escaping the earth's gravity, and perhaps here Lewis's influence on the genre's later development can be seen.[51] As Ransom's time in the spaceship elapses, he develops a more positive reaction to being in space, a response that marks the start of Ransom's 'three-volume transformation from terrified victim to anointed guardian of the planet.'[52] At one point he exposes his body, unclothed, to the changeless flood of sunlight entering the spaceship, which produced a sense of 'intense alacrity' in Ransom; now, 'he felt vigilant, courageous and magnanimous as he had seldom felt on Earth.' Ransom's initial conception of 'space,' it is explained, had been based on 'a nightmare, long engendered in the modern mind by the mythology that follows in the wake of science,' and now the nightmare was falling off, to the extent that, 'the very name "Space" seemed a blasphemous libel for this empyrean ocean of radiance in which they swam. [...] No: Space was the wrong name. Older thinkers had been wiser when they named it simply the heavens.'[53]

This vision of cosmic space as a benign heavenly ether is presented in opposition to the idea of 'space' as an empty, lifeless and cold medium, in a juxtaposition that draws directly from Lewis's scholarly work on the medieval model of the cosmos. For example, in *The Discarded Image* Lewis noted that in the 'cosmic imaginings of [the] modern,' space is understood as a hostile 'pitch black and dead-cold vacuity.'[54] In *Out of the Silent Planet* Lewis countered this view in two ways. Firstly, the darkness of outer space, as it is commonly perceived, is re-cast in the novel as a 'bath of pure ethereal colour,' in a reversal that works against the idea of the cosmos as a hostile space.[55] Second, Lewis criticizes the perceived emptiness of space, suggesting that, instead of being understood as a 'blank space' that might suggest occupation or exploration in the face of the unknown, the cosmos might in fact be instilled with a more spiritual presence.[56] Indeed, later in life Lewis wrote at length about the perception of God in the universe, and following reports that the Russians 'have not found God in outer space,' he remarked that if one were to 'send a saint up in a spaceship and he'll find God in space as he found God on Earth,' in an echo of Ransom's role in the novel.[57] In providing an opposition to the modern, secular understanding of the cosmos in this way, Lewis acknowledged the wisdom of 'older thinkers' in conceptualizing outer space as 'the heavens.' Indeed, this inversion of modern expectations in the novel was noted by one reader, who wrote to Lewis in 1938, stating that she was 'content with the fact that you have turned "empty space" into heaven!' In his response, Lewis remarked that 'the substitution of heaven for space' was his 'favourite idea in the book,' and it seems clear from this exchange that this was one of the key thematic messages of *Out of the Silent Planet*.[58]

When Ransom finally arrives on Mars, he has learned its "real" name in the language of 'Old Solar,' which is 'Malacandra.' The planet is described as having two distinct types of landscape – the barren, cold upland plains, and the densely vegetated, inhabited deep lowland valleys. The latter are explained as versions of the Martian 'canals' that were drawn by astronomers at the start of the twentieth century.[59] In this way Lewis established a believable world that fitted in with existing popular traditions, whilst also providing the imaginative space for his mythical interpretation of the cosmos. Or, as Lewis put it himself, 'I needed for my purpose just enough popular astronomy to create in the common reader a willing suspension of disbelief.'[60] Having established this scenario, the narrative sees Ransom escape his captors Weston and Devine, and befriend an indigenous population of Malacandrians, who have been interpreted as 'unfallen rational creatures free from the fears and temptations that plague our own wayward species.'[61] Ransom lives among them, learning about their culture and civilization, and upon discovering a series of stone carvings, identifies what appears to be an image of the solar system:

> The sun was there, unmistakably, at the centre of the disk: round this the concentric circles revolved. In the first and smallest of these was pictured a little ball, on which rode a winged figure. [...] In the next, a similar ball carried another of the flaming figures. [...] [The third ball] was there, but where the flame-like figure should have been, a deep depression of irregular shape had been cut as if to erase it.[62]

Whilst emulating diagrammatically the medieval model of the cosmos, albeit a heliocentric version as opposed to geocentric, this is the first direct indication in the novels that earth, the third ball in the carved image, appears to have fallen from grace in some way. This is seen in the erasure of the 'winged figure' that represents the spiritual guardian of each planet, and in the common refrain that earth is referred to as the 'silent planet' in an otherwise vibrant solar system. Ransom's initial terror at being in space is also attributed to earth's fallen status, with the 'mythology that follows in the wake of science' recognized as the precursor to Ransom's sense of estrangement. Indeed, the idea of earth as a 'fallen planet' is revisited in *Perelandra* and forms one of the central recurring motifs of the series.

The narrative of *Perelandra*, published five years after its prequel in the midst of the Second World War, starts with Ransom being taken to the planet Venus (known as 'Perelandra'), not by means of conventional spaceflight, but by a mystical creature known as 'Oyarsa.' 'I took a hero once to Mars in a space-ship,' Lewis later explained, 'but when I knew better I had angels convey him to Venus.'[63] As Lewis critic Jared Lobdell has pointed out, narrative conventions in science fiction sometimes act as a distraction, and for Lewis, 'the otherworldliness, not the science, is the point.'[64] As a result, a greater proportion of the novel is set on Perelandra itself, which is portrayed as a kind of utopia, a Garden of Eden before the Fall, its surface covered by vast oceans on the waves of which ride large mats of exotic, fruit-bearing vegetation. Here, Ransom was 'haunted, not by a feeling of guilt, but by surprise that he had no such feeling. There was an exuberance or prodigality of sweetness about the mere act of living which our race finds it difficult not to associate with forbidden and extravagant actions.'[65] In this way Lewis made the biblical allegory clearer, with references to forbidden actions and feelings of guilt or shame. Indeed, in a 1941 letter Lewis confirmed the meaning of his metaphor when he stated that 'Venus is at the Adam-and-Eve stage: i.e. the first two rational creatures have just appeared and are still innocent.'[66]

These passages demonstrate the ways in which Lewis succeeded in merging three types of mythology into his narrative: the space adventure, the medieval model of the cosmos, and the formative biblical story. Whereas some of the Christian undertones in *Perelandra* are certainly striking, as with the *Narnia* novels it would be a mistake to see the cosmic trilogy as having a primarily biblical message. As Lobdell suggests, the main subtext is the 'exploration of realms of the spirit,' which just happened to make use of the spiritual framework with which Lewis was most familiar.[67] In working these ideas into his revised cosmic spaces of 'the heavens,' 'Malacandra' and 'Perelandra,' the critique of the modern understanding of the cosmos is implicit – that a scientific understanding of outer space and the planets, and its manifestation in speculative science-fiction narratives, is lacking in an understanding of older models of the universe that embrace spiritual and ethical qualities. The significance of this re-casting of cosmic space becomes clear as Lewis articulates the threat to this sublime, ordered cosmos, manifested in what might be referred to as biological and spatial 'evolutionism.'

III Human bodies: cosmic evolution and corporeal violence

Imbedded in arguments about the exploration of space is the perceived relationship between the human body and the spaces of the interplanetary cosmos. In the works of Haldane, Stapledon and others, the human form was anticipated to undergo radical transformation in its propagation across the solar system – but how did Lewis envisage the human body in relation to cosmic spaces, and how was this related to his critique of the modern understanding of the cosmos? In the cosmic trilogy, rather than battling against the natural hazards of outer space and other planetary environments, which in the visions of Haldane and Stapledon would require the adaptation of the human form in the ultimate conquest of the cosmos, Ransom is left humbled at the benevolence of the cosmic realm, and it is in the actions of the novels' antagonists that we find the modern understanding of the cosmos pilloried and ridiculed by Lewis in his narrative structures.

In *Out of the Silent Planet* Ransom's view of space or 'the heavens' changed as he became aware of the benign influence of the cosmic ether on his body and mind. When Ransom is transported to Venus in *Perelandra* attention is drawn again to the effects that interplanetary travel have had on Ransom's body. Having arrived on the planet, Ransom realizes that his body had developed a 'parti-coloured appearance all over – no unnatural result of his one-sided exposure to the sun during the voyage.'[68] More than just an Englishman's suntan, this altered appearance represents a change that is coming over Ransom on a more fundamental level, whose full effect is only realized when he arrives back on earth. As he is helped out of his vessel by friends back home in England, they are faced with 'almost a new Ransom, glowing with health and rounded with muscle and seemingly ten years younger. In the old days he had been beginning to show a few grey hairs; but now the beard which swept his chest was pure gold.'[69] Far from being degraded by his encounters with outer space and the surface conditions of Venus, Ransom has, during the course of his adventures, been positively instilled with the beneficial effects of life in the cosmos.

This enlightened 'new Ransom' is contrasted in the cosmic trilogy with the malign effects that space travel brings out in Weston, the archetypal imperialistic scientist. Through these characters' interactions, Lewis confronted the idea that human life could eventually spread throughout the cosmos in a type of cosmic "seeding" propelled by advances in spaceflight technology. In *Out of the Silent Planet* the critique of Haldane and Stapledon's expansive vision of interplanetary posthumanism came to the fore, and Lewis parodied this view through the voice of Weston, who triumphantly declares on Malacandra the right of the human race to claim 'planet after planet, system after system, till our posterity – whatever strange form and mentality they have assumed – dwell in the universe wherever the universe is habitable.' Here the key element of Weston's vision of human life in the cosmos is its altered form and consciousness, or as Lewis later confirmed, the idea of 'perpetuating and improving the human species for the whole meaning of the universe.'[70] Although Lewis's fictional critique is one step removed when he has Ransom deride this philosophy of life as 'raving lunacy,' he made it clear in other accounts that 'the danger of "Westonism" I mean to be real,' and the reader should be left in no doubt as to the relevance of these passages to contemporary cultural discourse. In *Perelandra*, Weston returns, and having landed on Perelandra in his spaceship, again explains his motivations to Ransom whilst becoming increasingly animated, pronouncing that 'I became a convinced believer in emergent evolutionism. [...] Man in himself is nothing. The forward movements of Life – the growing spirituality – is everything. [...] I *am* the Universe.'[71]

Here Weston channels the French philosopher Henri Bergson's (1859–1951) 'creative evolution' and *élan vital*, which supposed that evolution had a kind of innate driving force. Lewis questioned Bergson's ideas in various accounts, particularly the notion of 'the "striving" or "purposiveness" of a Life-Force,' suggesting that, if such a life force existed, it would be analogous to God.[72] Whilst espousing these views in the novel, it becomes apparent that Weston is demonically possessed as he lurches into a disturbing seizure, his speech becoming increasingly muddled and manic. In Lewis's characterization, then, the weighty proclamations of Weston about eternal life and the universe become merged with the demonic utterances of the possessed Weston-Devil, finally confirming and at the same time undermining the malign influence of the modern understanding of the cosmos, taking this view to its implausible and absurd ends.

As the narrative of *Perelandra* transpires, it is up to Ransom to save the Eve-like 'Green Lady' from Weston's destructive influence. Thus, following a period of intense deliberation with his inner conscience, Ransom decides that 'Weston's body could be destroyed; and presumably that body was the Enemy's only foothold in Perelandra.'[73] Ransom echoes Lewis's address to the Oxford Pacifist Society, as he comes to the grim realization that violent conduct is the only morally just way in which the sanctity of Perelandra can be preserved. As has been pointed out by the theologian and ethicist Stanley Hauerwas, Lewis thought that 'war was a tribulation, but it was a tribulation that you could bear because it did not disguise itself as a pleasure.'[74] There then follows a lengthy account of the combat between the Weston-Devil and Ransom across the forests, oceans and caves of Perelandra, described in sometimes gruesome detail. When Ransom finally prevails, he becomes aware of his bodily injuries, the most severe of which is 'a wound in his heel [...] inflicted by human teeth [...] it was still bleeding.'[75] Here Ransom takes on an almost Christ-like status, as he sacrifices his body to help save Perelandra from Weston and the associated moral and spiritual threat of 'fallen' humanity. The historical context of the narrative is also relevant to the emerging theme of violence in the cosmos and its complex relationship with religious discourse that Lewis was developing in his written work. As Sanford Schwartz has pointed out, 'Lewis's fighting philologist was conceived in an atmosphere of looming international crisis [and] the issues surrounding the causes, conduct, and consequences of the Second World War are never far from the surface.'[76] In this context Ransom's combat with the Weston-Devil on Perelandra can be read as an echo of the Allied fight against a morally and spiritually defunct Nazi Germany, whose own ambitions of eugenic purity and geographical expansion have been well established.[77]

Perelandra ends with Ransom convening with the planet's spiritual guardians, whereupon he is sent back to earth. In this way, although Ransom has succeeded in saving Venus from 'falling' into a state of sin, earth remains as the 'silent planet,' whose fallen status in Lewis's analogy is manifested through certain types of morally and spiritually questionable activity. This is the focus of the final installment in the trilogy and, in a departure from the first two books, *That Hideous Strength* (1945) takes place entirely in and around the sleepy English university town of Edgestow, where the ominously-named 'National Institute for Co-Ordinated Experiments' (NICE) is spreading its malign influence. Some of the socio-scientific ambitions of this group are revealed by the returning character Devine, including 'sterilisation of the unfit, liquidation of backward races, selective breeding [and] biochemical conditioning.'[78] The experiments conducted by the NICE even include the animation of a severed head, which is disturbingly kept alive as 'the first of the New Men [...] the beginning of Man Immortal and Man Ubiquitous.'[79] Here Lewis echoes the language of Haldane and Stapledon, and positions the biochemical adaptation of the human form as a depraved and unethical practice that could only take place, in Lewis's cosmic context, on the fallen planet of Earth or 'Thulcandra.' Although not taking place in the interplanetary settings of the first two novels, Lewis's final installment drives deeper into the themes of corrupted scientific research and its association with earthly sin in its endeavor to produce new, purified and improved versions of humankind.

IV Spaceflight and the cosmos: the moral threat of human space exploration

A recurrent theme in Lewis's cosmic trilogy is the implied critique of space exploration, over and above the more general re-imagining of the science-fiction narrative that is infused with a medieval understanding of the cosmos. The most basic version of this critique can be found in *Out of the Silent Planet*, where a straightforward form of interplanetary imperialism is outlined. Indeed, in a 1939 letter Lewis revealed that one of his reasons for writing the novel was 'the discovery that a pupil of mine took all that dream of interplanetary colonisation quite seriously,' and as such, Weston and Devine arrive on Malacandra in an archetypal search of gold and glory.[80]

Weston states his intention to 'plant the flag of man on the soil of Malacandra [...] superseding, where necessary, the lower forms of life that we find,' whilst Ransom acts as the voice of reason and helps to prevent this imperialistic conquest.[81] However, the most overt criticism of space exploration, which at the time the book was written was anticipated by some but was by no means a reality, comes in *Perelandra* when Weston is derided as

> a man obsessed with the idea which is at this moment circulating all over our planet in obscure works of 'scientification,' in little Interplanetary Societies and Rocketry Clubs, and between the covers of monstrous magazines, ignored or mocked by the intellectuals, but ready, if ever the power is put into its hands, to open a new chapter of misery for the universe. It is the idea that humanity, having now sufficiently corrupted the planet where it arose, must at all costs contrive to seed itself over a larger area. [...] The destruction or enslavement of other species in the universe, if such there are, is to these minds a welcome corollary.[82]

Here a triumvirate of cultural trends is identified – science-fiction stories, interplanetary societies and scientific magazines – that would appear to represent a tangible threat to the ordered cosmos that Lewis established throughout the cosmic trilogy. It is telling that this description is provided not by Ransom or any other principal character in the novel, but by the 'narrator' himself, addressing the reader directly. The narrator plays a minor but important role, as he is tasked with witnessing and then telling Ransom's story, re-cast as 'fiction' lest people not believe Ransom's fantastic tales. The narrator is named on page 19 of *Perelandra* as 'Lewis' himself, not so much 'breaking the fourth wall' as reinforcing the possibility in the mind of the reader that this story originates in our world as we know it. This narrative construction allows the reader to understand the 'monstrous magazines' and 'little interplanetary societies' as not purely imaginative, but in reference to the thriving prewar groups that did indeed carry out rocket experiments, popularize the concept of spaceflight and produce some of the earliest design concepts of space hardware.[83]

Direct references such as these prompted Arthur Clarke to write to Lewis in 1943, as he evidently felt stung by Lewis's attack. Clarke stated that he wished 'to disagree, somewhat violently with you over a passage on p. 92 of *Perelandra*. [...] The interplanetary societies must be taken more seriously.'[84] In his response, Lewis confirmed his fear that 'a point of view not unlike Weston's is on the way' and expanded his critique, claiming that 'a race devoted to the increase of its own power by technology with complete indifference to ethics does seem to me a cancer in the universe.' Lewis cited Haldane and Stapledon as the chief purveyors of this view, whilst also parenthetically exclaiming, 'hang it all, I *live* among scientists!,' in reference to his residence at Magdalen College, Oxford.[85] This was the start of a correspondence of about a dozen letters between Clarke and Lewis, in which Lewis turned down invitations to speak to the BIS about his views and to attend Clarke's influential 1946 lecture on 'The Challenge of the Spaceship.'[86] The letters remained cordial but critical, and this tone is exemplified by Lewis's comment in his letter in September 1946: 'I wish your lecture every success *except* a practical realisation of space travel!'[87]

Lewis's critique of interplanetary travel is also evident in a number of poems and essays published towards the end of his career. In 'Prelude to Space,' Lewis deploys a rape metaphor when describing the launch of a space rocket and the inevitable, in the minds of those he detracts, spread of the human race throughout the universe:

> So Man, grown vigorous now,
> Holds himself ripe to breed,
> Daily devises how
> To ejaculate his seed
> And boldly fertilize
> The black womb of the unconsenting skies[88]

The poem goes on to list various human atrocities and diseases that would be spread throughout the cosmos should space exploration commence, including, 'Bombs, gallows, Belsen camp' in a crude but effective analogy explaining one aspect of Lewis's antipathy to this endeavor.[89] Moreover, in an essay entitled 'Religion and Rocketry,' Lewis provided the fullest explanation of his objection to space exploration, which can be read alongside his fictional critique of space exploration as, in the first instance, imperialistic, and, second, representative of a greater and more dangerous transgression, 'the fatal unilinear progression which so haunts our minds' or the obsession with never-ending scientific progress.[90] In his essay, Lewis explained that any future contact between humankind and alien species would be a 'calamity,' and based his objection around the understanding that humanity has 'fallen' in the moral and spiritual sense, and would only export a depraved ideology to other parts of the universe. 'We know what our race does to strangers,' he wrote. 'Man destroys or enslaves every species he can. [...] There are individuals who don't. But they are not the sort who are likely to be our pioneers in space. Our ambassador to new worlds will be the needy and greedy adventurer or the ruthless technical expert.'[91] Here we can see the shadow of Weston and Devine in Lewis's real fears of what the expansion of human influence to outer space might result in. He concludes with a statement of relief that we are 'still very far from travelling to other worlds' and further suggested that perhaps the 'vast astronomical distances' that get in the way of such conquests may in fact represent 'God's quarantine precautions,' preventing the 'spiritual infection' of one fallen species to another part of the cosmos.[92]

The onset of the Space Age drew forth further commentary from Lewis in his final years, and he commented in 1957 that 'I don't feel that "Sputnik" in itself is anything very dangerous,' whilst in a 1961 letter he noted that 'Yes, Gagarin is exciting.'[93] He also revealed in a 1963 essay that he was 'not much concerned about the "space race" between American and Russia,' suggesting that it was an 'excellent way of letting off steam' by the world superpowers.[94] Lewis seems to have modified his view slightly from earlier in his career, when he conflated spaceflight with a moral transgression into the cosmic realm. However, he went on to explain that there are 'three ways in which space-travel will bother me if it reaches the stage for which most people are hoping,' the first being a sense of disappointment that the poetic wonder of the moon would be aesthetically tarnished should humans set foot there, such that 'he who first reaches it steals something from us all.'[95] Second, Lewis repeated his earlier warning about the likelihood of humankind to destroy or enslave any rational alien species it may encounter. Finally, Lewis questioned the spiritual reasoning behind the exploration of space, stating that 'we are trying to cross a bridge, not only before we come to it, but even before we know that there is a river that needs bridging,' alluding to the lack of knowledge we have about life in the universe, and what that might state about our sense of spirituality.[96] What emerges, then, is a disparity between near-earth spaceflight and the expansive form of space exploration that would still have important moral and spiritual consequences for humanity in Lewis's view.

V Understanding Lewis as a critical myth-maker

By exploring C. S. Lewis's writings on outer space, it is possible to understand this writer as a critical myth-maker of outer space in the mid-twentieth century, whose works took into account spiritual and ethical dimensions in constructing a distinctive moral geography of outer space, and in doing so offered a warning against the anticipated militarization of outer space. Lewis's relevance in this historical framework derives from his substantial influence in the intellectual and cultural realms of the twentieth century, with his fictional and non-fictional works reaching millions of readers worldwide, and his intellectual achievements acclaimed to various degrees across theological, literary and historical registers.[97] Lewis deployed his critical interjection on humanity's relationship with the cosmos through a number of different written texts, including his letters, essays, academic work and poetry, but above all through his cosmic trilogy. Various aspects to Lewis's critique have been identified, including the nature of heavenly bodies, the role of the human body in interplanetary spaces and the motivations behind space exploration. As such, Lewis positioned a medieval understanding of an ordered, sublime and harmonious cosmos against the conception of interplanetary space as hostile, yet ripe for conquest and domination through harnessing the potential of science and technology.

In terms of a critique of the conquest of space, Lewis's work can be compared to the views of one other leading critic of space exploration in the twentieth century, the German-American political theorist Hannah Arendt (1906–75). In her 1963 essay 'The Conquest of Space and the Stature of Man' she postulated that 'the astronaut, shot into outer space [...] will be the less likely ever to meet anything but himself' in the search for a true 'Archimedean point' and that, ultimately, 'man can only get lost in the immensity of the Universe.'[98] Whilst it appears that Lewis and Arendt were not aware of each other's interventions on this topic and never crossed paths, together their writings amount to two separate strands of the philosophical critique of space exploration, Arendt's being a secular account that points towards the ultimate nihilism or narcissism of the prospect of space exploration, and Lewis's being a spiritual account that presents more of a positive understanding of humanity's relationship with the cosmos.

Lewis, however, as has been demonstrated in this chapter, offered this positive paradigm only having identified and criticized two forms of militarization that he saw as a moral threat: First, the threat of imperialism in outer space and its associated violence and hubris; and second, the threat of a bio-militaristic expansion of the very concept of humanity throughout the cosmos. More than just offering a utopian alternative to these ideas, Lewis's stories suggested that his view of a sublime and ordered cosmos would be worth fighting for, in the context of the existential struggles of the twentieth century, and that a positive vision of a benevolent cosmos could coexist with humankind regardless of the need for human space exploration. That Lewis's vision occurred in the overlapping categories of myth, narrative and scholarly work points towards the significance of engaging with a broad range of discursive texts, not just through channels typically associated with military agendas, in order to fully understand the militarization of outer space in the twentieth century.

Notes

1. Roger Lancelyn Green and Walter Hooper, *C. S. Lewis: A Biography*, London: Collins, 1974.
2. Meredith Veldman, *Fantasy, the Bomb and the Greening of Britain: Romantic Protest, 1945–1980*, Cambridge: Cambridge University Press, 1994, 54.
3. C. S. Lewis, *The Discarded Image: An Introduction to Medieval and Renaissance Literature*, Cambridge: Cambridge University Press, 1964.
4. Green and Hooper, *Lewis*, 55.
5. C. S. Lewis, 'Why I Am Not a Pacifist,' in idem, *The Weight of Glory and Other Addresses* [1949], New York: HarperCollins, 2001, 64–90; Stanley Hauerwas, 'On Violence,' in Robert MacSwain and Michael Ward, eds, *The Cambridge Companion to C. S. Lewis*, Cambridge: Cambridge University Press, 2010, 189–202, here 189.
6. Walter Hooper, ed., *C. S. Lewis: Collected Letters*, vol. 2: *Books, Broadcasts and War 1931–1949*, London: HarperCollins, 2004, 702 (emphasis in original).
7. C. S. Lewis, 'De Descriptione Temporum,' in Walter Hooper, ed., *Selected Literary Essays*, Cambridge: Cambridge University Press, 1969, 1–14, here 6.

8. Ibid., 8, 6.
9. Lewis, *Discarded Image*, 99.
10. David M. Smith, *Moral Geographies: Ethics in a World of Difference*, Edinburgh: Edinburgh University Press, 2000, 18.
11. See Fraser MacDonald, 'Anti-*Astropolitik*: Outer Space and the Orbit of Geography,' *Progress in Human Geography* 31.5 (October 2007), 592–615; Denis Cosgrove, 'Contested Global Visions: *One-World*, *Whole-Earth*, and the Apollo Space Photographs,' *Annals of the Association of American Geographers* 84.2 (June 1994), 270–94; K. Maria D. Lane, *Geographies of Mars: Seeing and Knowing the Red Planet*, Chicago: University of Chicago Press, 2011; and Rob Kitchin and James Kneale, eds, *Lost in Space: Geographies of Science Fiction*, London: Continuum, 2002.
12. Denis Cosgrove, *Geography and Vision*, London: I. B. Tauris, 2008.
13. Dennis Danielson, 'Intellectual Historian,' in MacSwain and Ward, *Cambridge Companion to C. S. Lewis*, 43–57, here 47.
14. Michael Ward, *Planet Narnia: The Seven Heavens in the Imagination of C. S. Lewis*, Oxford: Oxford University Press, 2008.
15. Walter Hooper, *C. S. Lewis: A Companion and Guide*, London: HarperCollins, 1996; Robert MacSwain, 'Introduction,' in idem and Ward, *Cambridge Companion to C. S. Lewis*, 1–15.
16. See Jared Lobdell, *The Scientifiction Novels of C. S. Lewis: Space and Time in the Ransom Stories*, Jefferson: McFarland, 2004; David Downing, *Planets in Peril: A Critical Study of C. S. Lewis's Ransom Trilogy*, Amherst: University of Massachusetts Press, 1992; T. A. Shippey, 'The Ransom Trilogy,' in MacSwain and Ward, *Cambridge Companion to C. S. Lewis*, 237–50; and Sanford Schwartz, *C. S. Lewis on the Final Frontier: Science and the Supernatural in the Space Trilogy*, Oxford: Oxford University Press, 2009.
17. See also Oliver Dunnett, 'Patrick Moore, Arthur C. Clarke and "British Outer Space" in the Mid-twentieth Century,' *Cultural Geographies* 19.4 (August 2012), 505–22; and William R. Macauley, 'Crafting the Future: Envisioning Space Exploration in Post-war Britain,' *History and Technology* 28.3 (October 2012), 281–309.
18. Downing, *Planets in Peril*, 35.
19. J. W. Lambert and Michael Ratcliffe, *The Bodley Head, 1887–1987*, London: Bodley Head, 1987.
20. Neil McAleer, *Odyssey: The Authorized Biography of Arthur C. Clarke*, London: Gollancz, 1992, 139.
21. Veldman, *Fantasy*, 46.
22. Hooper, *Books, Broadcasts and War*, 236; John Clute and Lee Weinstein, 'Lindsay, David,' in John Clute and Peter Nicholls, eds, *The Encyclopedia of Science Fiction*, London: Gollancz, 2015; http://www.sf-encyclopedia.com/entry/lindsay_david (accessed 15 July 2020).
23. Hooper, *Books, Broadcasts and War*, 753 (emphasis in original).
24. Green and Hooper, *Lewis*.
25. Hooper, *Books, Broadcasts and War*, 918, 236–7 (emphasis in original); Jeffrey Richards, '*Things to Come* and Science Fiction in the 1930s,' in Ian Hunter, ed., *British Science Fiction Cinema*, London: Routledge, 1999, 16–32, here 28.
26. C. S. Lewis, 'A Reply to Professor Haldane,' in Walter Hooper, ed., *On Stories and Other Essays on Literature*, New York: Harcourt Brace Jovanovich, 1982, 69–79, here 71.

27. Hooper, *Books, Broadcasts and War*, 236.
28. Frank Winter, *Prelude to the Space Age: The Rocket Societies, 1924–1940*, Washington, DC: Smithsonian Institution Press, 1983.
29. C. S. Lewis, 'On Science Fiction,' in Walter Hooper, ed., *Of Other Worlds: Essays and Stories*, London: Geoffrey Bles, 1966, 66; J. B. S. Haldane, *Possible Worlds and Other Essays*, London: Chatto & Windus, 1927.
30. Mark B. Adams, 'Last Judgment: The Visionary Biology of J. B. S. Haldane,' *Journal of the History of Biology* 33.3 (Winter 2000), 457–91, here 478.
31. Haldane, *Possible Worlds*, vi; Peter Bowler, 'Experts and Publishers: Writing Popular Science in Early-twentieth Century Britain, Writing Popular History of Science Now,' *British Journal for the History of Science* 39.2 (June 2006), 159–87, here 164.
32. Adams, 'Last Judgment.'
33. J. B. S. Haldane, 'Auld Hornie, FRS,' *The Modern Quarterly* 1.4 (Fall 1946), 32.
34. Idem, *Possible Worlds*, 304, 296, 302.
35. Geoffrey R. Searle, *Eugenics and Politics in Britain, 1900–1914*, Leyden: Noordhoff, 1976; Pauline Mazumdar, *Eugenics, Human Failings: The Eugenics Society, Its Sources and Its Critics in Britain*, London: Routledge, 1991.
36. Haldane, *Possible Worlds*, 304, 305.
37. Adams, 'Last Judgment.'
38. Mike Ashley and John Clute, 'Stapledon, Olaf,' in Clute and Nicholls, *Encyclopedia of Science Fiction*; Robert Crossley, 'Introduction,' in idem, ed., *An Olaf Stapledon Reader*, New York: Syracuse University Press, 1997, 2–5.
39. Olaf Stapledon, *Philosophy and Living*, vol. 2, Harmondsworth: Penguin, 1939, 305.
40. Idem, *Last and First Men* [1930], Harmondsworth: Penguin, 1987, 221.
41. Ibid., 252.
42. Lewis, 'De Descriptione Temporum,' 6.
43. See in this context also the contribution by Patrick Kilian, Chapter 8 in this volume.
44. Bob Parkinson, ed., *Interplanetary: A History of the British Interplanetary Society*, London: British Interplanetary Society, 2008.
45. Arthur C. Clarke to Olaf Stapledon, 19 April 1948, Stapledon Archive, University of Liverpool, H6.A.S.
46. Olaf Stapledon, 'Interplanetary Man?,' *Journal of the British Interplanetary Society* 7.6 (November 1948), 213–33, here 214.
47. Ibid., 215.
48. Ibid.
49. C. S. Lewis, 'The Planets,' in Walter Hooper, ed., *C. S. Lewis: Poems*, London: Geoffrey Bles, 1964, 12–15; Ward, *Planet Narnia*.
50. C. S. Lewis, *Out of the Silent Planet* [1938], London: Longmans, Green, 1966, 17.
51. See Thore Bjørnvig, 'Transcendence of Gravity: Arthur C. Clarke and the Apocalypse of Weightlessness,' in Alexander C. T. Geppert, ed., *Imagining Outer Space: European Astroculture in the Twentieth Century*, 2nd edn, London: Palgrave Macmillan, 2018, 141–62 (= *European Astroculture*, vol. 1).
52. Schwartz, *Final Frontier*, 6.
53. Lewis, *Out of the Silent Planet*, 26, 29.
54. Idem, *Discarded Image*, 111.

55. Idem, *Out of the Silent Planet*, 29.
56. For a critique of author Joseph Conrad's alluring 'blank spaces' on maps, see Felix Driver, *Geography Militant: Cultures of Exploration and Empire*, Oxford: Blackwell, 2001.
57. C. S. Lewis, 'The Seeing Eye' [1963], in Walter Hooper, ed., *Christian Reflections*, London: HarperCollins, 1998, 212–23, here 212, 217.
58. Hooper, *Books, Broadcasts and War*, 235.
59. K. Maria D. Lane, 'Mapping the Mars Canal Mania: Cartographic Projection and the Creation of a Popular Icon,' *Imago Mundi* 58.2 (July 2006), 198–211.
60. Lewis, 'Reply to Professor Haldane,' 71.
61. Schwartz, *Final Frontier*, 12.
62. Lewis, *Out of the Silent Planet*, 117.
63. Idem, 'On Science Fiction,' 69.
64. Lobdell, *Novels of C. S. Lewis*, 37.
65. C. S. Lewis, *Perelandra (Voyage to Venus)* [1943], London: Pan, 1983, 32.
66. Hooper, *Books, Broadcasts and War*, 504.
67. Lobdell, *Novels of C. S. Lewis*, 136.
68. Lewis, *Perelandra*, 49.
69. Ibid., 25.
70. Idem, *Out of the Silent Planet*, 145–6; Hooper, *Books, Broadcasts and War*, 262.
71. Lewis, *Out of the Silent Planet*, 24; Hooper, *Books, Broadcasts and War*, 262; Lewis, *Perelandra*, 81, 86 (emphasis in original).
72. Henri Bergson, *Creative Evolution*, New York: Henry Holt, 1911; C. S. Lewis, *Mere Christianity*, New York: HarperCollins, 1952, 22.
73. Idem, *Perelandra*, 133.
74. Hauerwas, 'On Violence,' 190.
75. Lewis, *Perelandra*, 173–4.
76. Schwartz, *Final Frontier*, 4.
77. See Richard J. Evans, *The Third Reich at War 1939–1945*, London: Allen Lane, 2008; and Michael Heffernan, *The Meaning of Europe: Geography and Geopolitics*, London: Arnold, 1998.
78. C. S. Lewis, *That Hideous Strength* [1945], London: HarperCollins, 2005, 44.
79. Ibid., 239.
80. Hooper, *Books, Broadcasts and War*, 262.
81. Lewis, *Out of the Silent Planet*, 145.
82. Idem, *Perelandra*, 73.
83. See, for instance, Alexander C. T. Geppert, 'Space *Personae*: Cosmopolitan Networks of Peripheral Knowledge, 1927–1957,' *Journal of Modern European History* 6.2 (2008), 262–86; Winter, *Prelude to the Space Age*; and Asif A. Siddiqi, *The Red Rockets' Glare: Spaceflight and the Soviet Imagination, 1857–1957*, Cambridge: Cambridge University Press, 2010.
84. Ryder W. Miller, *From Narnia to a Space Odyssey: The War of Ideas between Arthur C. Clarke and C. S. Lewis*, New York: iBooks, 2003, 36–8.
85. Hooper, *Books, Broadcasts and War*, 594 (emphasis in original).
86. On the latter, see Robert Poole, 'The Challenge of the Spaceship: Arthur C. Clarke and the History of the Future, 1930–1970,' *History and Technology* 28.3 (September 2012), 255–80.
87. Hooper, *Books, Broadcasts and War*, 741 (emphasis in original).

88. C. S. Lewis, 'Prelude to Space,' in Hooper, *C. S. Lewis: Poems*, 56–7, here 56.
89. Ibid.
90. C. S. Lewis, *The Abolition of Man*, London: Geoffrey Bles, 1943, 54.
91. Idem, *The World's Last Night and Other Essays*, New York: Harvest, 1960, 89.
92. Ibid., 91.
93. Walter Hooper, ed., *C. S. Lewis: Collected Letters*, vol. 3: *Narnia, Cambridge, and Joy, 1950–1963*, London: HarperCollins, 2004, 895, 1257.
94. Lewis, 'The Seeing Eye,' 219.
95. Ibid.
96. Ibid., 223.
97. MacSwain and Ward, *Cambridge Companion to C. S. Lewis*.
98. Hannah Arendt, 'The Conquest of Space and the Stature of Man' [1963], in idem, *Between Past and Future: Eight Exercises in Political Thought*, New York: Penguin, 2006, 260–74, here 272. On Arendt, see also Alexander Geppert's epilogue, Chapter 14 in this volume.

CHAPTER 7

One Nation, Two Astrocultures? Rocketry, Security and Dual Use in Divided Germany, 1949–61

Daniel Brandau

Over two decades ago, historian Christoph Kleßmann famously characterized the complicated relationship between East and West Germany as 'two states, one nation' ('Zwei Staaten, eine Nation'), promoting the idea of Germans having remained one nation after 1945. As this alludes to aspects of cultural or social unity, or some kind of continued entanglements at least, the notion has been controversially debated. As an 'array of images and artifacts, media and practices that all aim to ascribe meaning to outer space,' astroculture developed in very different ways on both sides of the Iron Curtain.[1] Fostered by spaceflight societies, lobby groups and government propaganda, it legitimized similar strategies towards rearmament in a world of divergent ideologies. It was not the peaceful visions of spaceflight that were at the center of attempts at legitimization and critique, but its aggressive, expansionist and militaristic undertones, tying it to another chief concern of both German states – 'Sicherheit,' or security. Sputnik 1, launched in October 1957, marked the advent of intercontinental ballistic missiles (ICBM) that had been anxiously anticipated since the end of the Korean War four years prior. After that technological breakthrough, the focus of the ensuing 'second Berlin crisis' of 1958 to 1961 shifted more and more towards technopolitics, ending with the construction of the Berlin Wall and a halt to the migration of experts and workers from the East to the West. The growing division dashed any hopes

Daniel Brandau (✉)
Freie Universität Berlin, Berlin, Germany
e-mail: daniel.brandau@fu-berlin.de

Alexander C. T. Geppert et al. (eds), *Militarizing Outer Space*
European Astroculture, vol. 3
https://doi.org/10.1057/978-1-349-95851-1_7

that postwar Europe could develop, through the exchange of knowledge, into a haven of international peace after two devastating wars. The reasoning was more than merely economic: in a world in which nuclear arms could strike any place on earth, international security became linked to security as a social process, all the way down to notions of individual safety.[2]

After the failures of technocracy and absolute military defeat, the legitimization of rocket technology seemed to be particularly complicated in postwar Germany. While the search for security characterized West German politics throughout the Cold War, debates about what assured or undermined security – military security, economic security, the security of the political system – took place in the many realms of popular culture.[3] Since the eighteenth century, the concept of security had carried spatial and temporal orientations and regarded technological innovation as a tool for its own enhancement – 'taming the future' was one of its strongest underlying principles. During the Cold War, 'security' consolidated its role as the centerpiece of all aspects of governance. The term invariably points to (and also discerns) threats and risks.[4] In the eyes of many critics, the risks of rocket technology seemed to facilitate the threats the Cold War produced and therefore required political oversight.

'Dual use' became increasingly important as an array of political strategies to separate civilian from military applications – not only within the West German government, but also the private sector.[5] It was a political construct with little basis in technological reality, but also a necessity due to the restrictions and popular skepticism towards military programs. The concept received a multi-layered meaning in the wake of the growing anti-militarist movements of the 1950s, negotiated and mediated through notions of outer space utopias and dystopias. Dual use highlighted the dangers of proliferation, but also seemed to offer a solution. An *Ersatzproduktion* – a term introduced by philosopher and activist Günther Anders (1902–92) – of spaceships instead of missiles was supposed to offer defense manufacturers a way out of fueling an arms race.[6] Despite its relevance in discussions on security, however, astroculture has rarely been acknowledged as more than a pop-cultural backdrop of the European political division. Even the more recent studies on postwar Germany have avoided comparative or entangled approaches, leaving Kleßmann's call for German-German perspectives unanswered – including, notably, Eckart Conze's landmark study on the Cold War security debate.[7] In light of this lacuna, space history may prove an unorthodox, but worthwhile approach, as long as it is not reduced to its positivist understanding of a history of technological realization.

One of the hitherto rather uncommon tracers of space-historical inquiry is the "dark" side of astroculture. Through shifting the focus to narratives, concepts and practices of militarization and notions of violence, this approach presents tools for understanding the complex asymmetries between East and West Germany in technopolitics and technoscience. 'Culture' also hints at the societal and political contexts of the production of spaceflight narratives, images or practices towards a 'social meaning of a life held in common.'[8] Those contexts were markedly diverging. In the West, enthusiasm was not just a state of mind or individual interest, but propelled a set of community-building practices that informed technopolitics. Private spaceflight societies were not independent, but mediated between industrial and government interests. By contrast, the East German government rarely relied on enthusiastic ventures originating from student circles or engineering industries, but sought to control and guide specific astrocultural activities and projects as sanctioned by the Sozialistische Einheitspartei Deutschlands (Socialist Unity Party, SED).

Futurism, or more specifically 'astrofuturism,' on the other hand, was as important to East German state propaganda as it was to Western enthusiast communities. It provided narratives and images of prospected changes over time that were anticipated to be mostly technological, but also social and political in nature.[9] For about a decade from the late 1940s to the late 1950s, a shared future seemed to outline a potential arena of European cooperation, possibly even bringing the East and West together in the long run. Technology and science seemed to uphold common truths regardless of ideological differences. The peaceful future in outer space seemed to be driven by a continent that had allegedly learned its lessons. It was no surprise that the International Astronautical Federation (IAF), founded in Paris and London in 1950 and 1951, closely mirrored the congresses of the Council of Europe that had been established in London in 1949 to jumpstart European political integration. West German spaceflight societies, sanctioned and promoted by Allied military authorities, played major roles in its foundation.[10] Such international cooperation was not driven by diplomats, but by engineers. Only a handful of them, including Wernher von Braun (1912–77), had been members of spaceflight societies during the late Weimar years.[11] Others had become involved in rocketry through V-2 and missile development at Peenemünde and the military industries of the Third Reich. The majority were young engineers and students at technical universities such as Stuttgart or Hanover. The latter profited from the expertise and reputation of their older colleagues, who in turn needed the students' idealistic enthusiasm to provide their societies democratic legitimacy.

The advent of spaceflight, with its ever-increasing series of Sputnik, Vanguard and Explorer satellite launches, and the coincident security crisis of 1957 and 1958, proved these idealists wrong. Spaceflight was technologically realized as a product of global conflicts, not as a political counteraction towards disarmament. Government institutions and military authorities, including NATO, became involved not to facilitate but rather to control and limit the exchange of knowledge. Rocketry promised great technoscientific progress but also apocalyptic destruction. West German engineers pioneered postwar European networks, but in order to participate in political decision-making processes and to acquire lucrative defense contracts they oriented themselves towards emerging West European political institutions and NATO. The East German government, on the other hand, actively promoted utopian and downplayed dystopian notions in order to differentiate spaceflight by communist countries from projects conducted in the West. Because there were many similarities in technology, communist propaganda focused on the alleged inherent motives that would present themselves in the long term.

These developments on either sides of the Inner German border were far from uncontested and stirred opposition even within the respective space enthusiast communities. Nevertheless, they had a considerable cultural impact and shaped the growing East-West divide, turning ideological theory into markers of the future that could be inscribed into exhibition objects as well as public debates. They centered on different understandings of rearmament, one explaining current militarisms through ideology (in the East) and the other focusing on projects instead of technologies and separating civilian from military uses through political institutions (in the West).

I Ideology and military innovation in a divided Europe

Astroculture had profound implications for both the legitimization and critique of defense and security in Central Europe, but its ideas, images, narratives and practices grew increasingly different. On a political level, the division of Germany in 1949 was motivated and explained as an economic and ideological one. The distinction between communism and capitalism was also believed to predetermine any chances of learning from the past and shaping the future. Both states tried to leave the Nazi past behind and presented themselves, in opposition to the respective other, as the legitimate "new" Germany.[12] When both states were remilitarized and integrated into defense alliances in the mid-1950s, technopolitics seemed to become a marker of that distinction. The idea of a joint spaceflight project, rediscovered as a peaceful future vision for the European continent during the intermediate decade of demilitarization, evolved into a catalyst on both popular and political levels. In the East, spaceflight imaginaries grew important in government propaganda to discern the long-term communist future from current militarism. In the West, astroculture became the basis for technoscientific networks where government programs were nonexistent.

At first it seemed that the international enthusiast networks of the interwar years could be revived. In the early to mid-1950s the European space *personae* claimed the 'non-political nature' of technology in general, and the alleged long-term peaceful goals of spaceflight in particular, to substantiate rhetorical acts of forgiveness. After the *Daily Mirror* denounced von Braun for his role in the V-2 development and rocket weapon attacks on London, Kenneth Gartland (1924–94) of the British Interplanetary Society (BIS) came to the defense of the German pioneer, '[because] he was no more morally responsible for civilian casualties in World War II than the designer of the Lancaster bomber.'[13] BIS President Val Cleaver (1917–77) proclaimed that there was an important long-term outcome of the German weapon program, namely the rocket as a potential 'deterrent of war' and guarantor of peace.[14]

Even before politicians could make up their minds about a European project, German engineers had restarted working in France, the United Kingdom, Italy, Spain and the Soviet Union. What began as the forced migration of expertise from defeated Germany to the industries of victorious Allied countries soon gave way to lucrative contracts for German engineers who chose to stay as workers or even managers – at least in Western Europe and the United States. Most German techno-experts who had worked in the Soviet Union, among them second-tier Peenemünders and specialists from universities and associated industries, were repatriated to East Germany after sharing their knowledge. Many immediately migrated to the West through Berlin, such as Helmut Gröttrup (1916–81) and Kurt Magnus (1912–2003).[15] Alternative networks were established in Eastern Europe, based on state-run science programs and the cooperation of nationalized companies, including those in East Germany.[16] In the West, the migration of experts boosted Western European technoscientific networks that became cornerstones of a new Europe amidst a new international order – the Cold War.

Articles and books on the V-2 rocket, circulating among American troops, astonished English-speaking Germans about how revolutionary the weapon had actually been, at first seeming to confirm war-time propaganda in retrospect.[17] Publications exhibiting a militaristic fascination with 'the weapon of the future' through the former enemies' eyes subverted the tight censorship rules applied by the Allied occupation authorities to military topics. They also gave a frame of reference around what became publishable to the large group of German engineers who had not migrated to Allied countries and tried to survive by writing fact or fiction themselves.[18] The dynamic between the longing to suppress and the wish to process the war into collective memory remained powerful throughout the 1950s and extended to the fascination for rockets. After former Peenemünde physicist Hans K. Kaiser (1911–85) had familiarized German readers with the rocket weapons of the war in his *Kleine Raketenkunde* published in 1949, it was Walter Dornberger (1895–1980), now working in the United States, who promoted the defense of Peenemünde as the 'cradle of spaceflight.' His apologetic portrayal in *V2: Der Schuß ins Weltall* was picked up and reiterated by his former colleagues, but not necessarily by those who had been involved in rocketry outside Peenemünde such as Eugen Sänger (1905–64).[19]

Despite Allied restrictions on German rearmament and manufacturing of aircraft and military technologies, and long before those restrictions were softened and loosened in 1952 and then lifted completely in 1955, the US Military Security Board in Wiesbaden encouraged the spin-off of a

Technische Universität Stuttgart student group, founded by the engineering student Heinz-Hermann Koelle (1925–2011) and soon joined by Sänger, into the Gesellschaft für Weltraumforschung (Society for Space Research, GfW, from 1956 Deutsche Gesellschaft für Raketentechnik und Raumfahrt, DGRR) in 1950 'to the fullest extent.'[20] The GfW, now with a license to engage in theoretical research, chose to use the same name as a Berlin spaceflight society of 1937–44. The latter had been student-run as well and tolerated by the authorities as a reservoir for German weapon development programs, with Peenemünde technical director von Braun and military director Dornberger enticing away its most capable members such as engineer Krafft Ehricke (1917–84) and the aforementioned Kaiser. With the postwar GfW, the prospect of creating a similar pool of future technoscientific experts still seemed far off, but the benefits nevertheless foreseeable as long as the students worked transparently and embraced international cooperation.[21]

Together with 31-year-old engineer Heinz Gartmann (1917–60), Koelle sent an open letter to fellow European spaceflight societies on 22 June 1949, in particular the BIS and the Groupement Astronautique Français. They stated that the rocket could be 'not only a weapon, but an instrument of peaceful research.' They also expressed their hope that networks of expertise across Western Europe were not just seen as a by-product of migration, but could 'foster friendly relations and a successful exchange of knowledge.'[22] Almost exactly one month after the proclamation of West Germany's new democratic constitution, the *Grundgesetz*, this was nothing short of a bold idea to politicize big technology. International technoscientific networks seemed to recover faster even than diplomatic channels and indicated a technocratic solution to the problem of century-old intra-European antagonisms. In a letter to von Braun, now working for the US Army, Koelle boasted:

> It is our great advantage that we are our own masters, there are no ministries or superior agencies that we need to ask for any kind of permission. However, we are bound to the rules and oversight of Allied research control of course, but it has become a form of art to interpret the restrictions in a reasonable way, which we are gladly doing with our own interests in mind. So far nobody has put obstacles in our way. The most important restriction is of course that we are not allowed to pursue any kind of actual rocketry, meaning we are not allowed to build rockets or even trial them on a test stand.[23]

Despite engineers having been integral to the war effort, there was a willingness by Allied authorities to approve of their self-proclaimed role as peacemakers. In turn, this offered chances to students, young professionals and non-professional enthusiasts who did not migrate or work overseas, but stayed in Germany and tried to establish new opportunities that were independent of the career networks of the war generation. Civilian spaceflight seemed to check all the boxes of a common goal – particularly when demonstrated through European collaboration.

Together with the BIS, the Groupement Astronautique Français and the American Rocket Society, the GfW played a major role in the foundation of the International Astronautical Federation in London in 1951.[24] In Germany, its underlying concept corresponded with Chancellor Konrad Adenauer's call for an 'overcoming of nationalism' in Europe after the catastrophes of two major wars and at a time when 'idealist internationalism' also faced skepticism. Political scientist John H. Herz (1908–2005), who had emigrated from Germany to the United States prior to the Second World War, argued that 'the security dilemma today is perhaps more clear-cut than it ever was before,' that it was at odds with utopianism and called for a new realpolitik.[25] Discussions on the European Defence Community, a treaty which was signed in 1952 but ultimately remained unratified in the French and Italian parliaments, seemed to conciliate internationalist visions with a pragmatic approach towards defense. The General Treaty of 1952 loosened most of German armament restrictions until 1955, and the Western European Union complemented the European Economic Community through a mutual assistance pact.[26]

Several years before the West German state became a member of NATO in 1955, the idea of a Western European alliance offered a practical political vision for the continent. In the wake of that introduction of defense as the primary driving force of European integration, the Peenemünders Karl Poggensee (1909–80) and Friedrich Staats (1913–2002) founded the Arbeitsgemeinschaft für Raketentechnik (Study Group for Rocketry, AFRA; from 1958 Deutsche Raketengesellschaft, German Rocket Society, DRG), in the North German city of Bremen in 1952. Cooperation with the GfW brought the AFRA IAF membership and members of the Stuttgart group, in turn, access to the old career networks that became relevant again due to the rebuilding of the West German defense industry from 1955.[27] Civilian application remained a utopian denominator at a time when the German spaceflight societies diversified politically. Vows to focus on civilian applications in the future, less and less clearly defined, planned or dated, merely covered up that the pacifists were losing the disagreement over the dual-use potentials of their areas of expertise.

GfW members were particularly divided about whether to re-allow militarism into their visions of spaceflight. Some of the 330 members (around 1,000 by the end of the decade) believed that rocketry had to be limited to civilian applications. Amongst them was Alfred Fritz (1911–2001), who had experienced the deployment of V-2s as a soldier and became an influential voice in the pacifist arm of the society. Eugen Sänger tried to mediate and argued that European international cooperation could automatically make missiles, the 'illegitimate child of spaceflight,' superfluous.[28] In hindsight, Sänger was careful not to rule out military options in the medium term; with the establishment of his Forschungsinstitut für Physik der Strahlantriebe (Research Institute for Reaction Propulsion Physics) in 1954, he became a defense contractor himself.[29] Nevertheless, the notion that civilian spaceflight and peace went hand in hand with post-nationalism was still predominant and influential among West German rocket and spaceflight enthusiasts. Both Stuttgart and Bremen were industrial centers, the former shaped by the local car industry and machine-building, the latter by aviation and shipbuilding. Since Bremen was located in the British sector of Lower Saxony in northwestern Germany, administrators and British companies, such as De Havilland, were willing to cooperate – despite vociferous V-2 remembrance and disdain in the London area.[30]

Since the GfW focused on the communication of pacifist visions and European techno-utopias, while its industrial partners were advocating for military contracts through transatlantic cooperation, the enthusiasts were able to separate applications that were actually closely related. This also meant that military aspects had to be downplayed in astroculture. Von Braun applauded the idea of a federation as early as 1949 since 'noble enthusiasm in the future of rocketry is stronger than national sentiments.'[31] Three years later, however, having started to develop the Redstone missile for the US Army, he accused the still-pacifist GfW both of hypocrisy or lack of discipline when its members seemed unable to constrain their 'self-righteousness' in light of the intensifying Cold War. Habits of 'solemnly proclaiming their abhorrence of any development dedicated to other than pacific purposes' just showed the new generation's inability to subordinate themselves to a military hierarchy, von Braun thought. He admitted that the rocket had 'two faces like Janus,' but its 'outstanding challenge to science and technology of the age in which we live' was the problem of diplomats, and it would take a post-Cold War world to abolish rocket weapons.[32]

Within the GfW not many agreed with von Braun. Sänger was still of the conviction that spaceflight had to be achieved through progress in astronautics, such as the photon drive, not conventional chemical rocketry.[33] When the third International Astronautical Congress (IAC) was held in Stuttgart in 1952, Alfred Fritz organized a concomitant exhibition that focused on civilian spaceflight projects. It juxtaposed the model of a V-2 missile with futuristic depictions and calculations, comparing the huge costs of the Second World War to the prospected (cheaper) costs of spaceflight. A poster extended this comparison into the Cold War by including contemporaneous American defense spending (Figure 7.1). The low figures for landings on the moon (10 billion DM = ca. $27 billion adjusted for inflation in 2020) and, by the year 2000, even Mars (20 billion DM = ca. $52 billion in 2020), in hindsight, significantly underestimated their true and later prospected costs.[34] They suggested that von Braun was wrong about the sheer necessity of dual use and that Sänger was right: a purely civilian realization of spaceflight seemed possible. The poster was shown on a wall, adjacent to a spaceship cockpit mock-up. Other illustrations were taken from the famous *Collier's* article series on the future of spaceflight that had started in March

Figure 7.1 In addition to technological objects, amongst them a full-size V-2 model, the 1952 spaceflight exhibition at the Landesgewerbemuseum in Stuttgart presented calculations about the projected costs of missions to the moon and beyond. It was to be cheaper than warfare, curator Alfred Fritz (1911–2001) was convinced.

Source: Deutsches Raketen- und Raumfahrt Museum e.V. records, 1942–1969, M0718/box 6. Courtesy of Stanford Libraries, Department of Special Collections.

1952, in particular Chesley Bonestell's (1888–1986) space station based on Wernher von Braun's concept.[35] Visual depictions and mathematical calculations pointed towards each other. If the themes and goals of future spaceflight transcended cultures, then its considerable, yet manageable, costs could be shared through international cooperation.

Alfred Fritz hoped to develop his exhibition into a permanent museum, the Raketen- und Raumfahrtmuseum Stuttgart. The project, which would swallow most of his energy for the twelve years that followed, received considerable political support. Most notably, minister president of Baden-Württemberg and later German Chancellor Kurt Georg Kiesinger (1904–88) endorsed Fritz's idea of an exhibition that did not shy away from debating the "dark side" of rocketry and spaceflight, as a mediator in the contested process of remilitarization. Fritz's GfW colleagues readily accepted his visions of a peaceful future in outer space. Nonetheless, the widespread self-designation of 'enthusiasts' obscured the fact that, despite common ground in the fascination for technology, there were deep political divisions within German society that began to penetrate the spaceflight clubs.[36]

The car city of Stuttgart remained the major hub for all matters space until 1955. Daimler-Benz hired young propulsion specialists, fully aware of the technology's dual-use potentials. The federal Ministry of Transport meanwhile assumed that spaceflight was naturally to become their political responsibility one day since they were responsible for meteorology and interested in observation through weather rockets and, perhaps, satellites in the future.[37] Long-term GfW supporter and minister of transport Hans-Christoph Seebohm (1903–67) did not foresee that Franz-Josef Strauß (1905–88) and his newly established Ministry of Defense would immediately acknowledge the profound implications of rocketry for German and European security, seek influence over the enthusiast networks and tie them to institutions and professional industrial control.[38] GfW members were in unison about embracing Western cooperation, but also reluctant to allow defense interests to dictate long-term goals. But while West German spaceflight enthusiasts were still trying to adapt to the realities of remilitarization, they largely failed to foresee, address or process the ramifications of the actual breakthroughs in spaceflight that originated in the East.

II Astrofuturism and socialist realism in East Germany

The authoritarian government of the German Democratic Republic (Deutsche Demokratische Republik, DDR) supported the production of astrofuturism in all its utopian and violent variants, but did not rely on a similar kind of astroculture originating from and giving 'social meaning' to the collaboration of private groups and industries. Over its first two decades East Germany developed into what historian Alf Lüdtke has called *durchherrschte Gesellschaft* – a state '"seeped through" the whole of society,' bringing together 'the elements of "politics", "society" and "everyday life, subjective experience and perception".'[39] Unlike West Germany, where the military, industry and the state developed an interest in networks evolving at engineering schools, the East German government sought immediate control over any aspects of technology enthusiasm or astroculture through its mass organizations. Precisely because astrofuturism gained paramount importance in official propaganda after Sputnik, its cultural aspects did not develop the same significance in technoscientific communities. Technology enthusiast societies were only supported as long as they followed guidelines or engaged in functions defined by the ruling party. These complications even intensified after Sputnik, and the sole East German society, the Deutsche Astronautische Gesellschaft (DAG), was tolerated, supervised and eventually integrated only because it gained influence within the IAF, therefore generating clout on behalf of East Germany in a largely Western organization at a time when Western countries refused to recognize the DDR diplomatically.

Among intellectuals and ideologues, initially there was widespread skepticism towards rocket technology and 'colonialist' spaceflight visions, both of which were considered tokens of the fascist past. This proved particularly controversial as the ruling SED used popular science as an ideological and educational tool.[40] Stalinist calls for socialist realism in the arts spilled over to East German popular literature, after an official decision by the SED in March 1951 against the voices of prominent socialist intellectuals Bertolt Brecht (1898–1956), Hanns Eisler (1898–1962) and Ernst Bloch (1885–1977).[41] Only a few spaceflight novels came to see the light of day, amongst them Ludwig Turek's (1898–1975) *Die goldene Kugel* (1949) about the arrival of a Venusian satellite and Arthur Bagemühl's (1891–1972) adventure *Das Weltraumschiff* (1952).[42] Furthermore, there was uncertainty about the modern rocket's fascist legacy. Both Peenemünde (the site where the V-2 rocket had been developed) and the former Mittelbau-Dora concentration camp (where the V-2 had been mass-produced in the Mittelwerk facilities) were on East German soil. Of the specialists who had been forced to work in the Soviet Union, some permanently returned to the DDR such as Werner Albring (1914–2007). These were uneasy complications for a state that tried to distinguish itself as antifascist.

The youth organization Freie Deutsche Jugend (FDJ), set up as the centerpiece of ideological education towards the new generation of Germans after the Nazis and the basis of the future communist state, controlled spaceflight discussions until 1955 through its publishing houses for youth literature and popular science for children and adolescents. By 1950 the FDJ was a mass organization of more than one million members between 13 and 25 years of age, with the aim to represent and educate the East German youth in accordance with party positions. Dissent was increasingly quashed under Stalinist guidelines, which worked so effectively that West German observers in 1950 thought the FDJ was the 'most reliable and powerful Soviet-German organization.'[43]

Even after Stalin's death in March 1953 the FDJ remained dedicated to socialist realism in literary depictions of the future, in particular themes related to building the 'Workers' and Peasants' state,' such as farming and advances in industrial production. Space-related topics were all but avoided while the party sought an official position.[44] The problem of how to interpret and manage the dual-use concept of rocketry became pressing in 1955. Amidst growing tensions over missile developments, the DDR engaged in establishing the Warsaw Pact in May 1955. At the same time, discussions of artificial satellites for science and military purposes culminated in American and Soviet public announcements of launches for the International Geophysical Year (IGY) of 1957/58. Philosopher Georg Klaus (1912–74), professor at Humboldt-Universität zu Berlin, proposed ideological distinction. While technologies initially developed to suppress could, according to Marxist theory, eventually be used to liberate, the problem of spaceflight was 'technological as well as societal,' and societies had to advance before they could deliberately decide to pursue peaceful spaceflight only.[45]

SED Central Committee Secretary Kurt Hager (1912–98) underscored the urgency of public engagement in the spaceflight topic, reflecting its relevance to Soviet and East German security interests. According to him, propaganda had to persuade young people that 'the future is in the East,' and that rearmament in the West should be read as the resurgence of fascism, allegedly ingrained in Western capitalism.[46] The FDJ, aware of the similarities between American and Soviet rocket developments and the employment of former Nazi experts, remained skeptical, so that Hager insisted and made no secret of what this propaganda was mainly about: keeping young engineers, especially the enthusiastic hopefuls, in East Germany, or even recruiting them from the West. They were desperately needed for the Second Five Year plan, designed for building high-tech infrastructures and making the state Cold War-ready. Instead, young people often migrated to the West through a still open Berlin (labelled *Republikflucht*) because they 'did not understand that the future is ours.' Hager called on the FDJ publishers to discuss futuristic topics and disseminate technological and scientific knowledge in any way possible.[47]

The SED's 'ideological offensive' of 1956 and 1957 presented opportunities for East German experts such as Heinz Mielke (1923–2013), a former Luftwaffe engineer-turned-astronomer, as well as science journalists Horst Hoffmann (1927–2005) and Peter Stache (1934–2010).[48] They not only received assignments to publish on the topic. In 1960 they also established the DAG in East Berlin.[49] Unlike its West German counterparts, however, engineers remained underrepresented. The location of its offices in East Berlin was politically motivated and sanctioned through the centralized administration of properties rather than driven by the needs or interests of industries, as Berlin was by no means an industrial hub, particularly not in aviation. In the West, spaceflight societies had been supported by industries and institutions when the latter foresaw their use as reservoirs of knowledge and human capital. East Germany, with its planned economy, only promoted the foundation of the DAG when there was a government need for spaceflight propaganda and youth education. That propaganda had to praise technology as an integral part of the social order dominated by communism, extending into both civilian and military accomplishments.[50]

The depiction of military use remained problematic. The military-owned Militärverlag published manuscripts that covered technological details, with an apologetic foreword pointing out young engineers (as quasi-'workers') having been misdirected and used by evil Nazi elites.[51] By contrast, any historical problematization of technology seemed 'liberal' and could lead to political troubles. For instance, the SED Central Committee reacted with outrage to the almost-publication of *Herz und Asche, Part II* by the FDJ publisher Neues Leben in 1958. In that novel, Latvian author Boris Djacenko (1917–75) depicted Red Army soldiers searching for and kidnapping V-2 engineers as well as raping a German woman. Hager criticized the 'liberal mindset' at the publishing house, transferred temporary control over day-to-day business to the Central Committee and replaced most of the editors. Those who had been involved in the publication process faced factional trials because of an alleged 'neglect of political vigilance.'[52]

The SED also commissioned propaganda artworks that hinted at dual-use capabilities without detracting from the vision of a peaceful future in the stars. The poster 'Die Sonne geht im Osten auf!' ('The Sun Rises in the East!'), for instance, was created in 1959 by the Deutsche Werbe- und Anzeigengesellschaft in Leipzig, an advertising agency owned by the party (Figure 7.2). It accentuated the sun as both a symbol and object of a realm that communist countries would soon be able to explore, indicating the 'rise' of the East through technoscience. The depiction of the rocket in the center implied potentials for military use, but its vertical ascent above the moon and into the stars emphatically highlighted the astrofuturist context.

When it came to the history of modern technology, differences between Eastern and Western rocketry remained much more difficult to identify. The SED newspaper *Das Neue Deutschland* depicted Wernher von Braun, an 'aristocrat collaborator,' as the personification of moral corruption and the alleged continuity of fascism in the West.[53] But when illustrators remained careful not to lose themselves in speculation or ideological exaltation, editors at Neues Leben warned that a 'general humanist tendency' no longer sufficed 'to use [illustrations] for the socialist education of children.' When the *Mosaik* magazine initiated a series on spaceflight, the editors warned artist Josef Hegenbarth (1894–1962), recipient of the prestigious 1954 East German Nationalpreis, that his illustrations had to do more than 'counter Western depictions,' but must also find new motifs to show the superiority of communism.[54] In a sign of growing confidence in its political system through the superiority in technoscience, the government began to embrace and instrumentalize the dual-use aspects of technology more aggressively.

Figure 7.2 The propaganda poster 'Die Sonne geht im Osten auf!' ('The Sun Rises in the East!') was issued by the Socialist Unity Party in early 1959, entwining military and civilian aspects within the motif of a rocket, the latter carefully outweighing the former through astrofuturistic imagery.
Source: Courtesy of Deutsches Historisches Museum.

III Propaganda and the problem of dual use

During the second half of the 1950s and particularly after 1957, it became obvious in both the East and West that the military dimension of rocketry was to become a far more pressing matter to European security than civilian spaceflight. Notions of utopia and dystopia remained more important in West German popular debates because a diverse number of actors, including enthusiast groups, were involved in the production of technoscientific networks. In turn, astroculture retained its functions in the discussion and legitimization of democratic governance both on a national level and in processes of West European integration.

The West German state had regained its sovereignty in 1955 and instituted the Bundeswehr as a new army under democratic control. Its East German equivalent, the Nationale Volksarmee (NVA) or National People's Army, followed just a few months later. While ICBMs were still in development in the United States and the Soviet Union, the deployment of intermediate-range systems such as the Soviet R-11/Scud-A and the American Corporal rendered the technology's implications for European security obvious, leading to concerns over the proliferation of knowledge. Both US and European military authorities became interested in bringing West German space enthusiast networks under military control. In East Germany, the state's grip over the radio and space enthusiast networks that would later form the DAG was tight to begin with, and there were no technoscientific networks truly independent from state institutions or the (compared to the West much smaller) aviation or defense industries.

After the remilitarization of the East German state, any claim of a fundamental East-West difference in defense policies rang duplicitous. In 1955 the popular science compilation *Weltall Erde Mensch* (Universe Earth Humankind), published by Neues Leben, had been chosen as the party's gift to receivers of the Jugendweihe, the SED-organized, all-but-mandatory initiation of young people at the age of 14 – making it the single most published book of its time. It was significantly revised through nine editions until 1960. Karl Böhm (1913–77), a civil servant within the ministry of culture, was influential in heavily expanding the chapters on outer space and spaceflight.[55] Together with fellow popular science writer Rolf Dörge, Böhm acknowledged in 1956 that 'spaceflight is no utopia anymore,' but again in a socialist realist sense, meaning that it was possible, but not necessarily sensible or urgent.[56] It was also an admission of the realities of technological applications: spaceflight was neither utopian or dystopian, but both civilian and military use stabilized the social, political and economic order in its development towards communism.

In 1959 and 1960, at the height of the 'second Berlin crisis,' even SED-commissioned illustrations started to highlight military over civilian use of rockets. The propaganda poster 'Zeitalter des siegenden Sozialismus' ('Age of Victorious Socialism') pointed to the military potentials of Soviet missiles, using imagery to reassure its citizens (and warn the West) of the defense capabilities of the Eastern bloc (Figure 7.3). Whether the goal was space exploration or warfare in or through outer space, the East was ready to defend its communist future. While the rocket was shot into space from a red-tinted Eastern bloc, it pointed to the West and was able to reach the other side of the globe. North America is clearly visible on the left.

Despite the division in ideological explanations – aggression, projected as a future outlook, always came from the other side of the Iron Curtain – there was no fundamental difference in technology. Yet the notion of technology itself being 'ambivalent' did not play a significant role in East German propaganda. This was due to the materialistic argument that technology's moral dimensions were bound to the ideology of a collective body, not individual choices. Notions of utopia and dystopia had no functions in a debate that was no longer taking place outside of ideological affirmations.

The West German government was worried about East German post-Sputnik propaganda influencing the Western anti-militarist or pacifist movements.[57] However, disagreements over the dual-use potentials of rocketry had erupted within West German enthusiast societies frequently since 1956, and had more to do with processes of West European integration and transatlantic military cooperation than with ideologies. The new Ministry of Defense readily cooperated with Theodore von Kármán's (1881–1963) NATO Advisory Group for Aerospace Research and Development (AGARD). The NATO agency was instituted in Neuilly-sur-Seine, France, in 1953 to strengthen the transatlantic alliance in key technologies of the future. At the same time, Eugen Sänger tried to align the GfW closer with the reemerging South German aviation industry, and he found a close ally in industrialist Ludwig Bölkow (1912–2003).[58]

When in April 1956 a group of former Third Reich rocketeers including Walter Dornberger and Helmut von Zborowski (1905–69) joined representatives of the newly-established West German Defense Ministry at an AGARD conference in Munich, this was met with irritation in Germany and abroad. Both Kármán and the West German government also extended invitations to young spaceflight enthusiasts including those who were not trained engineers or scientists. While this would remain their only invitation to NATO meetings, Alfred Fritz, Ludwig Neher (1896–1970) and Werner Büdeler (1928–2004) held specific value since they had acted as brokers in the networks through secretarial roles as well as extensive language and communication skills. They knew people and publications. Fritz was asked to turn over lists of books, partners and professionals in the field – and was himself no longer needed after that knowledge had been organized and actors acquired for AGARD.[59] The NATO agency's intent to use enthusiast groups as a military resource drew immediate backlash from other attendees of the Munich meeting. In the eyes of West German journalists, military networks, especially ones that were structured around former Nazi experts, carried dystopian

Figure 7.3 The East German government was more forthcoming than its West German counterpart in praising the military and civilian potentials of rocket technology. Instead of astrofuturist imagery, this 1960 propaganda poster 'Zeitalter des siegenden Sozialismus' ('Age of Victorious Socialism') highlighted planet Earth and a rocket steering towards space as well as North America.
Source: Courtesy of Deutsches Historisches Museum.

connotations. 'Slim, beautiful and deadly – that is, the rocket. The new sword of the powerful of today, and the carrier of a menacing intercontinental death of tomorrow,' journalist Adalbert Bärwolf (1921–95) warned readers of his new book depicting the new networks.[60] A British representative was vexed when former *Wunderwaffen* engineers such as V-1 developer Fritz Gosslau (1898–1965) proudly explained their pivotal contributions to the German war effort: 'Those of his audience who had experienced the V-1 at the receiving end wondered to what unpleasant consequences youthful obsession with aeromodelling may lead if exaggerated, as in the case of the young Gosslau.'[61] The student participant from Aachen (and future NASA engineer) Jesco von Puttkamer (1933–2012) was irritated that research and development had become intertwined with nostalgic views on militarism and the history of technology.[62]

At the IAC in Rome just a few months later in September 1956, the astronautical societies seemed to have arrived at their pinnacle and crossroads at the same time. At an audience Pope Pius XII (1876–1958) admonished that engineers could not simply relinquish the responsibility they carried. Pius's remarks indirectly opposed the notion promoted by former Peenemünders that their knowledge and idealism had merely been exploited by the Nazi leadership.[63] The Pope invoked philosopher Bertrand Russell (1872–1970), who in December 1954 had asked the British public in a BBC radio broadcast if humankind was indeed '[...] so destitute of wisdom, so incapable of impartial love, so blind even to the simplest dictates of self-preservation' to risk eliminating life on planet Earth through nuclear weapons.[64] By 1957 the connection between rockets and atomic warheads became widely acknowledged.

The opposition to West German rearmament inside the country had initially been driven by both pacifists and nationalists who rejected international military cooperation. Conservative supporters of the Adenauer government tended to see the prospect of rearmament as a necessary step in Western integration and security. By early 1955, around 58 percent of those asked in a nationwide poll agreed.[65] Fears of a war that would have pitted East and West Germans against each other nevertheless increased and peaked in late 1957 and 1958. While it took a few weeks for official East German propaganda to take off, the launch of the Soviet satellite Sputnik 1 was mostly conceived as an 'uncanny' machine in the Western press. In a December 1957 poll, only about 25 percent of Germans advocated spaceflight, around the same number were opposed, but half of those asked were still unsure about what to make of it.[66] The IGY was turned into a showcase of the technoscientific capabilities in Eastern and Western Cold War Europe, but few Germans bought into the idea of 'technocratic innocence.'[67]

In the spring and summer of 1958, the *Kampf dem Atomtod* (Fight Against Atomic Death) gained traction as a popular movement, led by Social Democrats, with huge demonstrations and a failed attempt to initiate a referendum over the ban of nuclear weapons on West German soil.[68] A central symbolic motif, the missile came to embody the ultimate carrier of atomic weapons threatening the European continent. It was depicted in posters and displays in several cities such as Hamburg and Frankfurt, and highlighted the "dark side" of that technology for every pedestrian to see. 'Miscarriages, leukemia and genetic malformations for entire generations!,' read a stand-up display by the German branch of the War Resisters' International in Frankfurt (Figure 7.4). With the aim of raising awareness, the sign associated the impressive and sleek model of a missile with both the uncertainties of technological dual use and the dangers of genetic disorders. Fears of extensive links between missile development, nuclear weapons and the effects of radiation on a genetic level had culminated when reports on the launch of Sputnik 1 coincided with a fire at the British Windscale nuclear facility on 10 October 1957.[69]

The movement was politically diverse and began to wane in 1959, losing most of its immediate political relevance. That was partly due to the Social Democrats' new party program. The *Godesberger Programm* prepared the party for a future government role, accepting the necessities of national defense but also promoting the idea of exerting control over the military industries.[70] If a 'belief in the security of growth and progress shaped the 1960s and early 1970s,' as Eckart Conze has argued, it was won through pragmatism and a sense of Cold War realities.[71] In politics and industrial manufacturing this translated to the acceptance of both military and civilian applications, which were nowhere as apparent as in rocketry. By the late 1950s even intellectuals such as Günther Anders began to accept the necessity of dual use. His concept of *Ersatzproduktion* did not translate to 'missiles instead of spaceships,' but the other way around. Because he feared that military industries might at some point start to pressure governments into conflicts in order to maintain the level of industrial production, Anders saw civilian applications as a way to liquidate dual-use machines.[72] Despite its Marxist undertones, this was also an acceptance of the Cold War status quo and the need for national defense, even if that acceptance involved the (re-)militarization of spaceflight utopias.

Despite such attempts at pragmatism, quarrels between idealists and opportunists continued to change Western spaceflight societies from within. A growing number of AFRA/DRG and GfW/DGRR members not only had professional backgrounds, but they wanted enthusiasm for space limited to outreach work and their societies to act as lobby groups promoting dual use to the government and investors. Alfred Fritz saw no choice but to support the DGRR loyally in its process of professionalization, even when that implied a departure from immediate spaceflight interests and a discrepancy between public communication and internal goals. Fritz, in his role of DGRR secretary, went as far as to shield Eugen Sänger from criticism by having critical letters 'vanish in the drawer.'[73] Yet when he and two colleagues requested permission from the DGRR board to set up a satellite tracking group – he

Figure 7.4 A giant missile mock-up associated nuclear angst with fears of disease and genetic damage on the Roßmarkt in Frankfurt am Main, West Germany, in March 1958.
Source: Photograph by Klaus Meier-Ude. Courtesy of Institut für Stadtgeschichte, Frankfurt am Main.

had already secured funding from the Max Planck Society – they were denied. Sponsor and businessman Ludwig Bölkow thought that actual space-related activities did not go well with an advocacy organization for industrial rocket development. The Max Planck Society would not have to wait long since its institutes participated in national and European initiatives such as ESRO from the early 1960s. By that time Alfred Fritz and his fellow enthusiasts were no longer involved.[74]

The professionalization of West German spaceflight societies did not cause its members to strictly adhere to government guidelines or only serve Western security interests. Hence dual use became even more of a political challenge on the federal level. Engineers from Sänger's own research institute accepted contracts to develop missiles for Egypt under Gamal Abdel Nasser (1918–70). Knowing full well the political risks, they arranged their employment through a shadow company in Switzerland.[75] Only later, when the Hamburg-based *Der Spiegel* magazine in 1963 uncovered and reported the activities and explicated the dangers that arose to the national security of Israel, this eventually led to the public condemnation of those involved.[76] The AFRA/DRG, by contrast, was much more straightforward in its development of actual rockets in student work groups to begin with. Its goal was not so much to attract industrial sponsors as to jumpstart developments and spin off businesses from within the society. Led by engineer Berthold Seliger (1928–2020), student groups built solid-propellant rockets and tested them in Cuxhaven at the North Sea coast.[77] Initially the Bundeswehr was interested in cooperating with the North German enthusiasts, and a local battalion assisted with a helicopter to retrieve used rockets from the sea.[78] Again Max Planck institutes sponsored the entirely research-focused experiments. After his two-stage 'Cirrus' rocket had achieved altitudes of 40 kilometers in vertical shots, however, Seliger founded the Waffen- und Luftrüstung AG (Weapon and Military Aviation) company in Hamburg and tried to entice DRG student members away. That barely concealed openness towards lucrative military contracts turned into a disaster for its parent enthusiast society when Seliger did not rule out selling weapons to non-NATO states.[79]

In 1963 and 1964 West German federal agencies, through drastic limitations to amateur rocketry and model aviation, put an end to the DRG's experiments. The government feared another international scandal and supported the continued aim of both the military industry and the Defense Ministry to gain control over all major developments in missiles and rocketry, overruling projects in the region for reasons of national security.[80] The South German GfW/DGRR became the space technology subdivision of the lobby group Wissenschaftliche Gesellschaft für Luft- und Raumfahrt (Scientific Society for Aeronautics and Spaceflight). Its political arm was spun off into the Gesellschaft für Weltraumforschung m.b.H. in Bonn, the first official German space agency, now part of the Deutsches Zentrum für Luft- und Raumfahrt (DLR).[81]

IV Cold War astroculture and the realities of dual-use development

In both postwar German states, astroculture grew from a niche interest to an experimental site of the imagination, with often productive and sometimes problematic relations to the actual sites of technological realization and use. Despite its extraterrestrial motifs, discussions on spaceflight were very much down to earth. They realigned interests, beliefs and hopes for the future with political and economic systems and the international order into which both states tried to integrate. Nevertheless, there was an increasing discrepancy in exactly *who* produced astroculture and to what end. This was due to the growing relevance of the dual use of rocketry and spaceflight technologies to European economic and military security in the latter half of the 1950s.

In the West, astroculture produced technoscientific communities through shared practices, interests and languages. That proved to be less important in the East, where the centralized government was quick to acknowledge the utility of astrofuturism in technopolitics and propaganda. The West German government tried, without success, to block the acceptance of the East Berlin DAG into the IAF in 1961.[82] That membership was interpreted as a significant political recognition of the communist state by a mostly Western organization. As soon as the IAF had made its decision, the party pushed for tighter control over the DAG, which now had to report to the Gesellschaft zur Verbreitung wissenschaftlicher Kenntnisse, the official state institution for the dissemination of scientific knowledge. Unlike astronomers who were usually allowed to attend conferences in the West, DAG members had to wait for the Belgrade IAC in 1967 to actually meet their international peers. The initial admission, decided at the IAC in Washington, DC, in October 1961, had to be celebrated *in absentia*. Despite these complications it was a political success, only a few months after Yury Gagarin's spaceflight, further demonstrating through technology that, perhaps, the future *was* in the East.[83]

To many observers, the West German emphasis on the utopia of civilian spaceflight seemed to risk hypocrisy as well as a blind eye towards the dangers of remilitarization. Utopian connotations remained prevalent and propelled a debate that changed, but still continued throughout the 1960s. By then, West German spaceflight societies had been transformed into industry lobby groups or government agencies. They had not so much given up their resistance to military use as they had simply excluded those members, such as Alfred Fritz, who had insisted on a critical and independent role. One reason for this was professionalization, another the failure of engineers to gain political influence on their own. Even in national or international technology programs, engineers rarely sat in on the early committees. It was academic scientists, their independence guaranteed by the constitution of 1949, who were trusted with a new function that became important in Cold War Western Europe: counter-expertise.[84]

The conflict between West German federal institutions, in particular the Ministries of Traffic, Defense and Nuclear Research/Science, over responsibility for rocketry as a dual-use business added to growing government control over technological development. It resulted in a split between developments that related to matters of proliferation (handled jointly by the Ministries of Defense and Economics) and civilian scientific research (coordinated by the Ministry of Science). For some, it was the second time that engineers were unable to reinvent their profession as a distinct pillar of the democratic state after the failure to establish a Weimar Ministry of Technology in 1919.[85] The utopia of spaceflight, however, remained an important argument in the debate over the military dimensions of technoscience, led by a heterogeneous group of actors and agencies. Since rocket development in a market-based economy constantly touched upon problems of proliferation, dual use continued to be a major concern throughout the Cold War.

In the 1960s narratives about spaceflight put forth by propagandists as well as technoscientific experts grew increasingly divergent in their inherent political and ideological strategies of legitimization. In East Germany, rocketry's ambivalent applications were acknowledged, personalized and explained through ideology. In the West, they were carefully separated through dual use. East German propaganda claimed that the peaceful future was in the East, and the dystopian, militaristic past was in the West, only to add that the Soviet Union possessed more powerful weapons than the United States and its allies, including ballistic missiles. The claim carried an acknowledgment that true differences in technological development or use were hard to find. Instead, the peaceful future in the stars was to be the outcome of ideological distinction.

Notes

1. Christoph Kleßmann, *Zwei Staaten, eine Nation: Deutsche Geschichte 1955–1970*, Göttingen: Vandenhoeck & Ruprecht, 1997. For the term 'astroculture,' see Alexander Geppert's introduction to the first volume of the trilogy; idem, 'European Astrofuturism, Cosmic Provincialism: Historicizing the Space Age,' in idem, ed., *Imagining Outer Space: European Astroculture in the Twentieth Century*, 2nd edn, London: Palgrave Macmillan, 2018, 3–28, here 8. The article is based on research undertaken for my book *Raketenträume: Raumfahrt- und Technikenthusiasmus in Deutschland 1923–1963*, Paderborn: Schöningh, 2019. I would like to thank Alexander Geppert, Tilmann Siebeneichner, Mike Neufeld and Josh Armstrong for their helpful comments on earlier drafts of this chapter. All Internet sources were last accessed on 15 July 2020.
2. On the 'second Berlin crisis,' see Matthias Uhl, *Krieg um Berlin? Die sowjetische Militär- und Sicherheitspolitik in der zweiten Berlin-Krise 1958 bis* 1962, Munich: Oldenbourg, 2008; and Richard Maxwell, 'Technologies of National Desire,' in Michael J. Shapiro and Hayward Alker, eds, *Challenging Boundaries – Global*

Flows, Territorial Identities, Minneapolis: University of Minnesota Press, 1996, 327–60.

3. Eckart Conze, 'Sicherheit als Kultur: Überlegungen zu einer "modernen Politikgeschichte der Bundesrepublik Deutschland",' *Vierteljahrshefte für Zeitgeschichte* 53.3 (July 2005), 357–80, here 380. On NATO's involvement in science politics, see John Krige, 'NATO and the Strengthening of Western Science in the Post-Sputnik Era,' *Minerva* 38.1 (January 2000), 81–108. See also Michael Sheehan's contribution, Chapter 4 in this volume.
4. Beatrice de Graaf and Cornel Zwierlein, 'Historicizing Security – Entering the Conspiracy Dispositive,' *Historical Social Research* 38.1 (February 2013), 46–64, here 53. The so-called Copenhagen School has been particularly influential in laying out methods to analyze modern processes of securitization. For an overview, see Thierry Balzacq, Sarah Léonard and Jan Ruzicka, '"Securitization" Revisited: Theory and Cases,' *International Relations* 30.4 (December 2016), 494–531.
5. On 'dual use' as a concept, see the respective contributions by Alexander Geppert and Tilmann Siebeneichner, Michael Neufeld and Regina Peldszus, Chapters 1, 2 and 11 in this volume.
6. See Günther Anders, 'Die Toten: Rede über die drei Weltkriege (1964),' in idem, ed., *Hiroshima ist überall*, Munich: C. H. Beck, 1982, 361–94, here 374.
7. Eckart Conze, *Die Suche nach Sicherheit: Eine Geschichte der Bundesrepublik Deutschland*, Berlin: Siedler, 2009. See also Edgar Wolfrum, *Geglückte Demokratie: Geschichte der Bundesrepublik Deutschland von ihren Anfängen bis zur Gegenwart*, Munich: Pantheon, 2007 and Axel Schildt and Detlef Siegfried, *Deutsche Kulturgeschichte: Die Bundesrepublik von 1945 bis zur Gegenwart*, Munich: Carl Hanser, 2009.
8. Paul James et al., *Urban Sustainability in Theory and Practice: Circles of Sustainability*, London: Routledge, 2015, 53.
9. The term 'astrofuturism' was coined by Douglas De Witt Kilgore; see idem, *Astrofuturism: Science, Race, and Visions of Utopia in Space*, Philadelphia: University of Pennsylvania Press, 2003.
10. On the history of the IAF, see Leslie Shepherd, 'Prelude and First Decade,' *Acta Astronautica* 32.7–8 (July–August 1994), 475–99; and idem, 'The Origin of the International Astronautical Congress and the Birth of the IAF,' *IAF Newsletter* 1 (March 2000), 9–11.
11. On the Weimar *Raketenrummel*, see Michael J. Neufeld, 'Weimar Culture and Futuristic Technology: The Rocketry and Spaceflight Fad in Germany, 1923–1933,' *Technology and Culture* 31.4 (October 1990), 725–52; and Alexander C. T. Geppert, 'Space *Personae*: Cosmopolitan Networks of Peripheral Knowledge, 1927–1957,' *Journal of Modern European History* 6.2 (September 2008), 262–86.
12. Carl H. Nordstrom, US Military Security Board Koblenz, letter to Heinz-Hermann Koelle, 27 February 1951, National Archives and Records Administration (hereafter NARA), 1410420, HMS: A1 140.
13. Kenneth Gatland, letter to the editor, *Daily Mirror*, 20 October 1952, US Space & Rocket Center Huntsville, Wernher von Braun Collection, British Interplanetary Society correspondence (hereafter USSRC/WvBC/BIS); copy courtesy of Michael J. Neufeld.

14. Val Cleaver, letter to Wernher von Braun, 27 May 1951, ibid.; copy courtesy of Michael J. Neufeld.
15. Werner Albring, *Gorodomlia: Deutsche Raketenforscher in Russland*, Munich: Luchterhand, 1991; André Steiner, 'The Return of German "Specialists" from the Soviet Union to the German Democratic Republic: Integration and Impact,' in Matthias Judt and Burghard Ciesla, eds, *Technology Transfer out of Germany after 1945*, Amsterdam: Harwood, 119–30.
16. See Matthias Judt, 'Exploitation by Integration? The Re-Orientation of the Two German Economies after 1945: The Impact of Scientific and Production Controls,' in idem and Ciesla, *Technology Transfer out of Germany after 1945*, 27–48.
17. See 'Guided Missiles – the Weapon of the Future,' *Intelligence Bulletin* 4.4 (April 1946), 1–30, here 1. The *Intelligence Bulletin*, issued by the US War Department, was available in circles that closely cooperated with American or British forces. More widely available was Willy Ley's *Rockets, Missiles and Space Travel*, New York: Viking, 1944.
18. On Allied censorship, see Mitchell G. Ash, 'Denazifying Science and Scientists,' in Judt and Ciesla, *Technology Transfer out of Germany after 1945*, 61–80.
19. Walter Dornberger, *V2: Der Schuß ins Weltall*, Esslingen: Bechtle, 1952. Sänger and von Braun had been at odds over how to reach space (via advancements in aviation versus large-scale rocketry) since the 1930s. After the war had ended, the former remembered the latter's willingness to cooperate with the military as a premature 'abandonment' of space enthusiasm; see Hermann Oberth, letter to Wernher von Braun, 12 October 1952, Smithsonian National Air and Space Museum Archives, Willy Ley Collection, 30/4; on the quarrel, see also Wernher von Braun, letter to Willy Ley, 6 February 1952, ibid.
20. See Carl H. Nordstrom, US Military Security Board Koblenz, letter to Heinz-Hermann Koelle, 27 February 1951, National Archives and Records Administration (hereafter NARA), 1410420, HMS: A1 140.
21. Koelle repeatedly assured the Security Board that US researchers were among the first to profit from any GfW research. He highlighted his exchange of ideas with ballistics expert C. J. Pierce at Ohio State University – and even worked with him as a student assistant in 1951. See Heinz-Hermann Koelle and Heinz Gartmann, letter to Carl H. Nordstrom, Military Security Board of the US Army, 30 November 1950, NARA, 1410420, HMS: A1 140. See also, in the same folder, eidem, letter to the Scientific Research Division of the Military Security Board of the US Army in Wiesbaden, 31 August 1950; and Heinz-Hermann Koelle, Anzeige über die Aufnahme von Forschungstätigkeit nach dem Militärregierungsgesetz Nr. 23, August 1950.
22. Heinz Gartmann and Heinz-Hermann Koelle, 'Resolution,' *Weltraumfahrt* 1 (1949), 14; Shepherd, 'Prelude,' 478.
23. Heinz-Hermann Koelle, letter to Wernher von Braun, 27 April 1950, Library of Congress, Wernher von Braun Collection (hereafter LoC/WvBC), Box 43, German correspondence, Gesellschaft für Weltraumforschung (GfW) 1950–53 (my translation).
24. Shepherd, 'Prelude,' 477–8.

25. John H. Herz, 'Idealist Internationalism and the Security Dilemma,' *World Politics* 2.2 (January 1950), 157–80, here 164. See also Depkat, *Politik der europäischen Integration*, 177–85.
26. Gabi Schlag, *Außenpolitik als Kultur: Diskurse und Praktiken der Europäischen Sicherheits- und Verteidigungspolitik*, Wiesbaden: Springer, 2016, 119–35; Lutz Köllner, Klaus A. Maier and Wilhelm Meier-Dörnberg, 'Die EVG-Phase,' in Militärgeschichtliches Forschungsamt Potsdam, ed., *Anfänge westdeutscher Sicherheitspolitik 1945–1956*, vol. 2: *Die EVG-Phase*, Munich: Oldenbourg, 1990.
27. Catrin Behlau, Severin Roeseling and Anja Seufert, *50 Jahre DLR Lampoldshausen 1959–2009*, Lampoldshausen: Deutsches Zentrum für Luft- und Raumfahrt, 2009, 32.
28. Eugen Sänger, *Forschung zwischen Luftfahrt und Raumfahrt*, Tittmoning: Pustet, 1954, 17; on the number of GfW members, see Heinz-Hermann Koelle, letter to Wernher von Braun, 27 April 1950, LoC/WvBC, Box 43, German correspondence, GfW 1950–53.
29. Johannes Weyer, *Akteurstrategien und strukturelle Eigendynamiken: Raumfahrt in Westdeutschland 1945–1965*, Göttingen: Otto Schwartz, 1994, 106–9.
30. British and Belgian newspaper debates over the involvement of former Nazi engineers rekindled irregularly after Wernher von Braun's successes in the Explorers program and, from 1959, at NASA. See, for example, 'Von Braun Film Banned,' *Washington Post* (24 September 1960), 6.
31. Wernher von Braun, letter to the British Interplanetary Society, 29 September 1949, USSRC/WvBC, BIS correspondence; copy courtesy of Michael J. Neufeld.
32. Ibid.; Wernher von Braun, 'Space Travel – Its Dependence upon International Scientific Cooperation,' translated by Henry J. White for the Department of Defense Office of Public Information, 1 August 1952, USSRC/WvBC, IAF 1952; copy courtesy of Michael J. Neufeld. On the traditional dichotomy between technocrats and servants of power, see Bernd W. Kubbig, *Wissen als Machtfaktor im Kalten Krieg: Naturwissenschaftler und die Raketenabwehr der USA*, Frankfurt am Main: Campus, 2004, 35.
33. Eugen Sänger, *Zur Mechanik der Photonen-Strahlantriebe*, Munich: Oldenbourg, 1956. On Sänger's concept of a photon drive, see Alexander C. T. Geppert, 'Die Zeit des Weltraumzeitalters, 1942–1972,' in idem and Till Kössler, eds, *Obsession der Gegenwart: Zeit im 20. Jahrhundert*, Göttingen: Vandenhoeck & Ruprecht, 2015, 218–50, here 233–43 (= *Geschichte und Gesellschaft*. Sonderheft 25).
34. The costs for the whole Apollo program have been estimated around $25.4 billion, or $152 billion adjusted for inflation; for a current estimation, see https://www.forbes.com/sites/alexknapp/2019/07/20/apollo-11-facts-figures-business/#4746b3a63377.
35. See Wernher von Braun, 'Crossing the Last Frontier,' *Collier's* (22 March 1952), 24–9, 72, 74, here 24–5.
36. On the politics behind the Raketen- und Raumfahrtmuseum Stuttgart and Kiesinger's involvement, see Alfred Fritz's reflections in idem, 'Vorstandsbericht DRRM,' 27 April 1962, Stanford University Libraries, Department of Special Collections (hereafter SSC), M718/2/4.
37. Niklas Reinke, *Geschichte der deutschen Raumfahrtpolitik: Konzepte, Einflußfaktoren und Interdependenzen 1923–2002*, Munich: Oldenbourg, 2004, 48–51. On (failed) GfW satellite projects, see Hans-Karl Paetzold, letter to Alfred Fritz, 11 October 1957, SSC, M718/1/12.

38. For Seebohm's reflections upon the early years, see Hans-Christoph Seebohm, 'Grußworte,' in Deutsche Gesellschaft für Raketentechnik und Raumfahrt (DGRR), ed., *10 Jahre: 1948–1958*, Stuttgart: DGRR, 1958, 3.
39. Alf Lüdtke, '"Helden der Arbeit" – Mühen beim Arbeiten: Zur missmutigen Loyalität von Industriearbeitern in der DDR,' in Hartmut Kaelble, Jürgen Kocka and Jürgen Zwahr, eds, *Sozialgeschichte der DDR*, Stuttgart: Klett-Cotta, 1994, 188–213, here 188; Mary Fulbrook, 'Retheorizing "State" and "Society" in the German Democratic Republic,' in Patrick Major and Jonathan Osmond, eds, *The Workers' and Peasants' State: Communism and Society in East Germany under Ulbricht 1945–71*, Manchester: Manchester University Press, 2002, 280–98, here 288.
40. On popular science in East Germany, see Igor J. Polianski, 'Das Rätsel DDR und die "Welträtsel": Wissenschaftlich-atheistische Aufklärung als propagandistisches Leitkonzept der SED,' *Potsdamer Bulletin für Zeithistorische Studien* 36/37 (June 2006), 15–23, here 21.
41. Sonja Fritzsche, *Science Fiction Literature in East Germany*, Oxford: Oxford University Press, 2006, 71–5. On Soviet space fiction, see Matthias Schwartz, 'A Dream Come True: Close Encounters with Outer Space in Soviet Popular Scientific Journals of the 1950s and 1960s,' in Monica Rüthers et al., eds, *Soviet Space Culture: Cosmic Enthusiasm in Socialist Societies*, Basingstoke: Palgrave Macmillan, 2011, 232–50, here 232–6.
42. Ludwig Turek, *Die goldene Kugel*, Berlin (Ost): Dietz, 1949; Arthur Bagemühl, *Das Weltraumschiff*, Berlin (Ost): Luscie Groszer, 1952. Bagemühl's novel was published by an independent publisher of fiction for youth. Both remained their only respective spaceflight novels, especially after the FDJ began controlling the market for books for young people.
43. Wolfgang Langenbucher, Ralf Rytlewski and Bernd Weyergraf, *Kulturpolitisches Wörterbuch: Bundesrepublik Deutschland/DDR im Vergleich*, Metzler: Stuttgart 1986, 296; Ulrich Mählert and Gerd-Rüdiger Stephan, *Blaue Hemden – Rote Fahnen: Die Geschichte der Freien Deutschen Jugend*, Opladen: Springer, 1996, 85, 101.
44. Mählert and Stephan have outlined the enormous efforts undertaken by the FDJ to acquire and protect influence within East German universities and scientific institutions since the late 1940s, supported by the USSR military police and the communist government; see ibid., 68–76.
45. See Asif A. Siddiqi, 'Korolev, Sputnik, and the International Geophysical Year,' in Roger D. Launius, John Logsdon and Robert Smith, eds, *Reconsidering Sputnik: Forty Years since the Soviet Satellite*, Amsterdam: Harwood, 2000, 43–72, here 47–8; Georg Klaus, 'Auf dem Wege zum Weltraumschiff,' *Urania-Universum* 1 (1955), 176–82, here 180–1; and Roger D. Launius, James Rodger Fleming and David H. DeVorkin, eds, *Globalizing Polar Science: Reconsidering the International Polar and Geophysical Years*, Basingstoke: Palgrave Macmillan, 2010.
46. Kurt Hager, 'Referat des Genossen Professor Kurt Hager,' in Zentralrat der Freien Deutschen Jugend, ed., *Seid Bahnbrecher des Neuen: Zentrale Propagandistenkonferenz der FDJ am 10. und 11. Dezember 1955*, Berlin (Ost): Junge Welt, 1956, 3–36, here 3; see also Peter Graf Kielmansegg, 'Konzeptionelle Überlegungen zur Geschichte des geteilten Deutschlands,' *Potsdamer Bulletin für Zeithistorische Studien* 7.23–24 (October 2001), 7–15, here 13.
47. Hager, 'Referat,' 18–19, 21.

48. On the 1956/57 'ideological offensive,' see Siegfried Lokatis, 'Die "ideologische Offensive der SED", die Krise des Literaturapparats 1957/58 und die Gründung der Abteilung Literatur und Buchwesen,' in idem, Simone Barck and Martina Langermann, eds, *'Jedes Buch ein Abenteuer': Zensur-System und literarische Öffentlichkeiten in der DDR bis Ende der sechziger Jahre*, Berlin: de Gruyter, 1997, 61–96, here 61–2.
49. Reinke, *Geschichte der deutschen Raumfahrtpolitik*, 322.
50. 'Information für Genosse Prof. Hager über die Deutsche Astronautische Gesellschaft,' 3 November 1960, Stiftung Archiv der Parteien und Massenorganisationen der DDR, Bundesarchiv, SED Zentralkomittee Wissenschaft (hereafter SAPMO-BArch, SED ZKW), DY 30/IV A2/9.04/243.
51. See, for example, Heinz Mielke, *Der Weg ins All: Tatsachen und Probleme des Weltraumfluges*, Berlin (Ost): Neues Leben, 1956.
52. 'Protokolle zu Aussprachen zum Fall Neues Leben,' 24 August–27 September 1957, SAPMO-BArch, SED ZKW, DY 30/IV A2/9.04/684; Kurt Hager, 'Mitteilung an Genossen Hörnig, Abteilung Wissenschaften,' 3 January 1958, ibid.
53. The initial idea was offered by Soviet marxist Ernst Kolman (1892–1979); see his 'Prahlhans am Pranger,' in Neues Deutschland, ed., *Die Zeit trägt einen roten Stern im Haar: Neues Deutschland berichtet über den Vorstoß der Sowjetunion in den Weltenraum*, Berlin (Ost): Neues Deutschland, 1957, 19. Attacks on Wernher von Braun increased during the 1960s; see Michael J. Neufeld, '"Smash the Myth of the Fascist Rocket Baron": East German Attacks on Wernher von Braun in the 1960s,' in Geppert, *Imagining Outer Space*, 117–40.
54. 'Information for the ZK Secretary concerning *Mosaik*, Neues Leben (Verlagsleitung),' July 1958, SAPMO-BArch, SED ZKW, DY 30/IV A2/9.04/684; Fritz Löffler, 'Josef Franz Hegenbarth,' in Historische Kommission bei der Bayerischen Akademie der Wissenschaften, ed., *Neue Deutsche Biographie*, vol. 8, Berlin: Duncker & Humblot, 1969, 226–7.
55. *Weltall Erde Mensch*, Berlin (Ost): Neues Leben, 1954; see also subsequent editions in 1955–60. 'Bericht über die naturwissenschaftliche Propaganda und die Herausgabe naturwissenschaftlicher Literatur 1954,' 18 February 1955, SAPMO-BArch, SED ZKW, DY 30/IV A2/9.04/684.
56. Karl Böhm and Rolf Dörge, *Gigant Atom*, Berlin (Ost): Neues Leben, 1957, 289.
57. The SED financially and organizationally supported communist parties in the West, such as the Bund der Deutschen in Hamburg, that tried to engage and gain from the pacifist movements. These attempts remained largely unsuccessful. See Stefan Appelius, *Pazifismus in Westdeutschland: Die Deutsche Friedensgesellschaft*, Aachen: G. Mainz, 1999, 335–8.
58. Theodor Benecke, 'Dankesschreiben an die Teilnehmer des AGARD-Seminars in München,' April 1956, SSC, M718/1/14; see idem and August Wilhelm Quick, *History of German Guided Missiles Development: AGARD First Guided Missiles Seminar, Munich, Germany, April 1956*, Braunschweig: Appelhans, 1957.
59. 'Letter from the AGARD Office in Paris to Alfred Fritz,' 25 March 1957, SSC, M718/1/14.
60. Adalbert Bärwolf, *Da hilft nur beten*, Düsseldorf: Muth, 1956, 103 (my translation). On Bärwolf, see Klaus Müller, 'Verdienstvoller Reporter der Wissenschaft gestorben,' *Die Welt* (21 November 1995), 4.

61. Alfred R. Weyl, 'The Truth about German Guided Weapons: Reflections on the Recent Munich Symposium,' *Flight International* (25 May 1956), 648–9, here 649.
62. Jesco von Puttkamer, 'Die alten Peenemünder: Heinkelwerke wollen Fliegerabwehrraketen bauen,' *Blick in die Zeit: Süddeutsche Zeitung* (27 April 1956), 3 (supplement).
63. 'Die Ansprache S. H. Papst Pius XII,' *GfW Mitteilungen* 35 (October 1956), found in: SSC, M718/2/1. Dornberger promoted the notion of exploitation after he had shown no qualms about the primacy of military rationale in his 1952 work *V2: Der Schuß ins Weltall* – probably a move to support his fellow Peenemünders who had remained or migrated back to West Germany and often faced tougher ethical scrutiny there.
64. See Bertrand Russell und Andrew G. Bone, *Collected Papers of Bertrand Russell*, vol. 28: *Man's Peril*, London: Routledge, 2003, 89; Ronald William Clark, *The Life of Bertrand Russell*, New York: Alfred A. Knopf, 1976, 536–8; and Lawrence S. Wittner, *Confronting the Bomb: A Short History of the World Nuclear Disarmament Movement*, Palo Alto: Stanford University Press, 2009, 5.
65. See Hans-Peter Schwarz, *Die Ära Adenauer: Gründerjahre der Republik, 1949–1957*, Stuttgart: DVA, 1981; Hendrik Träger, '1949–1962: Starke Landesfürsten, schwache Parteien,' in Sven Leunig, ed., *Parteipolitik und Landesinteressen: Der deutsche Bundesrat 1949–2009*, Münster: LIT, 2012, 39–78, here 60; and Michael Werner, *Die 'Ohne mich'-Bewegung: Die bundesdeutsche Friedensbewegung im deutsch-deutschen Kalten Krieg (1949–1955)*, Münster: Monsenstein und Vannerdat, 2006, 454, 458–9.
66. See 'Umfrage zur Wehrdebatte,' *EMNID-Informationen* (December 1957), 4; and Alfred Fritz, 'Die Welt ändert ihr Gesicht,' SSC, M718/3/8. On the 'uncanny' interpretation, see Alexander C. T. Geppert, 'Anfang – oder Ende des planetarischen Zeitalters? Der Sputnikschock als Realitätseffekt, 1945–1957,' in Igor J. Polianski and Matthias Schwartz, eds, *Die Spur des Sputnik: Kulturhistorische Expeditionen ins kosmische Zeitalter*, Frankfurt am Main: Campus, 2009, 74–94, here 84–5.
67. Ash, 'Denazifying Scientists and Science,' 61; and idem, 'Wissenschaftswandel in Zeiten politischer Umwälzungen,' *NTM – Internationale Zeitschrift für Geschichte der Naturwissenschaften, Technik und Medizin* 3.1 (December 1995), 1–21, here 14.
68. Axel Schildt, '"Atomzeitalter" – Gründe und Hintergründe der Proteste gegen die atomare Bewaffnung der Bundeswehr Ende der fünfziger Jahre,' in Forschungsstelle für Zeitgeschichte, ed., *'Kampf dem Atomtod!' Die Protestbewegung 1957/58 in zeithistorischer und gegenwärtiger Perspektive*, Hamburg: Dölling & Galitz, 2009, 39–56.
69. Soraya de Chadarevian, 'Mice and the Reactor: The "Genetics Experiment" in 1950s Britain,' *Journal of the History of Biology* 39.4 (November 2006), 707–35; John Beatty, 'Scientific Collaboration, Internationalism, and Diplomacy: The Case of the Atomic Bomb Casualty Commission,' *Journal of the History of Biology* 26.2 (June 1993), 205–31.
70. See Helga Grebing, 'Die theoretischen Grundlagen des Godesberger Programms,' in Sven Papcke and Karl Theodor Schuon, eds, *25 Jahre nach Godesberg: Braucht die SPD ein neues Grundsatzprogramm?*, Berlin: Europäische Perspektiven, 1984, 9–17.

71. Eckart Conze, 'Security as a Culture: Reflections on a "Modern Political History" of the Federal Republic of Germany,' *GHI London Bulletin* 28.1 (2006), 5–33, here 7.
72. Günther Anders, 'Die Toten: Rede über die drei Weltkriege (1964),' in idem, ed., *Hiroshima ist überall*, Munich: C. H. Beck, 1982, 361–94, here 374.
73. Alfred Fritz, letter to Walter Hecker, 12 October 1959, SSC, M718/3/5.
74. Fritz noted that Eugen Sänger tried to gradually exclude 'amateurs,' those who did not hold engineering or science degrees (about 25 percent of GfW/DGRR members), from the inner circle and influential positions. But while Sänger believed that spaceflight enthusiasm was still important, Ludwig Bölkow, sponsor of both the DGRR and DRG, demanded the dissolution of work groups with a spaceflight focus. Walter Stanner, letter to Alfred Fritz, 15 November 1958, SSC, M718/3/5; Alfred Fritz, 'Handwritten Notes,' 1 October 1959, ibid.; Reinke, *Geschichte der deutschen Raumfahrtpolitik*, 58.
75. Niels Hansen, *Aus dem Schatten der Katastrophe: Die deutsch-israelischen Beziehungen in der Ära Konrad Adenauer und David Ben-Gurion: Ein dokumentierter Bericht*, Düsseldorf: Droste, 2002, 640–1.
76. 'Deutsche Raketen für Nasser/Rüstung: 36, 135 und 333,' *Der Spiegel* 17.19 (8 May 1963), 56–71. Inge Deutschkron, *Israel und die Deutschen: Das besondere Verhältnis*, Cologne: Wissenschaft und Politik, 1983, 209; Michael J. Neufeld, 'Rolf Engel vs. the German Army: A Nazi Career in Rocketry and Repression,' *History and Technology* 13.1 (March 1996), 53–72.
77. 'Deutsche Raketengesellschaft schießt,' 24 January 1961, DM, LR 10192.
78. See Karl Poggensee, 'Starts der Deutschen Raketen-Gesellschaft in Sahlenburg,' 18 September 1961, DM, LR 10188.
79. Weyer, *Akteurstrategien*, 66.
80. Federal Government of Germany, Luftverkehrsgesetz in der Neufassung, 10 January 1959, in *Bundesgesetzblatt I* (1959), 9; idem, Luftverkehrsordnung, 10 August 1963, in *Bundesgesetzblatt I* (1963), 652. For reactions, see, for example, 'Niedersächsische Landesregierung befiehlt: Raketenforscher, kehrt marsch,' *Cuxhavener Zeitung* (6 June 1964), 7.
81. Thomas Bührke, 'Raumfahrt als Staatsaufgabe'; available at http://www.dlr.de/100Jahre/desktopde-fault.aspx/tabid-2565/4432_read-10545/.
82. Reinke, *Geschichte der deutschen Raumfahrtpolitik*, 322–3. The DRG had cooperated with the East German DAG and their representative Heinz Mielke; see 'Mitteilungen der Deutschen Raketengesellschaft e.V.,' Februar 1962, DM, LR 10192. On the DAG's inclusion into government (and international) spaceflight projects in the East, see Colin Burgess and Bert Vis, *Interkosmos: The Eastern Bloc's Early Space Program*, Heidelberg: Springer, 2015, 1–10.
83. Information for Genosse Professor Hager about the Deutsche Astronautische Gesellschaft (DAG), 3 November 1960; Ferdinand Ruhle, letter to Walter Ulbricht, 20 September 1961; Herbert Pfaffe, 'Report on Participation in the XVIII IAC,' 14 November 1967, all in SAPMO-BArch, SED ZKW, DY 30/IV A2/9.04/243.
84. On counter-expertise in the Cold War, see Ulrich Beck, 'The Reinvention of Politics: Towards a Theory of Reflexive Modernization,' in idem, ed., *Reflexive Modernization: Tradition and Aesthetics in the Modern*, Cambridge: Cambridge University Press, 1995, 1–55, here 13–20; Eckart Conze, 'Modernitätsskepsis

und die Utopie der Sicherheit: NATO-Nachrüstung und Friedensbewegung in der Geschichte der Bundesrepublik,' *Zeithistorische Forschungen/Studies in Contemporary History* 7.2 (November 2010), 220–39.

85. On the conflicts over authority on the federal level, see Reinke, *Geschichte der deutschen Raumfahrtpolitik*, 64–5; on the 1919 attempt by the Society of German Engineers, see Karl-Heinz Ludwig, *Technik und Ingenieure im Dritten Reich*, Düsseldorf: Droste, 1974, 31–2.

PART III

Armoring Minds and Bodies

CHAPTER 8

Participant Evolution: Cold War Space Medicine and the Militarization of the Cyborg Self

Patrick Kilian

I Cyborg genealogies

In 2006 the Central Intelligence Agency (CIA) released a previously classified report on *The Soviet Bioastronautics Research Program* dating back to 22 February 1962. Finally disclosed as a result of the Freedom of Information Act (FOIA) policy, this extensive study covered a wide range of topics such as the 'Training of Cosmonauts' and 'Human Factors Engineering' together with the 'Psychopolitical Implications of Soviet Biomedical Experiments in Space' and the 'Soviet Propaganda Drive.' The report concluded that the '1961 orbitings of Vostoks I and II enabled the Soviets to be the first to test human survival in space with life sustaining systems [...] and thus advanced Soviet bioastronautics research substantially beyond that of the United States.' According to the CIA's findings, this research program 'began about 1950 as part of a broad, carefully planned, high-priority program' that undoubtedly follows Russia's 'long-term objectives of interplanetary travel' and also 'includes the intensive study of man-machine relationships during extended time periods in flight.'

In order to fully integrate man into the space capsule and to make him function effectively inside the technological system, the Soviet scientists introduced 'biocybernetic controls,' electronics, automation and computer mechanisms into their human factors research program. The fields of application

Patrick Kilian (✉)
Universität Zürich, Zurich, Switzerland
e-mail: patrickandreas.kilian@uzh.ch

Alexander C. T. Geppert et al. (eds), *Militarizing Outer Space*
European Astroculture, vol. 3
https://doi.org/10.1057/978-1-349-95851-1_8

were manifold: the CIA report names the electronic regulation of brain functions to control states of wakefulness and sleep, but also much more fundamental features such as the control of behavior and consciousness as well as metabolic regulation to the changing climates of the space capsule. Quoting Andrey Ivanovich Prokhorov, member of the Presidium of the Scientific Council of Cybernetics of the Academy of Sciences, Soviet Union, the study predicted that 'the science of cybernetics will become increasingly important in the conquest of space.' Even more alarming from a political perspective, this cutting-edge research might introduce the 'present social period as the beginning of the era of Cosmic Man.'[1]

The vision and the fear of a blurring of the human-machine boundary with all its far-reaching consequences to the question of agency and the renegotiation of the human-nonhuman status was by no means an invention of the Cold War but goes back deep into early modern history. Radical Enlightenment thinkers such as Julien Offray de La Mettrie (1709–51) expressed ideas about man functioning like a machine. Under the intense pressure of the evolving Cold War confrontation and with new technologies and 'rationality' emerging from Second World War research, these automata discourses returned with an unprecedented degree of urgency.[2] This chapter situates the man-machine relationship and bio-cybernetic research within the culture of space medicine and explores the political impact of this discourse for the expected militarization of outer space as well as its effects on the idea of human nature and evolution. Conceptualized as a belligerent evolutionary struggle to adapt human beings to the hostile environments beyond the atmosphere, the militarization of outer space during the early Cold War was not solely about missile technology. It was also about engineering and reinventing human nature for the predicted future of 'mechanized men and humanized machines' in which 'the barriers separating living and non-living are being broken.'[3]

Historian Slava Gerovitch has shown that cybernetics, in fact, had a much more complicated history in the Soviet Union than the CIA report assumed. After being stigmatized as a capitalist pseudoscience in the 1950s, it successively became accepted and widely celebrated during the early 1960s; eventually it was adapted to space medicine and psychology. Encouraged to actualize the Communist Party's ideological concept of the 'new Soviet man' within the bodies of the cosmonauts, engineers struggled under immense pressure to adapt man to the hazards of the extraterrestrial environment in order to create a technologically and politically reliable Soviet self, able to reach new worlds.[4] These plans to reinvent and enhance human nature for spaceflight and to create cybernetically controlled man-machine systems were anything but limited to the Soviet Union and have a much better known counterpart in the United States, the cyborg. Since Donna Haraway promoted the cyborg as the iconic prototype of the coming 'post-gender world' in her famous 1985 'Cyborg Manifesto,' this phenomenon has continuously oscillated between its military genealogy and a postmodern/posthuman feminist utopia transcending fact and fiction. As Haraway commented, 'from one perspective, a cyborg world is about the final imposition of a grid of control over the planet, about the final abstraction embodied in a Star Wars apocalypse waged in the name of defense,' but she continued, 'from another perspective, a cyborg world might be about lived social and bodily realities in which people are not afraid of their joint kinship with animals and machines, not afraid of permanently partial identities and contradictory standpoints.'[5] Accordingly, cyborgs have not one but multiple histories and, presumably, they even have more than one possible future. As soldiers, they populate the digital battlefields of postmodern command-control-communication-intelligence (C^3I) technology with its satellite- and drone-controlled warfare. As metaphors, they inhabit the worlds of economics, literature and cyberspace; and, as a theoretical concept, they have migrated into the academic sphere to inform science studies, gender, automata, computer and artificial intelligence discourses.[6]

Exploring the origins of the cyborg in the works of MIT mathematician Norbert Wiener (1894–1964), such as Wiener's conception of the 'antiaircraft (AA) predictor' during the Second World War, Peter Galison has warned with a sharp critique of Haraway 'that the associations of cybernetics (and the cyborg) with weapons, oppositional tactics, and the black-box conception of human nature do not so simply melt away.'[7] During his wartime project to develop an anti-aircraft defense system that would predict future flight paths of German bomber pilots and fire accordingly, Wiener began to conceptualize the enemy as a closed control cycle in which human and airplane interacted as a servomechanical man-machine system. According to Galison, in 'this vision in which the enemy pilot was so merged with machinery that (his) human-nonhuman status was blurred [...], it was a short step [...] to a blurring of the human-machine boundary in general,' in such a sense that 'the servomechanical enemy became, in the cybernetic vision of the 1940s, the prototype for human physiology and, ultimately, for all of human nature.'[8] Envisioning the human self as being part of the servomechanical unit of a self-contained weapons system helped to shape space medicine's conceptions about man's place in the technological environment of the space capsule. Eventually, this mindset also laid the ground for the militarization of human nature.

II Bringing cyborg science to America

Shortly after the end of the Second World War, the conception of a fully integrated man-machine system made its way, first, into the vibrant worlds of postwar aviation medicine and, finally, into the debates of the newly established discipline of space medicine. In December 1949 flight surgeon and Director of Aeromedical Research at the Air Force School of Aviation Medicine (SAM), in Randolph Field, Texas, Paul A. Campbell (1902–82), published an article called 'Cybernetics and Aviation Medicine' in the *Journal of Aviation Medicine*. An earlier version of this paper had been previously presented at the annual meeting of the Aero Medical Association held in New York during August 1949.[9] At that time, Campbell was one of the first to anticipate the expansion of air medicine into space medicine and co-arranged one of the earliest symposiums exclusively dedicated to this new field of study at the University of Illinois in 1950. Years later, in 1958, he became director of the Space Medicine Department at SAM. After his retirement, Campbell published a book under the title *Earthman, Spaceman, Universal Man?*, addressed to a broad and non-scientific audience. In this book, he concluded that to withstand the threats and dangers of the outer space environment, man has 'to engineer himself around the slow process and the limitations of natural evolution. He must accept "black boxes" as aid but not replacement.'[10]

The only reference to this passage is a footnote to Norbert Wiener's cybernetic black-box ontology of human nature that, with biblical pathos, reminds the reader to 'render unto man the things that are man's and unto the computer only those things that are the computer's.' But Campbell also must have gotten in touch with another much lesser known – and much more troublesome – genealogy of the cyborg that predates the Space Age and harks back to military aviation medicine in Nazi Germany.[11] At an outpost of the Luftfahrtforschungsanstalt München (Aeronautical Research Institute) that was relocated to Garmisch-Partenkirchen right at the foot of the Bavarian Alps in August 1944, a group of young researchers dreamed dangerously of human nature and the integration of man and machine. Conducted in seclusion from the exploding world in the idyll of Garmisch but with decidedly military motivation, this research included aviation physiology, altitude acclimatization, acceleration, prostheses and radium therapy, as well as the use of radioactive rays as a possibility to deliberately accelerate and manipulate evolution.[12]

Right after capitulation, US Air Force Major General Harry G. Armstrong (1899–1983) conducted an expedition that went through Germany in search for promising scientists who worked in the field of military aviation medicine. The aim of this group – that also included Campbell – was to investigate German research that was conducted for the Luftwaffe in secrecy during the Second World War. Being part of the US Office of Strategic Services' (OSS) secret 'Operation Paperclip' program, which was initiated to bring German scientists, engineers and rocketeers to the United States, this mission was already motivated by the logics of the emerging Cold War and the aggravating tensions with the Soviet Union.[13] In Garmisch, Armstrong's expeditionary group found the young radiologist and head of research unit Ulrich K. Henschke (1914–80) and biomedical engineer Hans A. Mauch (1906–86) 'designing artificial limbs in a chicken coop.'[14] Both were immediately transferred to Heidelberg to meet with other Luftwaffe physicians at the US Air Force's newly founded Aero Medical Research Center. Among them was Nazi Germany's leading figure in aviation physiology, Hubertus Strughold

(1898–1986), the future 'father of space medicine' and former head of the Luftfahrtmedizinisches Forschungsinstitut (Research Institute for Aviation Medicine), a branch of the Reichsluftfahrtministerium (Ministry of Aviation) in Berlin.[15]

In Heidelberg, the German scientists were asked to reproduce all of their journal articles and classified research papers they had worked on for the Luftwaffe during the Second World War and translate them into English. After months of work, they completed their final report that would be published in 1950 as the comprehensive and nearly thirteen hundred pages long two-volume *German Aviation Medicine: World War II*.[16] This impressive study outdated the work of their American colleagues in many respects and laid the ground for the advancement of aviation medicine to space medicine; a progression that finally became institutionalized in February 1949 with the establishment of the first Department of Space Medicine at Air University's School of Aviation Medicine, Randolph Air Force Base, Texas, upon a directive by Armstrong, and with Strughold as its first director.[17] Not only did the Heidelberg study include a chapter on 'Man Under Gravity-Free Conditions' by physicist Heinz Haber (1913–90) and physiologist Otto Gauer (1909–79) that anticipated space medicine's future challenges, the collection also contained a report by Henschke and Mauch entitled 'How Man Controls' that reversed Wiener's conceptions from the perspective of the anti-aircraft gunner to the pilot who was thought of 'as part of a control cycle.' In their view, 'control' was a crucial feature of the servo-mechanical man-aircraft system and a key element to modern and future warfare: 'With a slight control movement,' they stated, 'man [...] can also cause more devastation than large armies of laborers or soldiers could have done a few centuries ago.' Thus, Henschke and Mauch reduced the human function within these control processes to the sensory 'efficiency of the human eye' and the decision-making capacities of the brain, 'which makes possible a logical discrimination of things which are seen.'[18] It was cyborg science from Nazi Germany that, although almost forgotten, belongs to the larger history of man-machine interaction in outer space, which very soon would become of strategic and military importance in the course of the Cold War Space Race, as indicated in the CIA report discussed above.

III Survival of the fittest

After the completion of the *German Aviation Medicine* study, the US Army closed the Heidelberg center in March 1947 and began to transfer more than 30 German air surgeons to the United States in a clandestine mission that would become famous as Operation Paperclip. Many of them, with Strughold leading the way, established themselves within the emerging field of space medicine and space psychology. In addition to Strughold, Haber, Henschke and Mauch, the list included high-altitude physiologist and mountaineer Bruno Balke (1907–99), surgeon Hermann Becker-Freyseng (1910–61), meteorologist Konrad Büttner (1903–70), aeromedical specialist Hans-Georg Clamann (1902–80), physiologist Ulrich C. Luft (1910–90) and Luftwaffe psychologist Siegfried J. Gerathewohl (1909–95). Most of them shared a dubious past in the Third Reich. While Nazi Germany's cutting-edge rocket engineering research project, conducted by Wernher von Braun (1912–77) and Walter Dornberger (1895–80), relied heavily on the manpower of prisoners of war and concentration camp detainees, it was researchers working in the field of aviation medicine who misused humans as test objects. One of the leading figures of these inhumane experiments was physician and close friend to Heinrich Himmler, Sigmund Rascher (1909–45), who exposed humans to extreme cold to develop medical techniques for the sea rescue of pilots.[19] Even if the later Paperclip émigrés did not actively participate, at least they knew of the deadly low-pressure chamber and hypothermia experimentations conducted on concentration camp inmates at Dachau. These experiments had been publicly presented to the convened aviation medicine community by physiologist Ernst Holzlöhner (1899–1945) at the 'Ärztliche Fragen bei Seenot und Wintertod' conference held from 26 to 27 October 1942 in Nuremberg.[20]

In a Faustian bargain to exchange knowledge for silence, these controversial pasts were all too often hidden by the American authorities that put the German scientists in positions to work on the adaptation of human nature for the hostility of outer space. But the researchers also had to adapt themselves to their new political environment in Cold War America. In Nazi Germany, their wartime laboratory work on human survival at the boundaries of life and death in the extreme and pathological environment of upper atmosphere air combat was perfectly adaptable to the broader discourse of the social Darwinist and eugenic ideology of a 'survival of the fittest.' As described by historian Karl Heinz Roth, the German aeromedical scientists 'contended that flying was the highest degree of evolution and that the "struggle for survival" was its driving force' – a struggle that went from biological adaptation to politics and ultimately turned into 'battles for national superiority.'[21]

But the Paperclip émigrés quickly came to terms with American liberal democracy, and their self-conception seemed quite compatible with the Cold War's antagonistic climate. Flexible enough to translate their social Darwinist epistemology into the Cold War rhetoric of an evolutionary struggle for survival, Strughold and his colleagues were grateful 'to the people of the United States for their hospitality' and the military-sponsored research opportunities.[22] Being promised the 'widest freedom' to continue their work, the Paperclip scientists were eager to benefit from Cold War America's 'big science' space program.[23] The Faustian bargainers kept their word and the Germans prospered within the culture of space medicine laboratories that mushroomed all over the United States. Balke and Büttner joined Strughold and Haber at the Space Medicine Department at Randolph Field; Luft went to work at the Lovelace

Clinic for Medical Education and Research in Albuquerque, New Mexico; Gauer and Henschke headed to the Aerospace Medical Laboratory at Wright-Patterson Air Force Base in Dayton, Ohio; and the former Luftwaffe psychologist Gerathewohl affiliated with Air Force's SAM before changing over to NASA in 1960, where he became the Director of the Biotechnology Division at NASA Ames Research Center at Moffett Field, California.

In 1963 Gerathewohl published an exhaustively comprehensive monograph on the *Principles of Bioastronautics* that documents the achievements of US space medicine and human factors engineering up to that point. This book made use of the cybernetic picture of a totally integrated man-machine system inside the space vehicle and also referred to evolutionary discourse and the conception of a 'struggle for existence' as a moving power for the Space Race, holding that 'man's desire to defy gravity, to explore the unknown, and to conquer new worlds' is at least partly 'spurred by a will to survive in a highly competitive world.' Continuing with this evolutionary argument that slightly oscillates between Cold War anxiety and biological discourse, Gerathewohl maintained that 'the venture into space is more revolutionary than the invasion of land by the aquatic animal [...] for these creatures were merely moving from one terrestrial habitat to another, having a hundred million years for adaptation.' The political urgency of the current 'competitive' situation left no time for this to happen through natural selection, in light of demands for advanced 'biotechnology' and the construction of 'the most effective and reliable man-machine system' by means of 'servo-mechanisms,' 'an integrated display and control system' and the utilization of cybernetic 'feedback loop' technology (Figure 8.1).[24]

This passage revisited Campbell's idea of man having 'to engineer himself around the slow process and the limitations of natural evolution,' and attempts to define concrete technologies that would bypass the processes of biological adaptation in outer space. Within this intellectual framework, 'evolution' was not only conceptualized as a process that could be controlled and accelerated by technological intervention; it was also envisioned as being part of the Space Race's struggle for national supremacy. But Gerathewohl was by no means the first to express concerns about human nature as a crucial factor for Cold War competition beyond the atmosphere.

In his foreword to the proceedings of what was one of space medicine's acts of foundation, the symposium on the Physics and Medicine of the Upper Atmosphere, held 6–9 November 1951 at the School of Aviation Medicine, Armstrong stated that 'the effectiveness of the United States Air Force [...] is measured in terms of the deterrent effect its combat striking power has on potential aggressors. This striking power, in turn, is dependent, to a considerable degree, on the most efficient utilization of the man-machine complex.' To meet these challenges, of medical and military concern alike, editor and commandant at SAM, Brigadier General Otis O. Benson (1902–82),

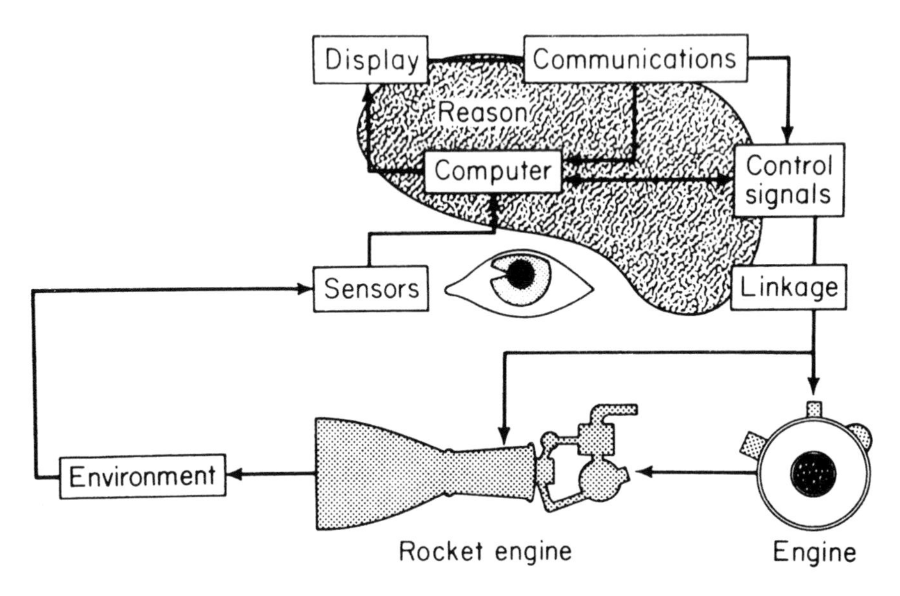

Figure 8.1 A 1960s schematic representation of the cybernetic man-machine relationship in a rocket vehicle. The human body is reduced to visual perception and brain activity and fully integrated in a closed-circuit control loop. Depicted as grey matter, the brain is expected to provide 'reason' to the man-machine system and to interact with the computer. Note that the 'reason' element is the only part of the system that is not enclosed in a black frame and not directly linked to the arrow diagram, which suggests that the human mind preserves some independence within the cybernetic feedback loop.
Source: Siegfried J. Gerathewohl, *Principles of Bioastronautics*, Englewood Cliffs: Prentice Hall, 1963, 14. Illustration by A. M. Mayo, ca. 1959/60.

explained that it would be helpful, 'if possible, to alter human physiology specifically by one means or another in order that man's tolerance may be increased.' Doing so would require an entirely new cybernetically inspired mindset that would think of 'a vehicle; a man or a crew; and perhaps a weapon or weapons' as being fully 'integrated into a mutually compatible unit,' added Clayton S. White (1912–2004), co-editor and Director of Research at the Lovelace Clinic in Albuquerque, New Mexico.[25]

In the following years, these ideas that linked the human factor and the astronaut's body directly to Cold War politics and regarded evolutionary mechanisms as central for the conquest of outer space and international competition in dangerous environments, spread within the realm of space medicine culture. For example, in 1959, Air Force General Thomas D. White (1901–65) remarked in his preface to the influential space medical anthology *Man in Space* that 'in the event of military operations in space, man's presence and ability to perform in space could spell the difference between defeat and victory' and therefore would be crucial to 'ensure national survival.'[26] Soon after, James H. Doolittle (1896–1993), a retired Air Force General and the chairman of Space Technology Laboratory, Inc., declared in his welcoming address to an audience of over 1,000 space scientists at the August 1960 Ballistic Missile and Space Technology symposium held at UCLA that 'entirely aside from the stockpile of space technology and hardware [...], the greatest stimulus of all to competition is the competition for survival.'[27] According to Doolittle, winning the Space Race against the Soviet Union depended on the hard science of missile technology *and* on the evolutionary survival instinct within human nature.

Embracing evolutionary thought and Darwinian discourse, space medicine's concerns also included the militarization and weaponization of the astronaut's technological environment in outer space. At the symposium on Psychophysiological Aspects of Space Flight held on 26 and 27 May 1960 at Brooks Air Force Base, Texas, the Chief of the Medical Psychology Section at SAM, Bryce O. Hartman (1924–98), emphasized that 'space flight is an extension of military aviation' and that 'vast technological, biological, and human resources are being thrown into the battle.' Envisioning 'a historical turning point in the evolution of the pilot's job,' Hartman stated that the 'role of the man in the air is becoming more like the task performed by ground-based operators in advanced weapons systems.'[28] To think of the ecological niche of outer space as being totally weaponized was not without consequences within the logics of a Cold War thought-style that included fantasies about the blurring boundaries of biology and technology, fully integrated control-loop cycles and the idea of a 'machines-manship.' This was because picturing the technological system of the spacecraft predominantly as a militarized environment was only one step away from thinking about the humans inside and interacting with these machines as becoming a part of the weapon themselves.

Hartman was in good company in dreaming about man-machine interaction at the symposium, which was convened to conceptualize 'man as a complex, constantly interacting psychophysiological unit' and as 'an integral part of the total machine.'[29] Including papers by prominent figures such as rocket expert Ernst Stuhlinger (1913–2008), balloonist Joseph Kittinger (1928–), sleep researcher Nathaniel Kleitman (1895–1999), chronobiologist Franz Halberg (1919–2013) and Hubertus Strughold, the conference covered a broad agenda of topics concerning man's adaptation to outer space, ranging from isolation and confinement to small group interaction and personnel selection, weightlessness, stress, circadian rhythms, sensory overloading and cognitive processes. A Cold War 'sense of urgency' pervaded the climate of the conference, demanding peak intellectual performance and an interdisciplinary effort in order to conceptualize and develop the feedback systems of man-machine interaction. As stated by Benson, 'man's behavior is a joint function of many kinds of influences arising both inside and outside the body, and [...] the understanding and control of his behavior can be achieved only through the joint efforts of all life sciences.'[30] Two of the presenters took this postulation to heart and boiled the interdisciplinary idea of cybernetic man-machine symbiosis by Wiener, Mauch and Henschke down to an essence: the *cyb*ernetic *org*anism, or: cyborg.

IV Participant evolution and the militarization of the self

The names of those two speakers were Manfred E. Clynes (1925–2020) and Nathan S. Kline (1916–83), and their paper ran under the extravagant title 'Drugs, Space, and Cybernetics: Evolution to Cyborgs,' of which an abridged version was published in the September 1960 issue of the scientific journal *Astronautics*.[31] Clynes and Kline were newcomers to the by then well-established circles of military-funded space medicine. They came from outside of this scientific culture, making their living as scientists at a psychiatric hospital. Both were based at the Rockland State Mental Hospital at Orangeburg, only a few kilometers from New York City. Since 1952 Kline had held the position of Director of Research; by 1960 he had become a renowned psychiatrist, especially famous for his use of psychopharmacological drugs on schizophrenics.[32] His collaborator, the Austrian-born Clynes, was an intellectual "whizz kid" – he was a brilliant pianist and graduate from the Juilliard School of Music as well as an ambitious engineer and computer scientist. At Rockland, he was in charge of the Dynamic Simulation Laboratory at the hospital's research department. In 1960 Clynes invented the so-called CAT computer (Computer of Average Transients), a machine capable of calculating and increasing the resolution of average signals of a repeated heart or brain activity, extracted from electrical noise (Figure 8.2). Only a few weeks before the Psychophysiological Aspects of Space Flight conference, a *New York Times* article featured Clynes's bio-cybernetic research with humans and computers. Introduced as 'boyish-looking,' he was praised for 'adapting missile-control concepts to biology' in order to explore 'the interactions of the unconscious controls that regulate the human body.'[33]

The military genealogy of cybernetics has already been mentioned, specifically with regard to Norbert Wiener's AA predictor and the German engineering of human factors during the Second World War. But now, with cybernetics and missile control technology being adapted to biology, the military-industrial complex threatened to literally invade the human body. By suggesting the application of cybernetic theory for the purposes of space medicine and the human engineering of astronauts, Clynes and Kline went one step further still. On the one hand, they intended to use cybernetics to understand the unconscious controls that regulate the human body, as said in the *New York Times* article. On the other hand, they wanted to apply this knowledge in order to change and manipulate these controls, optimizing man's biology for spaceflight and eventually changing the course of evolution. Still, what were the politics behind their military-inspired technologies

Figure 8.2 A 1960 *Life Magazine* picture of Manfred E. Clynes (1925–2020; left) and Nathan S. Kline (1916–83; right) facing a several meter long print-out of the Computer of Average Transients (CAT), 'which predicted a man's pulse rate from his breath rate.' With the cyborg they went one step further, using electronic measurements for the purpose of prediction as well as for cybernetically governed regulation. *Source*: 'Man Remade to Live in Space,' *Life Magazine* 49.2 (11 July 1960), 77–8, here 78. Courtesy of Getty Images.

and cybernetic artifacts that came from missile control and Cold War air defense?[34]

In the first lines of their paper, Clynes and Kline predicted that the dawning Space Age would open up an entirely new chapter in the history of mankind: 'The challenge of space travel to mankind is not only to his technological prowess, it is also a spiritual challenge to take an active part in his own biological evolution,' they declared.[35] In the vein of the previously mentioned James Doolittle, who emphasized the importance of the evolutionary human survival instinct for winning the Cold War in space over the impact of the large 'stockpile of space technology,' Clynes and Kline situated spaceflight within the broader context of biology and evolution. 'In the past,' they stated, 'the altering of bodily functions to suit different environments was accomplished through evolution. From now on, at least in some degree, this can be achieved without alteration of heredity by suitable biochemical, physiological, and electronic modification of man's existing modus vivendi.'[36]

To characterize this paradigm shift in human history, Clynes and Kline proposed the term 'participant evolution' for their plan to create 'self-regulating man-machine systems' with an 'artificially extended homeostatic control system' to be implemented into the human body.[37] As described by physiologist Walter B. Cannon (1871–1945), homeostasis is the property of self-regulation in open biological systems to maintain internal stability in response to changes in the external environment.[38] Clynes and Kline predicted that artificially regulating and supporting this mechanism in order for man to adapt to the hostile outer space environment could potentially solve a long list of different problems such as wakefulness, radiation effects, metabolic problems, oxygenation, fluid intake and output, as well as troubles with the enzyme system, the vestibular function, cardiovascular control, muscular maintenance, perceptual disturbances, pressure, gravitation, sensory deprivation, erotic and emotional satisfaction and, eventually, psychoses.

The material culture to implement the concept of cybernetic 'participant evolution' was rather mundane from today's perspective. The paper listed an extensive catalogue of drugs and medicines that was at least as long as that of the challenges to be met: psychic energizers, amphetamines, epinephrine, norepinephrine, digitalis, quinidine, apresoline, ephedrine and reserpine. To administer these medicines and create self-adapting, biologically extended and cybernetically controlled spacemen, the authors suggested using the so-called Nelson-Rose osmotic pressure pump. Assembled from cheap materials such as polythene, glass, rubber and a semipermeable cellophane membrane, filled with saturated Congo red solution, distilled water and the desired drug, this device seemed to constitute the perfect cyborg technology. Divided among three chambers, the water moved by osmotic pressure through the semipermeable membrane into the Congo red solution, creating a mechanical force that ejected the drug out of the machine directly into the human body. Small in size, the Nelson-Rose pump had already been implanted in rats – technically the first cyborgs – and could deliver medications at continuous and variable rates, depending on the thickness of the cellophane membrane.[39] In contrast to popular culture's fanciful visualizations, narrations and mystifications, which all too often obscured the cyborg's genealogy with futuristic robot-like imagery, the initial practices to bypass evolution and reinvent the human self were rather low- than high-tech attempts. Its material culture was far from science fiction and more closely related to the epistemic techniques of 'bricolage' and recombination.

Aside from these technical details, Clynes and Kline also anticipated the popular fears of Soviet supremacy in outer space that were expressed in the 1962 CIA bioastronautics report. With regard to the political implications of their project, they linked the cyborg to the Cold War's territorial rivalry and geostrategic confrontation. Speculating about the near-future creation of large space stations and the possibility of Soviet space colonization, they pointed to the geopolitical situation and reminded their audience 'that the Russians among others are already operating on such a space station – known as Earth.'[40] Partly in jest, partly serious, this effort to situate Spaceship Earth and biological adaptation to outer space within the logics of Cold War confrontation made the plan to advance participant evolution a project of highest political impact.[41] Clynes and Kline further stressed the Cold War's Space Race competition by stating multiple times that 'we do know that in the Soviet Union there is a considerable amount of research in this area.'[42] To adapt human nature to the remoteness of outer space, to take control over biological evolution, and to make the astronauts function more effectively within the man-machine systems were medical concerns but at the same time an objective of political interest.

The cyborg idea was taken very seriously within the space medicine community. Its consequences were accepted to be 'radical' but 'not fanciful' and certainly worth of receiving 'serious scientific consideration.'[43] Soon after the Psychophysiological Aspects of Space Flight conference and the follow-up paper in *Astronautics*, NASA's Biotechnology and Human Research Department in Washington, DC, assigned its contractor Hamilton Standards, a Division of United Aircraft, to conduct a 'Cyborg Study' for further research. Completed in May 1963 – and potentially still under the influence of the CIA's warning not to fall behind – this project came to the promising conclusion that 'the idea of modifying man is an advanced concept, which must supersede conventional thinking and which will, in the long run, provide us with basic research data in the fundamental physiology of man during the conditions of space travel.'[44] The idea to optimize man's evolutionary fitness and to change human nature was imagined to be crucial to Cold War competition. To bypass biological evolution by means of cybernetic control and to use military technology to adapt the human body to the hostility of the cosmos became part of the militarization of outer space and the competitive logics to ensure political supremacy via evolutionary survival, ultimately leading to the militarization of the self inside the space capsule (Figure 8.3).

Discussing the human body's 'automatic defense against the billions of germs' and the 'endless resilience of the mind and the emotions in absorbing uncounted strains, pressures, and anxieties,' a 1959 space medicine textbook concluded: 'Truly a nobly wrought weapon system, your healthy, thinking

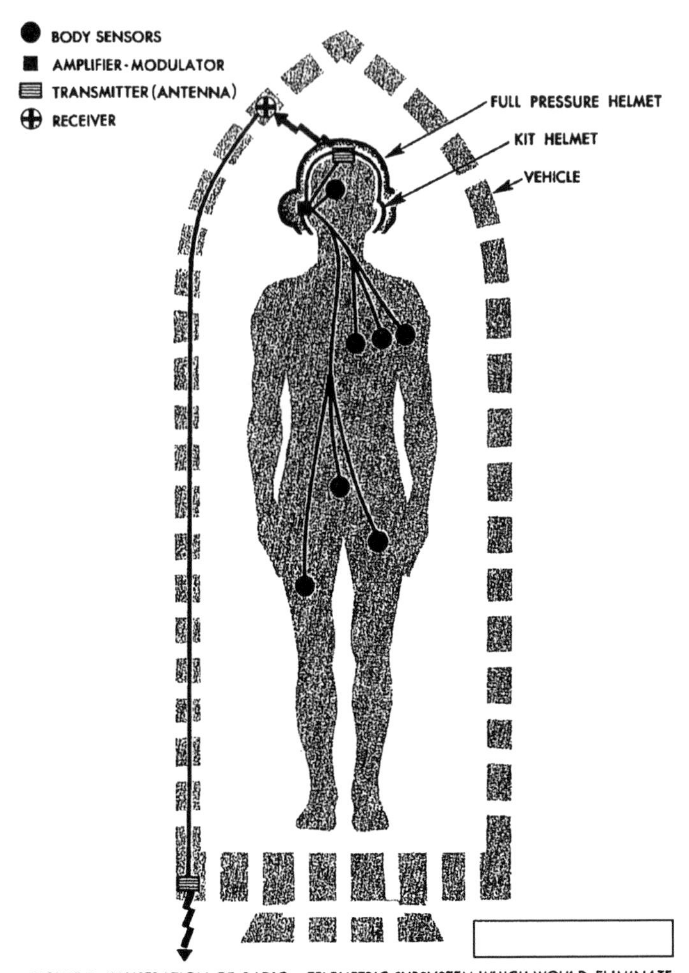

Figure 8.3 A schematic illustration from the 1962 Central Intelligence Agency report on the Soviet Bioastronautics Research Program, depicting the 'radio-telemetric subsystem' inside the space capsule that would measure and control the cosmonaut's bodily functions. Note that the cybernetically integrated man-machine system of the spacecraft is represented in the shape of a bullet.
Source: Central Intelligence Agency (CIA), Office of Scientific Intelligence, *The Soviet Bio-astronautics Research Program* OSI-SR/62-3 (22 February 1962), n.p. Courtesy of Central Intelligence Agency.

human being.'[45] As this association of humans with missiles signifies, and as the highly emphasized masculinity of the Cold War's predominantly male science community with all its gossip about missile sizes confirms, the cyborg was always, and without the shadow of a doubt, imagined as being male.

Although not discussing sex or gender specifically, Clynes and Kline consistently used the pronoun 'he' to talk about their cyborg spaceman. According to *his* masculinist and military genealogy, the cyborg was never intended to set the stage for Donna Haraway's feminist utopia of a coming post-gender world.[46] Her fear that the 'cyborg world is about the final imposition of a grid of control on the planet, about the final abstraction embodied in a Star Wars apocalypse waged in the name of defense,' manifested itself in these militaristic visions of a possible armament of the human body as a 'nobly wrought weapon system.'[47]

But even before *he* was presented to the elite of Air Force space medicine, the cyborg was introduced to the public sphere, with a *New York Times* article reporting on an advance copy of the paper just five days ahead of the conference. Under the headline 'Space Man Is Seen as Man-Machine,' the article informed America's space enthusiasts that 'a possible picture of the space man of the future has emerged from a radically new approach to the problems of space medicine. [...] He would not have to eat or to breathe. Those functions and many others would be taken care of automatically by drugs and battery-powered devices, some of which would be built directly into his body.'[48] Right from the beginning the cyborg was a cultural crossbreed, inhabiting both the worlds of popular imagination and scientific theory, and eluding binary distinctions between fact and fiction. In the following years the two researchers from Rockland received extensive media coverage, provoking authors, science journalists and science-fiction writers alike to comment on their brainchild.[49] Many of those referred explicitly or implicitly to the cyborg's military origins. For example, in the 1972 science-fiction novel *Cyborg* by Martin Caidin (1927–97), the former astronaut and Air Force test pilot Steve Austin was turned into a cyborg after experiencing a catastrophic air crash accident.[50] Being more adaptable and resilient to physical and psychological stresses after the operations, the cyborg was committed to fight the Soviet Union in underwater anti-submarine missions and in the Mars-like environment of the Israeli desert. During these operations, Austin not only performed as a spy but also as a human weapon.[51]

Not everyone was thrilled about the cyborg, though. A 1964 article in *Life Magazine* by science editor Albert Rosenfeld on the latest advances in space medicine, for example, named two firm opponents of the concept. Space medicine's grand seigneur Strughold and Toby Freedman (1924–2011), Director of NASA's Life Sciences Department, both opposed the cyborg which by then was fancifully characterized as consisting of a 'tiny, complex computer system constantly receiving and feeding back information to regulate the body to its changing environment.' Rosenfeld reported that these men found the idea 'unworkable' and even 'repugnant.'[52] Their articulated concerns were rather multi-layered – and perhaps maybe even superficial – ranging from technological to moral considerations. Was Strughold's dubious past in Nazi Germany and involvement in the gruesome human experimentations in the concentration camps responsible for now finding the vision of a 'superman' repugnant? Or was it the knowledge that Hitler's eugenic breeding programs did not succeed in creating a superhuman? Did his new home country cure him from his Second World War positivism? More likely, Strughold's and Freedman's dismissal might have been driven by professional interests and space medicine's competition for limited research funds. By that time, Freedman himself was developing an alternative concept for preparing the human body for spaceflight referred to as the 'Optiman.' Instead of seeking to change man's biological makeup, Freedman focused on more traditional techniques like physical education and training methods. A 1962 front-page article on Freedman's project in the *Los Angeles Times* entitled 'Physician Forecasts Space Age "Optiman" – Race on with Russia to Produce New Superman' underscores the argument that the Cold War's culture of confrontation in space was not only about missiles and satellites but also about human nature, biological and physical fitness and – last but not least – evolution.[53]

At the same time, the criticism of the cyborg was embedded in a larger discourse about the relationship of man and machine in spaceflight. This debate goes back to diverging paradigms proposed by Wernher von Braun, who imagined spaceflight as a manned endeavor, and the opposing unmanned satellite-system conceptions of Milton Rosen (1915–2014), President Dwight D. Eisenhower (1890–1969) and James Van Allen (1914–2006). Besides struggling with the technoscientific question whether humans could work as reliably as robotic machines or not, this discourse was also about political and military considerations.[54] Proponents of a manned space program such as the aforementioned Paul Campbell, for instance, argued that 'such traits as patriotism, courage and compassion will remain human heritages' and cannot be implemented into computers.[55] But there was also a more philosophical dimension to this debate relating to the consequences of cybernetic man-machine interaction in space that was especially lively in postwar German-speaking Europe.

In 1970 the Austrian philosopher Günther Anders (1902–92) published a sharp critique of human spaceflight under the title *Der Blick vom Mond* with special reference to the mechanization of man. Reversing Marshall McLuhan's (1911–80) famous philosophical dictum about the nature of media, Anders concluded that, with spaceflight, 'man has henceforth become a part, respectively, an extension of the machine' – a development with dire consequences for human agency.[56] Though not explicitly bearing upon the cyborg, Anders's critique strongly opposed spaceflight's goal of integrating man and machine. Following a similar line of reasoning as her former husband, Hannah Arendt (1906–75) associated spaceflight's technological endeavor to defy gravity with 'the attempt to create life in the test tube' in order to engineer and alter human nature 'to produce superior human beings.' She feared that both developments might introduce the coming of a dehumanized 'future man.' Cut off from his human and biological heritage – his *conditio humana* – as well as from his earthly habitat, this artificially enhanced 'future man' would be living in the dystopia of a technologically controlled outer space environment.[57] Another critic, the German cyberneticist and conservative publicist Karl Steinbuch (1917–2005), suspected in the third edition of his book *Automat und Mensch* that manned spaceflight and cybernetic man-machine systems might not be a 'human use of human beings.'[58] While the rather pragmatic critique of Strughold and Freedman was obviously motivated by self-interest and strategic considerations, those of Anders, Arendt and Steinbuch expressed more serious moral concerns that were inspired by an intellectual 'critique of technology.' What they shared was an understanding of the mechanization of man as a symptom of a typical Cold War dystopia – the post-humanistic and totally engineered world.[59]

V Medical utopia in the age of political dystopia

As highlighted by historians Michael D. Gordin, Helen Tilley and Gyan Prakash, 'much like utopia, dystopia, has found fruitful ground to blossom in the copious expanses of science fiction, but it has also flourished in political fiction [...]. Despite the name, dystopia is not simply the opposite of utopia [...]. It is a utopia that has gone wrong, or a utopia that functions only for a particular segment of society.' They propose 'to examine utopias (and dystopias) not for what they tell us about an intellectual construct in assorted individuals' heads, but rather for what they reveal about a set of abiding concerns and cultural formations that *generated* both the desire for utopian transcendence and the specific form that utopia/dystopia took.'[60] Outer space was *the* twentieth-century prototype of the entanglement of utopia and dystopia in Western popular imagination, representing Cold War Armageddon, extraterrestrial life, planned colonization and a new step for human civilization alike.[61] It developed into a symbolic geography onto which people projected the future of humanity and as such it was deeply inscribed into military contexts and the logic of rivalry. The deliberate or accidental reconfiguration of human nature in the course of spaceflight and its implications to future biological evolution were part of this discourse. Some of the ideas and imaginations that sprang from this mindset exceeded the cyborg vision of participant evolution by far.

One of those utopian (and at the same time also dystopian) conceptions about the reinvention of human nature in outer space was expressed at an interdisciplinary conference in London, far away from the military circles of US space medicine culture. In November 1962 a group of 27 well-known scientists came together for a symposium called 'The Future of Man' on the invitation of Gordon Wolstenholme (1913–2004), founding director of the Ciba Foundation, at the institute's new conference center at Portland Place. The contributors included six Nobel laureates, among them exobiologist Joshua Lederberg (1925–2008), biologist Peter Medawar (1915–87) and biochemist Francis Crick (1916–2004), who had discovered the double helix structure of DNA. The papers revolved around varying academic fields including demographic politics, genetics, eugenics as well as cybernetics and evolution. The latter two featured prominently in Julian Huxley's (1887–1975) opening lecture on 'The Future of Man – Evolutionary Aspects.'[62]

Another contributor went even further and ventured a glance into mankind's more distant future, discussing human evolution in outer space. In his paper on the 'Biological Possibilities for the Human Species in the Next Ten Thousand Years,' the geneticist and former communist J. B. S. Haldane (1892–1964) speculated about the evolutionary consequences of a possible atomic war, radiation damage and astronaut selection. He argued that, in case of 'a nuclear war, the survivors will have been heavily selected for radiation resistance, if such a selection is possible. If so they will be suited for astronautics.'[63] Apart from this involuntary biological conditioning for cosmic radiation, Haldane envisaged a rather exotic sounding set of genetic and medical practices to deliberately incorporate biological properties for the adaptation to outer space into human nature. For example, he proposed that 'gene grafting may make it possible to incorporate such features into the human stocks. The human legs and much of the pelvis are not wanted. Men who had lost their legs by accident or mutation would be specially qualified as astronauts.' Advancing this idea further, Haldane speculated that 'if a drug is discovered with an action like that of thalidomide, but on the leg rudiments only, not the arms, it may be useful to prepare the crew of the first spaceship to the *Alpha Centauri* system.'[64]

Seemingly unimpressed by the Thalidomide/Contergan scandal that shocked a whole generation in the late 1950s and confronted the public with the "dark" side of modern medicine, Haldane imagined future fields of application for substances like the one sold by the German pharmaceutical company Grünenthal.[65] Besides medication, he also considered manipulating human nature by the technique of 'gene grafting' to better equip man for his future life in outer space. Ruthlessly optimistic from a scientific viewpoint, Haldane expressed some reservations about the political implications of man's evolutionary makeover for service in outer space. Forecasting 'that even as soon as ten thousand years in the future there will be a real prospect of our species dividing into two or more branches, either through specialization for life on different stars or for the development of different human capacities,' Haldane concluded that evolutionary progress along such trajectories could become 'a terrible danger, as such species could fail to understand one another. [...] And such misunderstandings can generate quarrels and even war.'[66] What might happen if the two separate and competing space programs of the Soviet Union and the United States split up the evolutionary process of the human species into two different trajectories? Would this development finally transform the cold into a hot war? Within such a speculative framework, medical utopia and political dystopia seemed to go hand in hand.[67]

Haldane's idea that space war need not necessarily be the result of a struggle between rational actors, and that it could more likely be the accidental outcome of misunderstandings, communication difficulties and misperceptions of the 'other,' expressed an understanding that would later play a central role in Cold War historiography. Indeed, already by the 1970s – and still in the heat of the moment – the so-called post-revisionist school of historians began reevaluating the origins and dynamics of the East-West conflict as a series of mutual misperceptions and misunderstandings.[68] With his evolutionary approach, the geneticist Haldane seemed to sketch out this political interpretation in terms of biology, reiterating the logic of the Cold War with evolutionary vocabulary and applying it to human nature. As this discursive coincidence might outline, both interpretations relied on a similar Cold War mindset.

A few years later, the Ciba Foundation arranged another symposium entitled 'Ethics in Medical Progress with Special Reference to Transplantation.' One of the participants was Joseph E. Murray (1919–2012), a plastic surgeon and pioneer in the field of kidney transplantation who would be awarded the Nobel Prize in Medicine in 1990. Surprisingly, he concluded his paper on 'Organ Transplantation: The Practical Possibilities' with a surgical version of Haldane's ideas to alter man for outer space. He proposed 'that for the need of space travel, completely unanticipated physiological requirements may be met by the grafting of accessory organs, such as adrenal glands to overcome the stress of the environment of the moon, accessory lungs to accommodate the atmospheric conditions on Venus, or accessory extremities with which to crawl around Jupiter.'[69] As compared to this utopian vision, the cyborg almost appears 'all too human,' to borrow a line from Nietzsche.

VI Engineering humans

The militarization of outer space was as much about satellites and weapons as it was about human nature. Attempting to take control of human nature, space medicine experts found themselves confronted with a serious, almost Hamletian, question: 'Human engineering or engineering of the human being – which?'[70] But exercising control of the astronaut's biological makeup and the technological 'extensions of man' were much more than solely technical issues. These challenges were also entangled with the Cold War's geopolitical target to extend power over the cosmos. As noted by Odd Arne Westad, 'American and Soviet Cold War ideologies based an important part of their legitimacies on the control of nature, be it human nature or our physical surroundings.'[71] Yet how was this Cold War desire to control humans and their environments affected by the empty – and in many respects even placeless – topography of outer space? The French philosopher Paul Virilio supposed in 1993 that 'today, indeed, the place *where state-of-the-art technology occurs* is no longer so much the limitless space of an infinitely vast planetary or cosmic environment, but rather the infinitesimal space of our internal organs, of the cells that compose our organs' living matter [...]. Despite the grand illusion of the so-called conquest of space, [and] the implementation of absolute speed [... it is the] body that has become the ultimate planet.'[72] Armed with the seductive rationality of cybernetics that emerged from the military milieus of Wiener's AA predictor research and the improvised prosthesis workshop in Garmisch-Partenkirchen, the space medicine scientists did not hesitate in taking control over human nature and invading the human body.

The history of the cyborg as a utopian conception of the cybernetic control of human nature is inextricably linked with the Cold War dystopia of a potential loss of control – and of national power – in outer space. Making humans and machines meet was intended to increase the degree of control, reliability, effectiveness and alertness of the man-machine space system. It was about blurring the boundaries between the biological and the technological, and at times between utopia and dystopia. To 'alter human physiology' and to test how far 'man's tolerance may be increased' determined the effectiveness of the 'man-plane or man-plane-weapon combination.'[73] According to the military mindset among the space medicine elite, the idea of making humans and machines meet therefore also implied the dystopian vision of making humans and weapons meet. But the ambitious practices of a participant evolution also posed moral questions about the 'human use of human beings' and the consequences of eugenic interventions. Historian David Serlin has demonstrated in his insightful study *Replaceable You*, in which he describes 1950s reconstruction and prosthesis surgery as the 'other arms race,' that the Cold War struggle to mechanize the human body went far beyond the space medicine laboratories.[74] In fact, the human body was a permanent phantasm during the East-West conflict and, in 1961, even provoked the American writer and political activist Max Eastman (1883–1969) to proclaim a dangerous 'muscle gap' in analogy to the missile gap.[75]

But the vision to equip astronauts with cybernetic control-loop technology, to perform organ transplantations and to use practices of gene grafting in order to breed radiation-resistant spacemen was different from recovering amputated veterans and 'closing the muscle gap.' This vision did not rely on the epistemology of 'reconstruction,' but on a Cold War rationality of 'enhancement' and 'optimization.' It was about changing the course of evolution and the creation of a human prototype for mankind's future in outer space; it was about engineering humans in a time of urgency and confrontation. The technologies and material practices that should produce these resilient, reliable, adaptable and efficiently functioning spacemen were part of a scientific utopia, but also fragments of a paramilitary program to ensure national survival in a highly competitive climate. They were also effects of competing world views and contrasting ideologies about human nature and idealized virtues of the perfect citizen. Concluding that 'the purpose of the Cyborg is to provide an organizational system,' with which 'robot-like problems are taken care of automatically and unconsciously, thus freeing man to explore, to create, to think, and to feel,' Clynes and Kline promised nothing less than a cybernetic implementation of the Western Cold War ideals of freedom of thought, self-realization and creativity into the Space Age body.[76]

Notes

1. Central Intelligence Agency (CIA), Office of Scientific Intelligence, *The Soviet Bioastronautics Research Program* OSI-SR/62-3 (22 February 1962), 1, 2, 12, 13, 21, available at http://www.foia.cia.gov/sites/default/files/document_conversions/49/the_soviet_bio.pdf (accessed 15 July 2020). I am very grateful to Alexander Geppert, Daniel Brandau and Tilmann Siebeneichner for their invitation to the conference that preceded this book. Special thanks go to Bernard Heise and Melissa Marino for proofreading the manuscript. Research for this chapter was made possible by grants from the Swiss National Science Foundation.
2. On the Enlightenment automata discourse with respect to Cold War 'rationality,' see Paul Erickson et al., *How Reason Almost Lost Its Mind: The Strange Career of Cold War Rationality*, Chicago: University of Chicago Press, 2013, 34–8. On the history of the man-machine relationship, see Bruce Mazlish, *The Fourth Discontinuity: The Co-Evolution of Humans and Machines*, New Haven: Yale University Press, 1993; Anson Rabinbach, *The Human Motor: Energy, Fatigue, and the Origins of Modernity*, New York: Basic Books, 1990. For a historical-critical edition of La Mettrie's 1748 treatise, see Aram Vartanian, ed., *La Mettrie's* L'Homme machine*: A Study in the Origins of an Idea*, Princeton: Princeton University Press, 1960.
3. Richard R. Landers, *Man's Place in the Dybosphere*, Englewood Cliffs: Prentice-Hall, 1966, 3 and 17. On the history of 'human engineering' in the United States, see Rebecca Lemov, *World as Laboratory: Experiments with Mice, Mazes, and Men*, New York: Hill & Wang, 2005.
4. See Slava Gerovitch, *From Newspeak to Cyberspeak: A History of Soviet Cybernetics*, Cambridge, MA: MIT Press, 2002; and idem, '"New Soviet Man" Inside Machine: Human Engineering, Spacecraft Design, and the Construction of Communism,' *Osiris* 22.1 (January 2007), 135–57. See also Jay Bergman, 'Valerii Chkalov: Soviet Pilot as New Soviet Man,' *Journal of Contemporary History* 33.1 (January 1998), 135–52.
5. Donna Haraway, 'A Manifesto for Cyborgs: Science, Technology, and Socialist Feminism in the 1980s,' *Socialist Review* 80.2 (March–April 1985), 65–107, here 72.
6. See Les Levidow and Kevin Robins, eds, *Cyborg Worlds: The Military Information Society*, London: Free Association Books, 1989; Chris Hables Gray, ed., *The Cyborg Handbook*, New York: Routledge, 1995; Andrew Pickering, 'Cyborg History and the World War II Regime,' *Perspectives on Science* 3.1 (Spring 1995), 1–48; Anne Balsamo, *Technologies of the Gendered Body: Reading Cyborg Women*, Durham, NC: Duke University Press, 1996; Ian Hacking, 'Canguilhem Amid the Cyborgs,' *Economy and Society* 27.2–3 (May 1998), 202–16; N. Katherine Hayles, *How We Became Posthuman: Virtual Bodies in Cybernetics,*

Literature, and Informatics, Chicago: University of Chicago Press, 1999; Philip Mirowski, *Machine Dreams: Economics Becomes a Cyborg Science*, Cambridge: Cambridge University Press, 2001; Andrew J. Clark, *Natural-Born Cyborgs: Minds, Technology, and the Future of Human Intelligence*, Oxford: Oxford University Press, 2003; Stefano Franchi and Güven Güzeldere, eds, *Mechanical Bodies, Computational Minds: Artificial Intelligence from Automata to Cyborgs*, Cambridge, MA: MIT Press, 2005; and Ronald Kline, 'Where Are the Cyborgs in Cybernetics?,' *Social Studies of Science* 39.3 (June 2009), 331–62.

7. Peter Galison, 'The Ontology of the Enemy: Norbert Wiener and the Cybernetic Vision,' *Critical Inquiry* 21.1 (Fall 1994), 228–66, here 260. See also Norbert Wiener, *Cybernetics or Control and Communication in the Animal and the Machine*, Cambridge, MA: MIT Press, 1948.
8. Galison, 'Ontology of the Enemy,' 233.
9. Paul A. Campbell, 'Cybernetics and Aviation Medicine,' *Journal of Aviation Medicine* 20.6 (December 1949), 439–42.
10. Idem, *Earthman, Spaceman, Universal Man?*, New York: Pageant Press, 1965, 161. The proceedings of the space medicine symposium arranged by Campbell and held on 3 March 1950 at the University of Illinois, Chicago, were published as John P. Marbarger, ed., *Space Medicine: The Human Factor in Flights beyond the Earth*, Urbana: University of Illinois Press, 1951.
11. On the history of German aviation medicine under the swastika, see the works of Karl Heinz Roth, 'Flying Bodies: Enforcing States: German Aviation Medical Research from 1925 to 1975 and the Deutsche Forschungsgemeinschaft,' in Wolfgang U. Eckart, ed., *Man, Medicine, and the State: The Human Body as an Object of Government Sponsored Medical Research in the Twentieth Century*, Stuttgart: Franz Steiner, 2006, 107–37; and Karl Heinz Roth, 'Tödliche Höhen: Die Unterdruckkammer-Experimente im Konzentrationslager Dachau und ihre Bedeutung für die luftfahrtmedizinische Forschung des "Dritten Reiches",' in Angelika Ebbinghaus and Klaus Dörner, eds, *Vernichten und Heilen: Der Nürnberger Ärzteprozeß und seine Folgen*, Berlin: Aufbau, 2001, 110–51.
12. On the largely forgotten history of this 'medical research institute,' see Cornelius Borck, 'Das künstliche Auge: Zur Geburt des Cyborgs in der Sinnesprothesenforschung,' in Barbara Orland, ed., *Artifizielle Körper – Lebendige Technik: Technische Modellierungen des Körpers in historischer Perspektive*, Zurich: Chronos, 2005, 159–76, esp. 169–71.
13. On the history of Operation Paperclip, see, for instance, Annie Jacobsen, *Operation Paperclip: The Secret Intelligence Program that Brought Nazi Scientists to America*, New York: Little, Brown, 2014, esp. chapter 7, 108–32; Linda Hunt, *Secret Agenda: The United States Government, Nazi Scientists, and Project Paperclip, 1945–1990*, New York: St. Martin's Press, 1991; and Michael J. Neufeld, 'Overcast, Paperclip, Osoaviakhim: Looting and the Transfer of German Military Technology,' in Detlef Junker, ed., *The United States and Germany in the Era of the Cold War: A Handbook*, vol. 1: *1945–1968*, Cambridge: Cambridge University Press, 2004, 197–203. On the everyday life and memory culture of the German rocket specialists' community in Huntsville, Alabama, see Monique Laney, *German Rocketeers in the Heart of Dixie: Making Sense of the Nazi Past during the Civil Rights Era*, New Haven: Yale University Press, 2015.

14. Maura Phillips Mackowski, *Testing the Limits: Aviation Medicine and the Origins of Manned Space Flight*, College Station: Texas A&M University Press, 2006, 114 and 236, note 18. On the contributions of Harry G. Armstrong to the field of aviation and space medicine, see Harry G. Armstrong, *Principles and Practices of Aviation Medicine*, Baltimore: Williams & Wilkins, 1939; and idem, ed., *Aerospace Medicine*, Baltimore: Williams & Wilkins, 1961.
15. Mark R. Campbell et al., 'Hubertus Strughold: The "Father of Space Medicine",' *Aviation, Space, and Environmental Medicine* 78.7 (July 2007), 716–19; Shirley Thomas, 'Hubertus Strughold: The Father of Space Medicine Whose Dramatic Advanced Planning Encompasses the Universe,' in idem, ed., *Men of Space: Profiles of the Leaders in Space Research, Development, and Exploration*, vol. 4, Philadelphia: Chilton, 1962, 233–72.
16. US Air Force Surgeon General, ed., *German Aviation Medicine: World War II*, 2 vols, Washington, DC: Department of the Air Force, 1950. See also Mackowski, *Testing the Limits*, 115.
17. On the institutional origins of space medicine, see, for instance, Hubertus Strughold, 'From Aviation Medicine to Space Medicine,' in Kenneth F. Gantz, ed., *Man In Space: The United States Air Force Program for Developing the Spacecraft Crew*, New York: Duell, Sloan and Pearce, 1959, 7–18; Ursula T. Slager, *Space Medicine*, Englewood Cliffs: Prentice-Hall, 1962, esp. 13–17. A widespread outlook of the varying fields of research surrounding the sphere of space medicine is to be found in the proceedings of the Second International Symposium on Physics and Medicine of the Atmosphere and Space, sponsored by the School of Aviation Medicine, Texas, on 10–12 November 1958; see Otis O. Benson and Hubertus Strughold, eds, *Physics and Medicine of the Atmosphere and Space*, New York: John Wiley, 1960. For a detailed and valuable insight into the biographical and institutional history of space medicine, see John A. Pitts, *The Human Factor: Biomedicine in the Manned Space Program to 1980*, Washington, DC: NASA, 1985; Mackowski, *Testing the Limits.*
18. Ulrich K. Henschke and Hans A. Mauch, 'How Man Controls,' in US Air Force Surgeon General, *German Aviation Medicine*, vol. 1, 83–91, here 84, 86 and 90. Later on Mauch made a career in the United States and, in 1959, established his own research company called Mauch Laboratories that, from 1959 to 1964, served as a contractor for the Air Force and NASA, working on space suit designs; see Eugene Murphy, 'Hans Adolf Mauch,' in National Academy of Engineering, ed., *Memorial Tributes*, vol. 3, Washington, DC: National Academy Press, 1989, 259–65, esp. 262. See also Otto Gauer and Heinz Haber, 'Man under Gravity-Free Conditions,' in US Air Force Surgeon General, *German Aviation Medicine*, vol. 1, 641–4.
19. On the history of compulsory labor in the German rocket program, see Michael J. Neufeld, *The Rocket and the Reich: Peenemünde and the Coming of the Ballistic Missile Era*, New York: Free Press, 1995, esp. 167–96. For a detailed account of Sigmund Rascher's central role in the human experimentations at KZ Dachau and his personal relationship to Heinrich Himmler, see Leo Alexander, *The Treatment of Shock from Prolonged Exposure to Cold: Especially in Water*, Washington, DC: Office of the Publication Board, Department of Commerce, 1945.

20. For detailed accounts of the German doctors' dubious activities in Nazi Germany, see Ernst Klee, *Das Personenlexikon zum Dritten Reich: Wer war was vor und nach 1945?*, Frankfurt am Main: Fischer, 2003. For a brief discussion of Strughold's knowledge of the human experiments at Dachau concentration camp, see Mackowski, *Testing the Limits*, 92–7.
21. Roth, 'Flying Bodies – Enforcing States,' 123. On the (ab)use of Darwinian thought and evolutionary theory in Hitler Germany, see Richard Weikart, *From Darwin to Hitler: Evolutionary Ethics, Eugenics, and Racism in Germany*, Basingstoke: Palgrave Macmillan, 2004.
22. On the political readings of evolutionary theory during the Cold War, see Erickson et al., *How Reason Almost Lost Its Mind*, 150–7. These interpretations 'that restored biology over culture and put violence back into human nature' (150) were popularized by authors such as Robert Ardrey and Desmond Morris.
23. See Hubertus Strughold, *The Green and the Red Planet: A Physiological Study of the Possibility of Life on Mars*, London: Sidgwick & Jackson, 1954, xiii. On 'big science,' see Peter Galison and Bruce William Hevly, eds, *Big Science: The Growth of Large-Scale Research*, Stanford: Stanford University Press, 1992.
24. Siegfried J. Gerathewohl, *Principles of Bioastronautics*, Englewood Cliffs: Prentice-Hall, 1963, 9, 10, 13, 14. Gerathewohl also contributed a chapter to the *German Aviation Medicine* study entitled 'Psychological Examinations for Selection and Training of Fliers.' He knew Mauch and Henschke's proto-cybernetic work very well and quoted their paper 'How Man Controls' in his German postwar study on aviation medicine; see Siegfried J. Gerathewohl, *Die Psychologie des Menschen im Flugzeug*, Munich: Johann Ambrosius Barth, 1953, 39–40. For another volume on bioastronautics that details the interrelated problems of spaceflight, biological evolution and cybernetics, see Karl E. Schaefer, ed., *Bioastronautics*, New York: Macmillan, 1964.
25. See Harry G. Armstrong, 'Foreword,' xiii–xv, here xiii; Otis O. Benson, 'Preface,' xvii–xviii, here xvii; and Clayton S. White, 'Introduction,' in Otis O. Benson and Clayton S. White, eds, *Physics and Medicine of the Upper Atmosphere: A Study of the Aeropause*, Albuquerque: University of New Mexico Press, 1952, 1–5, here 1.
26. Thomas D. White, 'The Inevitable Climb to Space,' in Gantz, *Man in Space*, xiii–xv, here xv.
27. James H. Doolittle, 'Competition for Survival,' in Donald P. LeGalley, ed., *Ballistic Missile and Space Technology*, vol. 1: *Bioastronautics and Electronics and Invited Addresses*, New York: Academic Press, 1960, 3–6, here 4.
28. Bryce O. Hartman, 'Time and Load Factors in Astronaut Proficiency,' in Bernard E. Flaherty, ed., *Psychophysiological Aspects of Space Flight*, New York: Columbia University Press, 1961, 278–308, here 278 and 279.
29. Bernard E. Flaherty, 'Introduction,' in idem, *Psychophysiological Aspects*, 1–5, here 1.
30. Ibid., 3; Otis O. Benson, 'Preface,' in Flaherty, *Psychophysiological Aspects*, v–vi, here vi. On 'interdisciplinarity' as a constant and pervasive ideal within the scientific culture of Cold War America, see Jamie Cohen-Cole, *The Open Mind: Cold War Politics and the Sciences of Human Nature*, Chicago: University of Chicago Press, 2014, esp. chapter 3 'Interdisciplinarity as a Virtue.'
31. Manfred E. Clynes and Nathan S. Kline, 'Drugs, Space, and Cybernetics: Evolution to Cyborgs,' in Flaherty, *Psychophysiological Aspects*, 345–71; Manfred

E. Clynes and Nathan S. Kline, 'Cyborgs and Space,' *Astronautics* 9 (September 1960), 26–7 and 74–6; for a reprint of this article, see Gray, *Cyborg Handbook*, 29–33.

32. See, for instance, Nathan S. Kline, 'Use of Rauwolfia Serpentina Benth: In Neuropsychiatric Conditions,' *Annals of the New York Academy of Sciences* 59 (April 1954), 107–32; see also Nicholas de Monchaux, *Spacesuit: Fashioning Apollo*, Cambridge, MA: MIT Press, 2011, 70.
33. Geoffrey Ponds, 'Young Scientist Leads Two Lives,' *New York Times* (20 March 1960), 83. For a concise summary of Clynes's and Kline's parallel lives as well as their collaboration on the cyborg concept, see Monchaux, *Spacesuit*, 67–78.
34. Langdon Winner, 'Do Artifacts Have Politics?,' *Daedalus* 109.1 (Winter 1980), 121–36.
35. Clynes and Kline, 'Drugs, Space, and Cybernetics,' 345.
36. Ibid., 346.
37. Ibid.
38. Walter B. Cannon, *The Wisdom of the Body*, New York: Norton, 1932. On the adaptation of this concept to human societies, see Jakob Tanner, '"Weisheit des Körpers" und soziale Homöostase: Physiologie und das Konzept der Selbstregulation,' in idem and Philipp Sarasin, eds, *Physiologie und industrielle Gesellschaft: Studien zur Verwissenschaftlichung des Körpers im 19. und 20. Jahrhundert*, Frankfurt am Main: Suhrkamp, 1998, 129–69.
39. J. F. Nelson and S. Rose, 'A Continuous Long-term Injector,' *Australian Journal of Experimental Biology* 33.4 (August 1955), 415–20; for a detailed discussion of the structure and mode of operation of the Nelson-Rose pump, see Clynes and Kline, 'Drugs, Space, and Cybernetics,' 348–51.
40. Idem, 352.
41. See Richard Buckminster Fuller, *Operating Manual for Spaceship Earth*, Carbondale: Southern Illinois Press, 1968. On the military origins of space cabin ecology and colonization, see Peder Anker, 'The Ecological Colonization of Space,' *Environmental History* 10.2 (April 2005), 239–68.
42. Clynes and Kline, 'Drugs, Space, and Cybernetics,' 362 and 366; see also eidem, 'Cyborgs and Space,' 76.
43. Frederick I. Ordway, James Patrick Gardner and Mitchell R. Sharpe Jr., *Basic Astronautics: An Introduction to Space Science, Engineering, and Medicine*, Englewood Cliffs: Prentice-Hall, 1962, 523–6, here 526.
44. Robert W. Driscoll, 'Engineering Man for Space: The Cyborg Study' (15 May 1963), in Gray, *Cyborg Handbook*, 75–81, here 76.
45. William A. Kinney, *Medical Science and Space Travel*, New York: Franklin Watts, 1959, 71.
46. On the culture of masculinity within the military-industrial complex, see, for instance, Brian Easlea, *Fathering the Unthinkable: Masculinity, Scientists and the Nuclear Arms Race*, London: Pluto, 1983; and Carol Cohn, 'Sex and Death in the Rational World of Defense Intellectuals,' *Signs* 12.4 (Summer 1987), 687–718.
47. Haraway, 'Manifesto for Cyborg,' 72.
48. 'Space Man Is Seen as Man-Machine,' *New York Times* (22 May 1960), 31.
49. For newspaper and magazine articles, see, for instance, 'Man Remade to Live in Space,' *Life Magazine* 49.2 (11 July 1960), 77–8; Jot Neri, 'My Wife and the Cyborgs,' ibid. 54.4 (25 January 1963), 19; David Lucas, 'Life Support

Systems,' *The Michigan Technic* 82.3 (December 1963), 10–13; 'Auf Brautschau,' *Der Spiegel* 19.1 (6 January 1965), 71–2; 'Todlos glücklich,' ibid. 20.53 (26 December 1966), 89–101; and Theo Sommer, 'Die Sucht nach dem Mond,' *Die Zeit* 22 (30 May 1969), 1. Popular non-fiction books include Kurt W. Marek, *Yestermorrow: Notes on Man's Progress*, London: Andre Deutsch, 1961, esp. 50; Daniel S. Halacy, *Cyborg: Evolution of the Superman*, New York: Harper & Row, 1965; David M. Rorvik, *As Man Becomes Machine: Evolution of the Cyborg*, New York: Doubleday, 1970; Victor C. Ferkiss, *Technological Man: The Myth and the Reality*, New York: George Braziller, 1969; and Alvin Toffler, *Future Shock*, New York: Random House, 1970, esp. the chapter 'The Cyborg Among Us,' 209–14.

50. Martin Caidin, *Cyborg*, New York: Arbor House, 1972; for a German translation, see idem, *Der korrigierte Mensch: Cyborg*, Munich: Goldmann, 1974. Caidin, who served as an Air Force Sergeant from 1947 to 1950, was very interested in aeronautics and human spaceflight and published multiple non-fiction books on this topic. For a contribution on space medicine, see idem and Grace Caidin, *Aviation and Space Medicine: Man Conquers the Vertical Frontier*, New York: Dutton, 1962. The *Cyborg* novel also became the blueprint for three television movies in 1973 and the popular television series *The Six Million Dollar Man* aired on the ABC network from 1974 to 1978.
51. On the intersections of military submarine medicine and space medicine, see Karl E. Schaefer, ed., *Environmental Effects on Consciousness: Proceedings of the First International Symposium on Submarine and Space Medicine*, New York: Macmillan, 1962. And on the use of deserts and other extreme terrestrial environments as 'earth analogs' for space medical and psychological research, see Sheryl L. Bishop, 'From Earth Analogs to Space: Getting There from Here,' in Douglas A. Vakoch, ed., *Psychology of Space Exploration: Contemporary Research in Historical Perspective*, Washington, DC: NASA, 2011, 47–77.
52. Albert Rosenfeld, 'Pitfalls and Perils Out There,' *Life Magazine* 57.14 (2 October 1964), 112–24, here 124.
53. Marvin Miles, 'Physician Forecasts Space Age "Optiman": Race on with Russia to Produce New Superman,' *Los Angeles Times* (14 November 1962), 1–2.
54. Roger D. Launius and Howard E. McCurdy, 'Robots and Humans in Space Flight: Technology, Evolution, and Interplanetary Travel,' *Technology in Society* 29.3 (August 2007), 271–82, esp. 275–7; for a more detailed discussion, see eidem, *Robots in Space: Technology, Evolution, and Interplanetary Travel*, Baltimore: Johns Hopkins University Press, 2008.
55. Campbell, *Earthman, Spaceman, Universal Man?*, 158.
56. Günther Anders, *Der Blick vom Mond: Reflexionen über Weltraumflüge*, Munich: C. H. Beck, 1970, 13: 'Während bis vor kurzem das Instrument als die "Verlängerung" des Menschen gegolten hatte, [...] ist nunmehr der Mensch zum Stück bzw. zur Verlängerung des Instruments geworden.'
57. Hannah Arendt, *The Human Condition* [1958], 2nd edn, Chicago: University of Chicago Press, 1998, 2.
58. Karl Steinbuch, *Automat und Mensch: Kybernetische Tatsachen und Hypothesen*, 3rd edn, Berlin: Springer, 1965, 310: 'Ob es jedoch ein humaner Gebrauch menschlicher Wesen ist, sie in den Weltraum hinauszuschießen, scheint zweifelhaft.' With this passage, Steinbuch referred to Norbert Wiener, *The Human Use of Human Beings: Cybernetics and Society*, Boston: Houghton Mifflin, 1950.

59. On the moral implications of space exploration, see also Oliver Dunnett's contribution, Chapter 6 in this volume.
60. Michael D. Gordin, Helen Tilley and Gyan Prakash, 'Introduction,' in eidem, eds, *Utopia/Dystopia: Conditions of Historical Possibility*, Princeton: Princeton University Press, 2010, 1–17, here 1, 4 (emphasis in original).
61. See, for instance, Howard E. McCurdy, *Space and the American Imagination* [1997], 2nd edn, Baltimore: Johns Hopkins University Press, 2011.
62. Julian Huxley, 'The Future of Man – Evolutionary Aspects,' in Gordon Wolstenholme, ed., *Man and His Future: A Ciba Foundation Volume*, Boston: Little, Brown, 1963, 1–22. For an analysis of the controversial reception of this conference in West Germany, see Ina Heumann, 'Wissenschaftliche Phantasmagorien: Die Poetik des Wissens in "Man and His Future" und ihre Rezeption in der Bundesrepublik,' in Dirk Rupnow et al., eds, *Pseudowissenschaft: Konzeptionen von Nichtwissenschaftlichkeit in der Wissenschaftsgeschichte*, Frankfurt am Main: Suhrkamp, 2008, 343–70.
63. J. B. S. Haldane, 'Biological Possibilities for the Human Species in the Next Ten Thousand Years,' in Wolstenholme, *Man and His Future*, 337–61, here 355. For a brief discussion of this conception, see Alexander von Lünen, 'The Perfect Astronaut Would be a Human without Legs: J.B.S. Haldane and "Positive Eugenics",' in Regina Wecker et al., eds, *Wie nationalsozialistisch ist die Eugenik?/ What is National Socialist About Eugenics? Beitrag zur Geschichte der Eugenik im 20. Jahrhundert/Contributions to the History of Eugenics in the Twentieth Century*, Vienna: Böhlau, 2009, 127–38.
64. Haldane, 'Biological Possibilities,' 354. A similar idea of gene grafting and the creation of 'polyploid men' by breeding 'radiation-resistant space personnel with many times more than the normal heredity units' is also to be found in Kinney, *Medical Science and Space Travel*, 135.
65. On the history of the Contergan scandal, see Rock Brynner and Trent Stephens, *Dark Remedy: The Impact of Thalidomide and Its Revival as a Vital Medicine*, New York: Basic Books, 2001.
66. Haldane, 'Biological Possibilities,' 359; for a similar line of thought, see Carsbie C. Adams, *Space Flight: Satellites, Spaceships, Space Stations, and Space Travel Explained*, New York: McGraw-Hill, 1958, 362.
67. For a sharp critic against Haldane's vision that already anticipates the dystopian quality of his ideas in its title, see Caryl Rivers, 'Grave New World,' in Miguel A. Santos, ed., *Readings in Biology and Man*, New York: MSS Information Corporation, 1973, 131–7; versions of this article also appeared in the *Saturday Review* (8 April 1972) and the *Chicago Tribune* (30 July 1972). For a German critique, see, for instance, Richard Kaufmann, *Die Menschenmacher: Die Zukunft des Menschen in einer biologisch gesteuerten Welt*, Frankfurt am Main: Fischer, 1964, 117.
68. For an early and influential study developing the concept of 'post-revisionism,' see Daniel Yergin, *Shattered Peace: The Origins of the Cold War and the National Security State*, New York: Houghton Mifflin, 1977. For a discussion, see, for instance, John Lewis Gaddis, 'The Emerging Post-Revisionist Synthesis on the Origins of the Cold War,' *Diplomatic History* 7.3 (July 1983), 171–91. On the 1970s, see the second volume in this trilogy, Alexander C. T. Geppert, ed.,

Limiting Outer Space: Astroculture after Apollo, London: Palgrave Macmillan, 2018 (= *European Astroculture*, vol. 2).

69. Joseph E. Murray, 'Organ Transplantation: The Practical Possibilities,' in Gordon Wolstenholme and Maeve O'Connor, eds, *Ethics in Medical Progress with Special Reverence to Transplantation: A Ciba Foundation Symposium*, Boston: Little, Brown, 1966, 54–65, here 65. This passage was also quoted with reference to the Cyborg in 'Tod überlebt,' *Der Spiegel* 22.3 (15 January 1968), 89–101, here 100.
70. Robert S. Pogrund, 'Human Engineering or the Engineering of the Human Being – Which?,' *Aerospace Medicine* 32.4 (April 1961), 300–15.
71. Odd Arne Westad, 'The New International History of the Cold War: Three (Possible) Paradigms,' *Diplomatic History* 24.4 (Fall 2000), 551–65, here 556; for a detailed discussion of Cold War concerns over human nature, see Mark Solovey and Hamilton Cravens, eds, *Cold War Social Science: Knowledge Production, Liberal Democracy, and Human Nature*, Basingstoke: Palgrave Macmillan, 2012.
72. Paul Virilio, *The Art of the Motor*, Minneapolis: University of Minnesota Press, 1995, 100, 109 (emphasis in original); originally published as *L'Art du moteur*, Paris: Galilée, 1993.
73. Benson, 'Preface,' xviii.
74. David Serlin, *Replaceable You: Engineering the Body in Postwar America*, Chicago: University of Chicago Press, 2004, 21–56.
75. Max Eastman, 'Let's Close the "Muscle Gap",' *Reader's Digest* 79 (November 1961), 122–5. On the Cold War body, see also Jane Pavitt, *Fear and Fashion in the Cold War*, London: V&A Publishing, 2008, esp. 18–39.
76. Clynes and Kline, 'Drugs, Space, and Cybernetics,' 348. On the virtue of 'creativity' as an American Cold War conception, see Jamie Cohen-Cole, 'The Creative American: Cold War Salons, Social Science, and the Cure for Modern Society,' *Isis* 100.2 (June 2009), 219–62.

CHAPTER 9

Starship Troopers: The Shaping of the Space Warrior in Cold War Astroculture, 1950–80

Philipp Theisohn

This chapter discusses the significance of the space suit to the history of the militarization of outer space. Yet its relevance might not be self-evident at first. The functionality of the space suit does not necessarily place it in a military context but rather first defines it as a shelter for the human body in space. In the period when spaceflight had already become a scientific enterprise but was not yet practiced – that is, from the beginning of research into liquid-fuel rockets in the 1920s to the first human spaceflight in 1961 – the functions of the space suit were already quite clearly defined. In 1936 Philip E. Cleator (1908–94), founder and president of the British Interplanetary Society, wrote that according to 'the conception of what has been termed a space-suit,' it should protect astronauts 'from the vacuum of space and the emanations of the sun' in their activities outside the spacecraft. 'Not unlike a diver's dress in appearance, the garment will completely cover the wearer. Its equipment will include [...] a miniature radio receiving and transmitting set. It has even been suggested that each suit be fitted with a tiny rocket motor, thus enabling the wearer to propel himself through space' (Figure 9.1).[1] In fact, except for the miniature rocket motor, such suits were already in development at the time. Designed as 'full-pressure suits' rather than 'space suits,' they were not made to be used in space but 'for pilots of balloons and aircraft with open cockpits.'[2] In 1931 Yevgeny Chertovsky (1902–61) developed the first of these full-pressure suits, the CH-1, in Leningrad;

Philipp Theisohn (✉)
Universität Zürich, Zurich, Switzerland
e-mail: philipp.theisohn@ds.uzh.ch

Alexander C. T. Geppert et al. (eds), *Militarizing Outer Space*
European Astroculture, vol. 3
https://doi.org/10.1057/978-1-349-95851-1_9

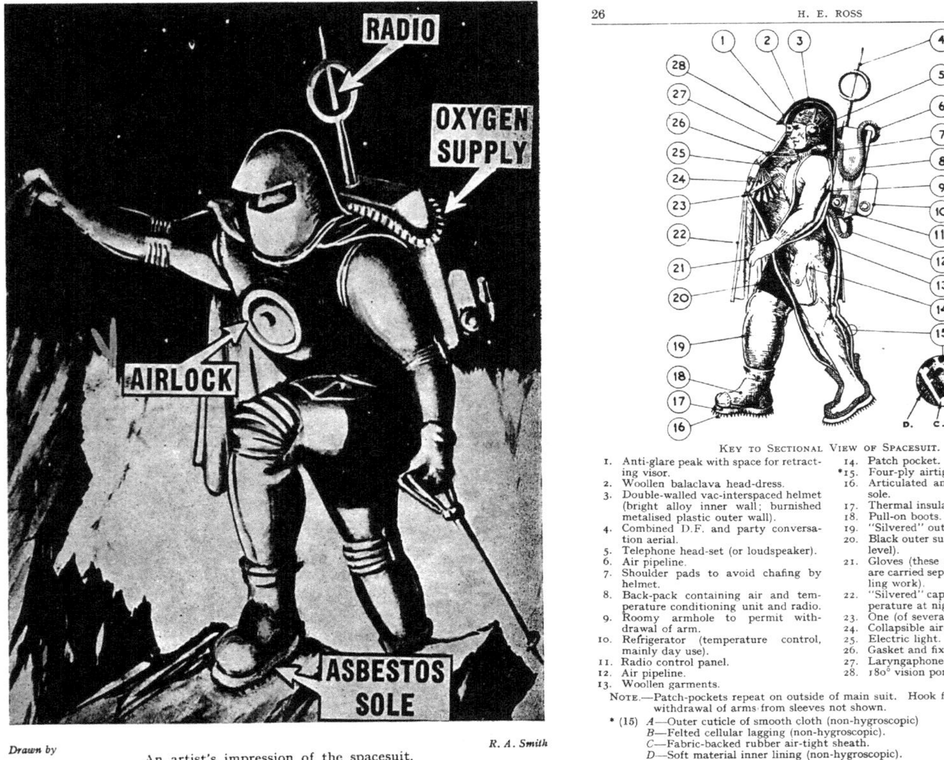

Figure 9.1 An artist's impression of a space suit (left) and a sectional view (right). Both graphics were drawn by Ralph Andrew Smith (1905–59), the British Interplanetary Society's 'chief designer.'
Source: H. E. Ross, 'Lunar Spacesuit,' *Journal of the British Interplanetary Society* 9.1 (January 1950), 23–37, here 24 and 26. Courtesy of British Interplanetary Society, London.

and in his remarks, Cleator refers to another prototype developed by J. B. S. Haldane (1860–1936) and Sir Robert Henry Davis (1870–1965).

It was not until later that full-pressure suits became associated with military purposes. While they were not put into use during the Second World War, they were further developed over the course of the war, both by the Soviets and the Americans. They then attained their military function during the Cold War, first in the context of surveillance flights, which were carried out at high altitudes to avoid missile-defense fire.[3] The suit enables its wearer to travel through spaces that would otherwise be lethal. By creating a passage through an impassable territory, the suit serves as a medium. This medial quality is first recognized when war bars entry to certain territories and the strategy for coping with enemies requires overcoming the human body. This is particularly the case for surveillance airplanes such as the Lockheed U-2 (developed in 1945), which had to fly at high altitudes to evade missile fire. High-pressure suits were also used with this aircraft, namely the Goodrich Mark III and IV models.[4]

Only with regard to this medial aspect does the space suit become relevant to military considerations. Or vice versa: the moment outer space transformed into a war zone, the 'space suit' actually became a space suit (namely with the SK-1, which Yury Gagarin [1934–68] wore in the Vostok 1 spacecraft).[5] At first, one may suspect that the military significance of space suits can at most possess a secondary, accidental character; space suits are worn where twentieth-century militarization unfolds.

All the same, this chapter argues that the militarization of outer space is actually discursively conceived with the space suit and therefore also emerges at a very specific point in time. Historically, one can very well both isolate and explain it. On the one hand, infusing the space suit with military potential required something like space anthropology. Already by the 1950s the anthropology of space had reached an adequate level of technological reflection. As Patrick Kilian's contribution to this volume shows, the development of biocybernetic controls – advanced decisively by the Soviets, viewed skeptically by the Americans – belongs constitutively to the Space Race.[6] 'The era of cosmic man' was also the era of the cosmic body. On the other hand, the excessive corporeal fantasies associated with military space exploration were curbed by the reality of human spaceflight, which began just in 1961. Adapting humans to the extraterrestrial biosphere turned out to be far less spectacular than technological fiction had dared to dream. But in a narrow window of time during the 1950s, the idea of the space suit as a corporeal adaptation to militarized space came to life. Its most extreme and revealing form can be found in Robert A. Heinlein's 1959 novel *Starship Troopers*. While literary fiction had already made space into a theater of armed conflict before Heinlein (with the narrative of an alien invasion having proven to be completely dominant), *Starship Troopers* fundamentally changed the basis of the discussion.[7] The novel depicted war in space as premised on the amalgamation of media technologies, political order and corporeality – an amalgamation that, in this form and around 1960, could solely be imagined in space. At the center of this amalgamation stood the suit.

Except for Nicholas de Monchaux's meticulously compiled documentation of the technological, material and fashion history of the space suit, recent engagements with the phenomenon in cultural studies are characterized by reflections on the body in space from the perspective of identity politics. Based on Donna Haraway's 'Cyborg Manifesto' (1985), both Megan Stern's thoughts on the 'inhuman gaze' (2000) and Debra Benita Shaw's remarks on the 'space suit as cultural icon' (2004) focus on the fusion of man and machine. According to Shaw, the space suit must be read as an icon of postmodernism. It 'has provided a potent symbolism for contemporary fusions with digital technology where simulations, data processing and prosthetics both describe and supplement the body.'[8] Adapting the human body to the conditions in space leads directly to the concept of a hybrid anthropology that was already associated with the term 'cyborg' in 1960.[9] It should be noted, however, that this currently popular postmodern interpretation of the cyborg astronaut – which aims in essence at undermining the categories of nature/culture, man/woman, subject/object – did not yet form part of space suit debates during the blossoming of spaceflight. Stern located her media-historical origin in the photographs of the Apollo 11 crew after the July 1969 moon landing, but for a long time, understanding the astronaut as a cyborg remained affirmatively connected to the concept of a martial order of the cosmic body both in factual space exploration and, especially, in literary space fiction.

And this is how the space suit became readable as armor in the sense of identity politics, which seems absurd considering the circumstances. Body armor is not at all necessary for space warfare, thus space combat suits have almost never been developed or constructed.[10] To this date, war in space has not required any fights, battles or close combat. Instead, it has been designed and simulated in the laboratories of military think tanks. The forces engaged in controlling extraterrestrial space (and thereby also space on earth) far surpass the power of human veins and muscles, which means that current space technology does not have any motive to deal with bodies at war.[11]

Yet the literary imagination of galactic warfare remains bound to the idea of body combat. According to literary scholar Steffen Hantke, in the age of mass destruction, science fiction has paradoxically promoted surgically enhancing the individual 'in a strangely anachronistic move,' which turns the individual in outer space into a hybrid object whose performance and imperfections are improved by technological extensions and substitutions.[12] While the reception of cyborg culture is typically linked to 1980s cyberpunk and gender discourse, those familiar with the narrative tradition of space warfare are more inclined to identify the origins of the cyborg in the technological enhancement and replacement of body parts so that humans are fit to enter into space combat. Darth Vader, the gloomy villain of *Star Wars* (1977), soon became the cipher for the desire to anatomically transform the human in space. There is a fundamental assumption behind Vader's mask: ruling a galaxy in the late twentieth century requires that the body be shaped anew. And this conviction is not only found in space fiction. It is, rather, the starting point of speculation in both military strategy and astrobiology as pursued by J. B. S. Haldane in the early 1960s.[13] The practice of corporeal enhancement calls for an act of engineering that adequately transforms the political idea of the cosmos into human flesh, reflecting how our understanding of the cosmic order influences our concepts of the body. A cosmos that is subjected to the idea of planetary evolution logically becomes populated with humanoid species that share our own desires and weaknesses and, in turn, act as reflections of our cosmic selves. And a galaxy that is hostile and threatening generates revisions to our body politics, which is primarily a question of representation and not of survival. To enter the galactic battlefield, humans must represent the martial order and its social, technological and ideological structures. From this perspective, the militarization of outer space corresponds to the code of war inscribed into bodies crossing the galaxy. The history and mutations of this code define one aspect of what can be called the *conditio extraterrestris* of the modern age. The analysis of how bodies are clothed in space fiction then uncovers the unformulated rules of the embattled heavens.

I Heinlein, *Starship Troopers* and the martial order of the cosmic body

Robert A. Heinlein (1907–88) was one of the individuals primarily responsible for fusing the discourses that transformed space suits into a medium of militarization around 1960. Heinlein, a four-time Hugo Award winner and one of the most influential space-fiction authors of the twentieth century – and with regard to his political views, also most controversial – did not approach the topic of the space suit unprepared. After time as a naval officer and an unfinished degree in mathematics and physics at UCLA, Heinlein became, starting in May 1942, a civilian engineer and personnel director at the Aeronautical Materials Lab of the Naval Aircraft Factory in Philadelphia, where he was partly responsible for the development of a 'high-altitude pressure suit.'[14] While Heinlein thus possessed firsthand technical knowledge of such suits, for decades this knowledge remained silent in his texts. It first resurfaced with the publication of his novel *Have Space Suit – Will Travel* in the *Magazine of Fantasy and Science Fiction* in 1958.

The novel is an action-packed piece of space fiction: an alien abduction with a subsequent chase scene, an intergalactic tribunal and a happy homecoming from planet Vega 5. The central role of the space suit is remarkable. It is not only stressed in the title but also actually staged as a medium throughout the text. The decisive detail of this suit is its miniature radio (already mentioned above with reference to Cleator), which connects the wearer of the suit to a galactic communication network.[15] When the high-school student Kip puts on the suit, he immediately becomes part of a story that exceeds the earth. The call for help he receives and answers is only the starting signal of an incipient assimilation to space. In this assimilation, the suit not only transmits signals through space, it is also a transmitter itself, the medium that makes the journey through space possible. Whoever is wearing a space suit is, as Kip learns, 'determined to space': he immediately communicates and operates on another frequency.[16] But the space suit, made by the protagonist while still on earth, proves inadequate for the fight for survival in outer space. The narrative primarily serves to evidence the suit's technical deficiencies. Time and again, the text constructs contingencies that ascertain the suit's shortcomings: from faulty air-conditioning to the lack of a water tank, exterior mirror and internal clock in the helmet. Then, at the end of the space adventure, one factor is especially decisive: the space traveler returns from alien planets with knowledge that enables him to continue studying mechanical engineering with a scholarship at MIT – in order 'to make space suits that work better.'[17]

At first glance, this may look like a naïve avowal of technological progress. But if one considers the central medial role of the suit – which, in Heinlein's novel, is not just part of the extraterrestrial world but rather produces that world and allows it to become narratable – then this comment should not be passed over so quickly. According to the logic of the text, the suit is the actual protagonist. Upon its first use, it immediately reveals space as hostile, and it enables its wearer to travel across this world so as to make possible, in the end, its own further development. While media history ordinarily takes the perfectibility of man as a matrix for describing technological developments, *Have Space Suit – Will Travel* conversely stages the human as a medium that the space suit uses to perfect itself. Only one year later, in 1959, Heinlein's follow-up novel, *Starship Troopers*, showed where to search for this perfection by converting the connection of space suit, space exploration and extraterrestrial communication into a logic that conceives of the militarization of society as having intrinsic value, a value from which political and moral solutions are to be derived.

Starship Troopers displays a devotion to military libertarianism.[18] The preservation and safety of the future global state is solely based on the free will of its citizens. No one is forced into military service, and everyone has a free choice between becoming a 'citizen' or remaining a 'civilian.' The single privilege that citizens acquire through their service in arms is political participation in democratic processes that civilians remain excluded from. *Starship Troopers* drew its ideological position from the line of thought Heinlein later adopted when he became involved – alongside science-fiction authors Gregory Benford (1941–), Poul Anderson (1926–2001) and Larry Niven (1938–) – in the Citizens Advisory Council on National Space Policy, consulted by the Reagan administration during the development of the SDI program.[19] The novel and Heinlein's intellectual and professional biography are thus related, yielding a constant interaction between fiction and politics. *Starship Troopers* adopts the political situation of the time, which witnessed two prevailing powers attempting to expand their reach. This goal necessitated threatening the other power with extermination. Hence, there is sufficient ground to approach the novel as an allegory for the conflict between the United States and the Soviet Union, which it transforms into a conflict between humanity and the 'Bugs' or 'Pseudo-Arachnids.' This species is characterized less by its appearance than its political organization, which classifies it as the absolute 'other.' The novel's conflict scenario offers numerous other historical analogies to its time. For instance, the disputed war zones 'Planet P' and 'Klendathu' could translate into combat fields on earth such as Korea, just as the Arachnid's destruction of Buenos Aires alludes to the permanent threat of nuclear war.[20] Pursuing this interpretation to its end would entail turning Heinlein's fictionalized universe into an allegorical space. Astroculture in general and Heinlein's novel in particular would be reduced to the invention of bizarre, arbitrary and substantially meaningless combat sites. Such a focus on military and political allegories would render the fictional quality of science fiction insignificant.

But *Starship Troopers* introduces into these political allegories an entity that cannot be deciphered allegorically. It places the space war on an entirely new foundation – the suit, or more precisely, the space combat suit that the Terran Mobile Infantry wear in the novel:

> No need to describe what it looks like, since it has been pictured so often. Suited up, you look like a big steel gorilla, armed with gorilla-sized weapons. [...] But the suits are considerably stronger than a gorilla. If an M.I. [Mobile Infantry] in a suit swapped hugs with a gorilla, the gorilla would be dead, crushed; the M.I. and the suit wouldn't be mussed.[21]

But there is also more to the suit than just an increase in strength:

> The 'muscles,' the pseudo-musculature, get all the publicity but it's the control of all that power which merits it. The real genius in the design is that you *don't* have to control the suit; you just wear it, like your clothes, like skin. Any sort of ship you have to learn to pilot; it takes a long time, a new full set of reflexes, a different and artificial way of thinking. [...] But a suit you just wear.[22]

The suit may at first appear to be nothing more than a gadget created by a speculative fantasy of a technological future, but the extensive information about its design reveals its structural significance. It is *not* a gadget or device that requires specific technical skills. While it does have a technological frame that lies in 'negative feedback and amplification,' it is described more or less as a creature, and one might even venture to say that the suit is a parasite. It does not break or get damaged, but it does become 'sick,' in which case it needs its own physician, 'a doctor of science (electromechanical engineering).' Although the space suit is designed to improve the citizen's body since it 'takes orders directly from your muscles and does for you what your muscles are trying to do,' the suit is defined as a '[c]ontrolled force [...] force controlled without your having to think about it.'[23]

'Controlled force': behind that expression rests a hidden truth since the control woven into the suit works in a twofold manner. It permits a soldier of the Mobile Infantry to control otherworldly forces, and, at the same time, it forces him to be under the control of a civilization that understands outer space as a zone that admits no civilians. The clothing of those who fight for terrestrial affairs in space not only strengthens their physical skills and connects them to channels of information, it also 'rig[s]' their 'eyes' and 'ears.'[24] The suit focuses the soldier's attention on both the battlefield and the network as it inscribes principles of sovereignty, equality and efficiency. It thereby reprograms and resets its wearer's understanding of culture, sex and morals. Shaw's rereading of space suits, which was mentioned above and is based on the methodological instruments of gender studies, later expands on exactly this point.[25] Although Heinlein's novel seems to seek the radical affirmation of the armored body in exact opposition to Shaw's theses, by doing so it prepares Shaw's argumentation. Even when Heinlein very clearly seems to frame the identity-politics dimension of the suit with the American 'frontier myth,' he still certainly recognizes that the dichotomies of subject/object, man/machine, individual/society, man/woman begin to dissolve under the influence of the suit. Replacing them with another dichotomy, of citizens/civilians – that is, the differentiation between those who wear the suit and those who do not – belongs to the political calculation of the text.[26]

The suit is viewed as a medium that transforms its wearer by subjecting him or her to what the novel calls 'force.' Although the actual meaning of this assertion is quite incomprehensible, Heinlein's combat suit does materialize the notion that, in space, armor and media is one and the same thing. This applies to the fictional representation of the 'Space Wars' since it seems at first that alien civilizations (such as the Bugs) are the main war targets. As a matter of fact, the most important agenda of the war is to undermine a serious threat: the limitless expansion to potential habitats. Thus, at stake is the loss of a connected human community, which would render individuals alienated from their earthly origins. By expanding to other planets, humanity is in danger of losing a global, operative and all-inclusive society, or, to be more precise, spaces may emerge where people are disconnected from the rest of humanity. *Starship Troopers* stages the moment of this loss with the narration of 'the drop,' namely the interval between soldiers leaving the ship and their arrival on a hostile planet:

> It's better after you unload. Until you do, you sit there in total darkness, wrapped like a mummy against the acceleration, barely able to breathe [...]. It's that endless wait in the dark that causes the shakes – thinking that they've forgotten you [...].
>
> Then suddenly nothing.
>
> Nothing at all. No sound, no pressure, no weight. Floating in darkness [...] free fall, maybe thirty miles up, above the effective atmosphere, falling weightlessly toward the surface of a planet you've never seen. But I'm not shaking now; it's the wait beforehand that wears.[27]

'The wait beforehand' – that is, the lapse of time in which the individual does not receive any response from the system – is more terrifying than the encounter with alien forces. Back on the battlefield, the soldier is reconnected with a power network that allows him to overcome the threat of a chaotic, senseless and unstructured cosmos. While the novel seems to neglect combat scenes including the use of arms, killings etc., it meticulously describes the modes and techniques of communication. To fight means to be located and traced, to be addressed through multiple audio circuits and to acquire control over frequencies. Again, Heinlein translates battles into media triumphs, which entail recapturing the network after a brief moment of being disconnected.

Heinlein's novel thus shows space combat to be an act of unending transmission. The main purpose of space war is to prepare the galactic soldier's body for transmission by equipping it with a marauder suit that is, in truth, a communication switchboard. Humanity's superiority, with its position of power in the universe, is not based on possessing more efficient weapons. Rather, it must be implemented through an ability to disseminate signals through outer space while exploiting the potential of a control system for communication that is – thanks to the suit – an extension of the human body. Heinlein's novel suggests that the power of such extensions can only be achieved in combat scenarios and, foremost, by citizens. They volunteer for military service because only as soldiers can they take part in the crucial network that transcends all planets. Volunteering makes them, unlike ordinary civilians, active members of the universal state. The emergence of the interstellar soldier, the 'Starship Trooper,' is strictly related to the necessity and feasibility of a communication network.

II The medium of force

To gain further insight into the mental horizon of galactic warfare, two issues must be considered: first, the effect that this type of military communication system has on the values it transmits, in particular with reference to human and terrestrial identity; and second, the reason why this network is strictly bound to military activity.

In 1950, nearly a decade before Heinlein published his novel, Harold A. Innis (1894–1952), a Canadian professor of political economy, argued that empires always depend on the balance of their media. His study *Empire and Communication* advanced an approach based on differentiating between 'space-biased' and 'time-biased' media.[28] At first sight, a direct line of thought connects Innis's theory to Heinlein's fictional representation of the relation between power and the media. In Heinlein's novel, the wisdom of the human empire is preserved in pseudo-decentralized storage. Remote harbors for encyclopedic knowledge are well known in space fiction, the most prominent being planet Terminus, which is chosen for securing the galaxy's knowledge during the collapse of the Galactic Empire in Isaac Asimov's *Foundation* series (1942–93).[29] In *Starship Troopers*, this role is filled by the 'Officer Candidates School.' Johnny Rico, the novel's protagonist, explains:

> I am not going to describe Officer Candidates School. It's like Basic, but squared and cubed with books added. In the mornings we behaved like privates, doing the same old things we had done in Basic and in combat and being chewed out for the way we did them – by sergeants. In the afternoons we were cadets and 'gentlemen,' and recited on and were lectured concerning an endless list of subjects: math, science, galactography, xenology, hypnopedia, logistics, strategy and tactics, communications, military law, terrain reading, special weapons, psychology of leadership, anything from the care and feeding of privates to why Xerxes lost the big one.[30]

Xerxes is probably a topic in 'History and Moral Philosophy,' the only course the narrator cares to provide detailed information about. It is not exactly what one would usually expect to find in the humanities since the course fosters adjusting history and moral philosophy to the logic of efficiency. They are transformed into an 'exact science.' The military perspective lays bare the complex history of human interaction – from Plato to the Korean War, from the 'absolute monarch to utter anarch' – so as to implement governmental codes, equations and 'mathematically verifiable moral reasons' in society.[31]

The Officer Candidates School thus initiates chosen individuals into a field of knowledge that constitutes the wisdom of citizens, the ones who have deliberately chosen to defend the global state in space suits. It seems logical to consider this institution – where we witness an elitist, cryptic and localized transmission of knowledge – as the enduring and complex counterpart to the audio circuits in the combat suits, which work as an egalitarian and space-transcending medium that is accessible to anyone who volunteers to use them. The role of media within this enclosed community generates a global state characterized by detailed ranks and hierarchies, grades of responsibilities, abilities, skills and knowledge – a state that is organized and directed by a very efficient administration, which stretches from the government on earth to the desert sands of a distant planet inhabited by hostile aliens.

A closer inspection, however, reveals that essential changes follow once Innis's paradigm is applied to outer space. Innis argues that the great civilizations of antiquity were bound to collapse due to an overemphasis on either time-biased or space-biased media. When empires begin to collapse, their media are typically imbalanced; they are either out of space or out of time. When space-biased media are overemphasized, nobody can spread information because decryption has become an arcane discipline. When time-biased media are overemphasized, media have become monopolized by a few who use information as a good to trade in their own interest; the preservation of knowledge is far from being one of their goals.

This argument is justified when it is applied to the role of parchment, stone or paper in ancient empires. Innis's theory was not, however, inclusive of the Space Age, in which media became global and invisible. But there is still a connection between political control and media: concepts of political control can encompass the vastness of the galaxy. The Galactic Empire broadcasts universal signals that everyone can receive instantly and simultaneously. In wars fought in outer space distance has a different value and is overcome by connectivity. Limitless expansion then involves thinking about outer space in terms other than size. In space, connectivity and communicative expansion have their own purposes.

Yet the question remains whether there is a connection between this paradigmatic change and the militarization of outer space as imagined in Heinlein's novel. Militarization goes a step further than clothing and training soldiers since it is an epistemological concept. The reader learns of this concept through the narrative of the Officer Candidates School which explains the militarizing process. This process is politically justified by the expansive urge of all life: all societies find themselves in a constant fight for survival and thus, just as morality secures internal social stability, war must be understood as a normal condition of external security (read: in outer space): 'war and moral perfection derive from the same genetic inheritance.'[32] The explosiveness of Heinlein's novel also consists in how it coolly postulates the inextricability of inward-directed morality and outward-directed aggression. In this context, the militarization of outer space entails integrating a terrestrial social order that is based on the principle of equality into the reality of a fundamentally belligerent galaxy, which humans have to adapt to technologically, politically and mentally:

> Nevertheless, let's assume that the human race manages to balance birth and death, just right to fit its own planets, and thereby becomes peaceful. What happens?
>
> Soon (about next Wednesday) the Bugs move in, kill off this breed which 'ain't gonna study war no more' and the universe forgets us.[33]

The political legitimation of militarization is now also joined by its performative implementation. In a society governed by the logic of space war, there is no place for discursive debates or reasoning. The didactics at

Officer Candidates School call, in the end, for always formalizing the content of the courses, for breaking it down into mathematical proofs. Morality is programmable since it is based on the aforementioned 'mathematically verifiable moral reasons.'[34] In this reduction of content to formulas, the novel follows the necessity of efficient communication. Militarization also implies that transmission absorbs tradition. In outer space, the epistemological core of civilization must be converted into a transmittable logic due to the necessities of an a priori hostile environment. From today's perspective, such a crucial transformation is perceived in the prospects of the digital, but in terms of representing humans in space, this transformation takes on the form of a military code. The media employed to transmit this code not only affect its content, they also shape the receiver and sender. Heinlein's concern with the theory of general semantics would have certainly provided more than fertile ground for this line of thought.[35] In his *Science and Sanity* (1933), Alfred Korzybski, the main exponent of general semantics, posits that it is possible to find connections – just as in *Starship Troopers* – between the body of the soldier, its reflexes, the semantic encodings it enables and the system of transmission that the soldier participates in.[36]

'The sovereign franchise' is the transmission system in Heinlein's novel. It erases personal characteristics such as 'place of birth, family of birth, race, sex, property, education, age, religion, et cetera.'[37] Social distinctions are managed by the system, whose only category of differentiation is 'responsibility.' Human beings are classified into a group that assumes social duties and a group that does not. The core of this ideology lies in Heinlein's libertarian beliefs, but the logic of the novel suggests that this explanation is not entirely sufficient. Instead, the origins of the depersonalization process are attributable to the notion of the individual as a potential agent of disturbance or resistance to communication in outer space, which makes any person a source of misunderstanding, questioning and unpredictable desires. Physical enhancement is the solution. In space, the human being is modified, rebuilt and remodeled into the Starship Trooper, who is a permeable medium for messages that are, in turn, nothing but force: 'force, naked and raw, the Power of the Rods and the Ax. Whether it is exerted by ten men or by ten billion, political authority is *force*.'[38] In short, force is both the message sent through the vastness of the universe by the sovereign franchise *and* the medium that forms the new individual, the citizen of outer space. The act of transmission is thus combat, fought in an armor that empowers humans to invade the galaxy and, at the same time, invades humans by programming altered identities.

III Utopian and dystopian military networks

The most striking aspect of the space war in Heinlein's novel is the physical absence of an enemy. The text justifies military intervention in global policy with the expedient of the alien civilization of the Bugs, whom the inhabitants of earth must combat. But the enemy remains almost invisible throughout the book. While Paul Verhoeven's 1997 film adaptation overemphasizes the physical appearance of the alien insect, Heinlein's novel refrains from giving much attention to direct encounters and close combat.[39] Rather, it focuses on the Bugs' alternative model of social organization, and the contours of this society become more blurry the more they are described. The Bugs are an evolutionary threat, born with skills and intelligence. Their civilization has a way of breeding information, of transmitting strategic knowledge through a kind of biological preprogramming that cannot be reduced to instincts:

> Those Bugs lay eggs. They not only lay them, they hold them in reserve, hatch them as needed. If we killed a warrior – or a thousand, or ten thousand – his or their replacements were hatched and on duty almost before we could get back to base. You can imagine, if you like, some Bug supervisor of population flashing a phone to somewhere down inside and saying, 'Joe, warm up ten thousand warriors and have 'em ready by Wednesday… and tell engineering to activate reserve incubators N, O, P, Q, and R; the demand is picking up.'[40]

While humans can still only imagine interaction by mobile telephone, the Bugs' communication network is genetically implemented and cannot be compared to anything human. It is an inbred system: '[…] their actions were as intelligent as ours (stupid races don't build spaceships!) and were much better coordinated. It takes a minimum of a year to train a private to fight and to mesh his fighting in with his mates; a Bug warrior is *hatched* able to do this.' As a species, the Bugs prove 'how efficient a total communism can be when used by a people actually adapted to it by evolution.' In the novel the Bugs are an antagonist who threatens humanity by natural superiority alone. They may be defeated every so often in battle – but as a species they will prevail. Actual combat is inefficient and insignificant since 'every time we killed a thousand Bugs at a cost of one M.I. it was a net victory for the Bugs.'[41] The real war is an inner war that is fought in order to overcome humanity's socio-biological deficiencies in the Space Age.

The imago of the alien aggressor serves as a model that helps recognize this deficiency. It is not the first time in Heinlein's work that humanity's opponent is a creature superior to man by virtue of its connectivity. In *The Puppet Masters* (1951) he introduced a species of space slugs from Titan that invades earth by tapping into the human nervous system. They melt human minds into a gigantic coordinated network that extends across the globe. In light of this phenomenon, the later description of the combat suit as a 'creature' that melds together with the soldier's body must be read as a compensatory analogy. The suit mimics the technique of the alien by taking control of the individual's body and mind. The enemy, be it bug or slug, assumes the role of an ideal community, which is envisioned as a utopian state of the swarm where the group directly carries out every action by means of the individual. This state proves to be a unique organism that spreads through space while embracing other forms of life and modifying their inner behavioral code.

Space war, as imagined here, can be read with Marshall McLuhan, namely 'as a process of achieving equilibrium among unequal technologies,' in other words, as approximating the standards of extraterrestrial communication.[42] The balance is accomplished by contrasting the communicative superiority of alien civilizations with the evocation of blunt force. The novel *The Puppet Masters* is written from the perspective of those who volunteer to take revenge on the slug invaders and to 'clean up Titan.' In their perspective, 'the human race has got to keep up its well-earned reputation for ferocity. The price of freedom is the willingness to do sudden battle, anywhere, any time, and with utter recklessness. [...] For who knows what dirty tricks may be lurking around this universe?'[43]

With regard to the suit policy that *Starship Troopers* established only eight years later, this statement reveals both a superficial and a deeper hidden logic. On the surface there is the conviction to countering control with raw strength while the deeper logic aims at improving man, at attempting to transform the human being into a new form of life in space. This process presupposes that the galaxy warrior is in permanent combat mode. Against the background of the evolution of man in combat, it is possible to reverse the assumption that the suit is an effect of a threat from beyond the planet. Instead, a universe filled with war-mongering aliens is the necessary consequence of humanity's reeducation program. It is not so much the alien body that forces humanity to adapt but rather humankind's corporeal adaptation to extraterrestrial life that summons the scheme of the Bugs. Their scheme reveals another agenda. Entering the Space Age is perceived as integral to a new cosmological, social, political and technological order, regardless of the potential enemies. On a figurative level Heinlein's texts tend to abhor the effigy of the alien and suffuse it with a repellent aura. Repulsion arises from the image of the 'masses,' the 'zillions' of fighters in the army of the Bugs and from their soft, naked bodies.[44] These are easily identified as those patterns encountered in twentieth-century totalitarian imageries: the masses, the flood, soft shapes, unclosed or open structures. All these aspects convey the importance of plasticity to sexuality, which, in turn, is the real threat to the human combatant.[45]

The combat suit allows for a twofold move, that is, 'a progressive dissociation of the individual from the group of which he is a putative part, and a domestication of combat as it moves ever closer to the interior of the defended territory and the interior of the defender's psyche.'[46] Thus while the battlefield serves as the stage for the lonesome fighter – an obsolete role in space – this illusion is used to promote the inner transformation of the space warrior. The suit can be read as a device that protects against the deformation of the humanist image of man as the last remaining humanist trait of the posthuman age. More specifically, the combat suit is the reconstruction of man in space, a shell that reassembles the shattered fragments of the human body and mind after 'the drop' into the new and greater cosmos. *Starship*

Troopers unveils this act of restoration. It would therefore be shortsighted to label this text 'revisionist.' While Heinlein's political agenda seems to be quite dominant within the plot, the novel exposes the futility of such revisionism. After all, it offers no prospects for humans beyond the galactic battleground: as soon as the war is over, the body falls apart. So war must prevail. The scenario of the alien enemy's absolute destruction is thus clearly impossible considering the realities of the narrated world. It is for this reason that both Heinlein's novel and Verhoeven's movie stage the war plot as an open-ended epic narrative.[47]

IV The space suit as postmodern icon

Situating the space suit in the cultural history of the twentieth-century militarization of outer space, *Starship Troopers* serves as an interface between technological fictions of the body, astroculture and ideologically supported militarism. One may consider Heinlein's testimony to be a bizarre snapshot of its time. It was only possible at this specific moment, after the beginning of biocybernetic speculation in the 1940s, but prior to the beginning of human spaceflight, which would then stage and code the space suit in a new way.[48] At the same time, it is a figure of thought that necessarily presupposes the American 'frontier myth,' as a look at how contemporary Asian and European films staged the space suit shows.[49] Nevertheless, Heinlein's combat suit still attained a noteworthy afterlife: noteworthy, because it took a global detour for it to evolve from merely space-bound clothing into a shell that fully absorbs the human leftovers of its wearers, transforming them into something else. To comprehend Verhoeven's 1997 film adaption of Heinlein's novel, it is necessary to trace the beginnings of this evolution of the space suit, and they are to be found in the Japanese adaptation of the material. That began with Testu Yano's (1923–2004) 1967 translation of the novel and was at first determined by a discussion of the text's fascistic dispositive. The actual cultural reassessment of *Starship Troopers* then occurred – as Takayuki Tatsumi has shown – about ten years later and took place not on a textual level but on a pictorial one. Kazutaka Miyatake's and Naoyuki Katoh's illustrations in the Japanese paperback edition removed the suit from the context of the novel and iconized it in an entirely new way (Figure 9.2).[50]

Accordingly, one sees a nighttime cityscape, not of this earth, with a gigantic angular greenish steel figure hovering in front of it. Human traits are no longer recognizable except for the basic anatomy of its extremities – two legs, two arms. Instead, over its right shoulder, one sees a rocket launcher integrated into the torso, its arms reminiscent of the barrels of a firearm. The pure functionality, the military purpose of the suit, has become dominant. All human characteristics have vanished. They have been coalesced and transformed into a body of combat. In comparison, the architecture in the background looks downright organic. At the same time, the illustrator captures

Figure 9.2 An illustration by Kazutaka Miyatake (1949–) and Naoyuki Katoh (1952–) taken from the Japanese 1977 paperback edition of *Starship Troopers*. Note how the pure functionality depicted here removes the suit from the context of the novel and iconizes it in an entirely new way.

Source: Courtesy of Kazutaka Miyatake and Naoyuki Katoh, Studio Nue.

the contrast of the extremely heavy armor and the seemingly weightless movement. This is to be read allegorically. Materialized in Heinlein's space suit is the ideology of an earth-centered and hierarchically organized chain of command that can only function if a person's individuality is suspended in outer space. Out of this ideology rises an icon that can actually no longer be ascribed to any human society. The cover illustration follows this analysis to establish the iconographic tradition of Japanese robots, which live from the idea of an oversized, armored flying suit. The first was *Mobile Suit Gundam*, an animated series which debuted in 1979. Following the show's cancellation, the first *Mobile Suit Gundam* toys were then launched in 1980 by the Japanese company Bandai.[51]

But the iconic isolation of the suit still remains decisive from the perspective of cultural history. Tatsumi has pointed out that the 'paradigm shift between the late 1960s ideological debate on *Starship Troopers* and the late 1970s visual representation of the same novel is symptomatic of the postmodern deconstruction of hierarchical relations between entity and function, ontology and aesthetic, existentialism and structuralism, symbol and allegory, that is to say, signified and signifier.'[52] Even if this basic assessment can be absolutely agreed with, the development may still be conceived less as break and more as a continual transformation. The fusion of body, communication and identity, which is concealed in the space suit, is already introduced in Heinlein's novel. It may be true that the novel depicts it as a future technology that at first assimilates those who are operating on the frontier, but it actually targets the whole collective of society. That the space suit, as a posthuman figuration, also dissolves the very dichotomies that were responsible for the military localization of the space suit on the boundaries between civilization and wilderness has been a discovery of postmodern cyborg theory, from Haraway to Stern and Shaw, and it would hardly have been conceivable without the aestheticization of the space suit by Japanese science-fiction iconography.

Paul Verhoeven's (1938–) 1997 film adaptation of the novel draws its particular appeal from the cultural displacement of the space suit from being an object of political and military projections to becoming a postmodern icon (Figure 9.3). The film stages the process of replacing the old narrative, which first created the space suit, with a new narrative that is now carried by the space suit. The opening plot is not only determined by the agenda of 1950s America but also stylizes this agenda in prudery, in a high-school setting and the invocation of a community's common fate against an enemy from afar who strikes mercilessly and with whom there cannot be any understanding beyond 'shock and awe.'

In its narrative structure, Verhoeven's film, even far more than Heinlein's novel, is a coming-of-age film that tells the story of the ascent of three schoolmates through the galactic war. The interplay between a community that has withstood 'millions of lightyears' – the 'Federation' – and, at the

Figure 9.3 This still from Paul Verhoeven's (1938–) 1997 film *Starship Troopers* depicts the scene in which a Federation desk clerk who only possesses metal prostheses in place of his legs and right arm welcomes Rico into the ranks of the Mobile Infantry, illustrating the Federation's order of bodies and clothing.
Source: *Starship Troopers*, directed by Paul Verhoeven, USA 1997. Courtesy of Touchstone Pictures.

same time, deliberate individual differentiation according to abilities, succeeds above all through the order of bodies and clothing. This interplay is depicted through different characters: Carl, assigned to the division of 'Military Intelligence,' is soon seen in a Gestapo coat; Carmen, who serves as a pilot, in a pilot's uniform; and Johnny Rico, who is assigned to the 'Mobile Infantry,' is already confronted at the information desk with the fact that an existence as a cyborg awaits him at the end of his career: 'The Mobile Infantry made me the man I am today.' With these words, Rico is welcomed by a desk clerk who only possesses metal prostheses in place of his legs and right arm. This confirms a hierarchical system in which the individual can only be successful by obeying the Federation's body politics. The suit is coded as a medium of the underprivileged, who, based on their poor high-school grades, cannot be used as 'mindworks' and therefore can only survive in the community by putting their bodies into action.

But at the same time, the suit proves itself to be the basis of the narrated world. While it does not protect its wearers from being torn into pieces by arachnoid aliens every now and then, it still helps them become a smooth, opaque surface that no one can penetrate. Visually, the space suit contrasts naked, raw sexuality – as Verhoeven's film displays it in the form of the Brain Bug – with an asexual, indecipherable figure, as is illustrated when Federation

Figure 9.4 This *Starship Troopers* still shows how Federation troops examine a dead bug, visualizing how the space suit contrasts naked, raw sexuality – displayed here in the form of the Brain Bug – with an asexual, indecipherable figure.
Source: *Starship Troopers*. Courtesy of Touchstone Pictures.

troops examine a dead bug (Figure 9.4). While the higher-ranking pilot Zander Barcalow – who does not wear a suit – is 'picked out' by the Brain Bug in the truest sense of the word (his brain is sucked out), the suit makes penetration impossible. The Mobile Infantry of the galactic war no longer possess any sexual identity. More precisely: they no longer have any identity that would allow them to be classified within a binary system – man/woman, man/machine, humanoid/arachnoid. They are not only indecipherable to the Arachnids but also to Heinlein's contemporaries. And when it is announced at the end of the film to the cheers of the troops assembling on a distant desert planet that the captured Brain Bug is 'afraid,' then that is only half true: it is not fear of humanity that makes this creature tremble. It is fear of what comes after humanity.

Notes

1. Philip E. Cleator, *Rockets Through Space: or The Dawn of Interplanetary Travel*, London: George Allen & Unwin, 1936, 189.
2. Isaac Abramov and Å. Ingemar Skoog, *Russian Spacesuits*, Berlin: Springer, 2003, 5.
3. Kenneth S. Thomas and Harold J. McMann, *U.S. Spacesuits*, New York: Springer, 2012, 8–9.
4. Ibid., 9–10.

5. *SK* stands for *skafandr kosmicheskiy*. The connotation of the diving suit is present as well since *skafandr* is a diving suit.
6. See Chapter 8 in this volume.
7. On the history of alien invasions in literature and film, to which Heinlein also contributed with *The Puppet Masters* (1956), see Heather Urbanski, *Plagues, Apocalypses and Bug-Eyed Monsters: How Speculative Fiction Shows Us Our Nightmares*, Jefferson: McFarland, 2007, 156–68.
8. See Nicholas de Monchaux, *Spacesuit: Fashioning Apollo,* Cambridge, MA: MIT Press, 2011; Donna Haraway, 'A Manifesto for Cyborgs: Science, Technology, and Socialist Feminism in the 1980s,' *Socialist Review* 80.2 (March–April 1985), 65–107; and Megan Stern, 'Imagining Space Through the Inhuman Gaze,' in Scott Brewster et al., eds, *Inhuman Reflections: Thinking the Limits of the Human*, Manchester: Manchester University Press, 2000, 203–16. Stern's central argument is that although the 'astronaut's visor [...], gleaming and impenetrable, recalling the helmets of armoured soldiers, [...] signifies masculine authority, progress and conquest,' the photographic portrait on the lunar surface 'enables "those previously invisible" to become "visible as subjects of their own representations" and "produces them as objects of that visibility"'; see ibid., 208. Debra Benita Shaw, 'Bodies out of this World: The Space Suit as Cultural Icon,' *Science as Culture* 13.1 (March 2004), 123–44, here 127.
9. The concept and term of the cyborg were introduced by Manfred E. Clynes and Nathan S. Kline, 'Cyborgs and Space,' *Astronautics* 9 (September 1960), 26–7, 74–6; followed by Robert W. Driscoll, 'Engineering Man for Space: The Cyborg Study' [May 1963], reprinted in Chris Hables Gray, ed., *The Cyborg Handbook*, New York: Routledge, 1995, 75–81. See also Patrick Killian's contribution, Chapter 8 in this volume.
10. Throughout the history of technology one may find such a combat suit only a single time, namely at the end of the 1980s with the Ames company's AX-5 model, which was actually conceived to allow astronauts to carry out military operations in space. This suit was entirely based on a mechanically strengthened 'hard suit.' See Monchaux, *Spacesuit*, 267–8.
11. Science fiction has, however, inspired the fashion of combat suits on earth. The most recent example so far is the TALOS (Tactical Assault Light Operator Suit). Designed for the US Special Operations Command and first presented in 2013, this suit copies the style of *Iron Man*'s armor as introduced in the Marvel film of the same name (2008), while actually realizing the technological program of twentieth-century military science fiction, including a powered exoskeleton and situational-awareness displays; see *Iron Man*, directed by Jon Favreau, USA 2008 (Paramount Pictures).
12. Steffen Hantke, 'Surgical Strikes and Prosthetic Warriors: The Soldier's Body in Contemporary Science Fiction,' *Science Fiction Studies* 25.3 (November 1998), 495–509, here 495.
13. On Haldane's paper 'Biological Possibilities for the Human Species in the Next Ten Thousand Years,' see Patrick Kilian's contribution, Chapter 8 in this volume.
14. William H. Patterson, *Robert A. Heinlein: In Dialogue with His Century*, vol. 1: *1907–1948: Learning Curve*, New York: Tom Doherty, 2010, 307. In addition to Patterson's monumental two-volume biography, H. Bruce Franklin's œuvre-based biography, *Robert A. Heinlein: America as Science Fiction*, Oxford: Oxford

University Press, 1980, which appeared during Heinlein's lifetime, is still highly informative for historically contextualizing Heinlein's fiction.

15. The novel emphasizes that the space-suit radio is not simply a combination of two independent technologies but rather a substantial fusion. It accomplishes this by making the protagonist's eventual survival dependent on activating a 'communicator-beacon' that, in turn, can only be reached with the help of a space suit; see Robert A. Heinlein, *Have Space Suit – Will Travel*, New York: Ace, 1958, 150–1.
16. Ibid., 250.
17. Ibid., 247.
18. It is well known that Heinlein believed in libertarianism. See William H. Patterson, *Robert A. Heinlein in Dialogue with His Century*, vol. 2: *1948–1988: The Man Who Learned Better*, New York: Tom Doherty, 2014, 247–60, on political libertarianism in general and Heinlein's connection to Barry Goldwater in particular.
19. Ibid., 444–6. For a more detailed analysis, see Peter J. Westwick, 'From the Club of Rome to Star Wars: The Era of Limits, Space Colonization and the Origins of SDI,' in Alexander C. T. Geppert, ed., *Limiting Outer Space: Astroculture after Apollo*, London: Palgrave Macmillan, 2018, 283–302, here 291–4 (= *European Astroculture*, vol. 2); and Alexander Geppert and Tilmann Siebeneichner's introduction, Chapter 1 in this volume.
20. See Franklin, *Robert A. Heinlein*, 116.
21. Robert A. Heinlein, *Starship Troopers* [1959], New York: Ace, 2006, 105.
22. Ibid.
23. Ibid., 106–7.
24. Ibid., 107.
25. Shaw, 'Bodies out of this World.'
26. *Starship Troopers* engages in an aggressive neutralization of the sexes in space combat; Verhoeven's 1997 film adaptation (where, for instance, showers are not gender-segregated) extrapolates this even more strongly. See Jamie King, 'Bug Planet: Frontier Myth in *Starship Troopers*,' *Futures* 30.10 (December 1998), 1017–26; and, above all, De Witt Douglas Kilgore, *Astrofuturism: Science, Race, and Visions of Utopia in Space*, Philadelphia: University of Pennsylvania Press, 2003, 82–110.
27. Heinlein, *Starship Troopers*, 5, 7.
28. Harold A. Innis, *Empire and Communications*, Oxford: Clarendon Press, 1950, 7–8.
29. The Encyclopedia Galactica – the official name for the project on the planet Terminus – can be read as yet further evidence of the identity of media and weapons in outer space. As an archive threatened by anarchic cosmic warfare, the Encyclopedia Galactica must become 'active' to protect itself. It must inscribe its contents into the galaxy, including into diplomatic negotiations, trade management, cultic and religious institutions, acts of manipulation and sabotage and, of course, warfare. This applies at least to the first *Foundation* trilogy, consisting of *Foundation* (1951), *Foundation and Empire* (1952) and *Second Foundation* (1953); see Isaac Asimov, *Foundation; Foundation and Empire; Second Foundation*, New York: Alfred A. Knopf, 2010.
30. Heinlein, *Starship Troopers*, 182.

31. Ibid., 189, 193.
32. Ibid., 195.
33. Ibid., 196.
34. Ibid., 193.
35. Heinlein was interested in the theory of general semantics and attended two of Korzybski's seminars in 1939; see Kate Gladstone, 'Words, Words, Words: Robert Heinlein and General Semantics,' *The Heinlein Journal* 11 (July 2002), 4–6. See also David E. Wright, 'General Semantics as Source Material in the Works of Robert A. Heinlein,' *ETC: A Review of General Semantics* 68.1 (January 2011), 92–109.
36. The main problem in Alfred Korzybski's theory is the difference between Aristotelian language (which separates factors that are actually not separable, such as *matter*, *space*, *time*, *body*, *soul* and *mind*) and non-Aristotelian language (which Korzybski views as provided by Albert Einstein and Hermann Minkowski); see idem, *Science and Sanity: An Introduction to Non-Aristotelian Systems and General Semantics*, 4th edn, Lakeville: Institute of General Semantics, 1958, 315–71. See also Anthony Enns's contribution, Chapter 10 in this volume.
37. Heinlein, *Starship Troopers*, 192.
38. Ibid., 66, 193. As Sergeant Zim reveals, 'We supply the violence; other people – "older and wiser heads," as they say – supply the control. Which is as it should be.' Wearing the suit thus entails entering a field of governmental control that cannot be questioned.
39. As King has pointed out, this overemphasis is a consequence of Verhoeven's attempt to stage 'a subversion of semiotic carriage through multi-level discourses which hystericise the polemics of Heinlein's novel into grotesques of American wartime propaganda material'; see King, 'Bug Planet,' 1019.
40. Heinlein, *Starship Troopers*, 160.
41. Ibid., 161.
42. Marshall McLuhan, *Understanding Media: The Extensions of Man* [1964], London: Routledge, 2001, 375.
43. Robert A. Heinlein, *The Puppet Masters* [1951], London: Hodder & Stoughton, 1987, 223.
44. Idem, *Starship Troopers*, 263. This is quite explicit throughout the novel. The threat of the Bugs is mirrored in the everyday horror of spiders, which pushes Johnny Rico to confess: 'I have never liked spiders, poisonous or otherwise; a common house spider in my bed can give me the creeps. Tarantulas are simply unthinkable, and I can't eat lobster, crab, or anything of that sort.' Ibid., 143. Furthermore, Rico's history teacher at the Officer Candidates School rejects the Platonic idea of the state by categorizing it as 'antlike communism.' Ibid., 191.
45. The first analysis of these structures was provided by Klaus Theweleit, *Männerphantasien*, vol. 1: *Frauen, Fluten, Körper, Geschichte*, Frankfurt am Main: Roter Stern, 1977; they are also evoked in Shaw, 'Bodies out of this World.'
46. Eric S. Rabkin, 'Reimagining War,' in idem and George E. Slusser, eds, *Fights of Fancy: Armed Conflict in Science Fiction and Fantasy*, Athens: University of Georgia Press, 1993, 12–25, here 19.
47. See Neil Badmington, *Alien Chic: Posthumanism and the Other Within*, London: Routledge, 2004, 61–2.

48. The linking of biocybernetics and the Space Race occurred especially through NASA's Division of Biotechnology and Human Research, located at NASA headquarters in Washington, DC. Starting in 1962, Robert Driscoll and his team began working on the concept of a cyborg capable of living in space – an optimization of the human for spaceflight that particularly included five aspects: 'artificial organs, hypothermia, drugs, sensory deprivation, and cardiovascular models.' See Ronald R. Kline, *The Cybernetics Moment: Or Why We Call Our Age the Information Age*, Baltimore: Johns Hopkins University Press, 2015, 174–5.
49. King, 'Bug Planet'; see also Gary Westfahl, *The Spacesuit Film: A History, 1918–1969*, Jefferson: McFarland, 2012, 220–89.
50. Takayuki Tatsumi, 'Postmodern Japan and Global Visual Culture,' in Brian McHale and Len Platt, eds, *The Cambridge History of Postmodern Literature*, Cambridge: Cambridge University Press, 2016, 405–18, here 407–8.
51. On the influence of the suits from *Starship Troopers* on the conception of the Mobile Suit Gundam, see Yoshiyuki Tomino, *Mobile Suit Gundam: Awakening, Escalation, Confrontation*, Berkeley: Stone Bridge Press, 2004, 8.
52. Tatsumi, 'Postmodern Japan and Global Visual Culture,' 409.

CHAPTER 10

Satellites and Psychics: The Militarization of Outer and Inner Space, 1960–95

Anthony Enns

The development of satellite technology was partly driven by utopian narratives of universal harmony, as many scientists and engineers saw it as a global endeavor that would unite the nations of the earth and lead to a more peaceful future. In a talk presented at the 12th International Astronautical Congress in 1961, for example, British engineer and science-fiction author Arthur C. Clarke (1917–2008) predicted that the global satellite communication system he had proposed in 1945 would eventually eliminate national and linguistic barriers and thereby 'link together the whole human race [...] in a unity which no earlier age could have imagined.' The first color photograph of the earth, captured by the ATS-3 satellite in 1967, was also featured on the cover of Stewart Brand's (1938–) *Whole Earth Catalog* to convey the idea of planetary wholeness, and such images inspired British environmentalist James Lovelock's (1919–) 'Gaia hypothesis,' which postulated that the entire planet was a single, living organism. Jody Berland thus argues that satellite photography served to promote world peace by representing 'the promise of earth without its wars.' Similarly, Lisa Parks describes this idea as a 'fantasy of global presence' that served 'to synthesize, contain, and transform the world's irreducibility into an iconic expression of global totality.'[1]

The utopian promise of space exploration was also associated with the exploration of psychic or inner space. The concept of 'consciousness expansion'

Anthony Enns (✉)
Department of English, Dalhousie University, Halifax, NS, Canada
e-mail: Anthony.Enns@Dal.Ca

Alexander C. T. Geppert et al. (eds), *Militarizing Outer Space*
European Astroculture, vol. 3
https://doi.org/10.1057/978-1-349-95851-1_10

was first introduced by American psychologist Timothy Leary (1920–96) and his associates at Harvard University to describe the effects of hallucinogenic substances like LSD and psilocybin, which were later termed 'psychedelic' or 'mind manifesting.' In the 1960s NASA scientists reportedly gave LSD to monkeys to replicate the disorienting effects of space travel, which suggested that there was a fundamental connection between outer and inner space. Leary subsequently argued that there was no difference between astronautics and psychedelics, as hallucinogenic drugs detached the mind from any fixed terrestrial location, which helped to prepare users for extraterrestrial migration. American ethnobotanist and self-proclaimed 'psychonaut' Terence McKenna (1946–2000) also claimed that psilocybin-containing mushrooms were the organic spores of an alien species seeking to quicken the process of evolution through alien-human symbiosis. The connections between astroculture and the newly emerging counterculture became particularly evident with the release of Stanley Kubrick's (1928–99) film *2001: A Space Odyssey* (1968), co-written by Clarke, whose 'total effect,' as described by one reviewer, 'so closely parallels that of LSD visions as to restimulate such experience in some viewers.' When marketing director Mike Kaplan became aware of this phenomenon he even created new posters that promoted the film as 'the ultimate trip.' The perceived similarities between astronautics and psychedelics eventually inspired an entirely new sub-genre of popular music, which was referred to as either 'space rock' or 'psychedelic rock.'[2]

The assumed connections between outer and inner space travel also led to a resurgence of interest in psychic phenomena. Although psychical research began in the nineteenth century, when it was often associated with religious experiences, trance states and occult manifestations, it was more often seen in the twentieth century as a pioneering effort to explore the extraterrestrial future of mankind. Like the development of satellite technology, the history of psychical research was driven by utopian narratives of universal harmony, as psychic abilities were understood as a means of expanding consciousness and achieving a sense of cosmic equilibrium. For instance, Clarke's prediction that communications satellites would foster a sense of global unity clearly echoed French philosopher Pierre Teilhard de Chardin's (1881–1955) concept of the 'noosphere,' the collective consciousness of mankind made possible by communication networks, 'which, perhaps anticipating the syntonization of brains through the mysterious power of telepathy, already links us all in a sort of "etherized" universal consciousness' – an idea that led him to support extraterrestrial migration as well as psychical research. Leary also predicted that the mind would be capable of telepathy, extrasensory perception and psychokinesis as soon as humans 'leave the surface of the planet and live in free space' – a theory that was allegedly proven by astronaut Edgar Mitchell (1930–2016), who claimed to have conducted a successful extrasensory perception (ESP) test from the Apollo 14 space capsule. If the history of astroculture is not limited to 'alien life or extraterrestrial technology,' but instead 'comprises a wider range of images, artifacts and activities conducted by a broader range of expert and amateur actants,' then the history of psychical research can be seen as an important yet largely unacknowledged aspect of astroculture.[3]

Unlike psychical research, however, satellite research was initially driven by military interests, as satellite reconnaissance was seen as a logical extension of earlier forms of aerial reconnaissance. Clarke's initial proposal was even inspired by the science-fiction stories of George O. Smith (1911–81), a radio engineer who worked on radar proximity fuses for the US military and was keenly aware of the potential military applications of space technologies.[4] Over time, psychical research also came to be seen as a potential tool for gathering military intelligence. Just as the history of astroculture was informed by both utopian narratives of universal harmony and imperialist narratives of military conquest, so too was the history of psychical research informed by the twin goals of consciousness expansion and territorial expansion.

This chapter examines the military deployment of satellites and psychics by showing how they were seen as complementary methods of intelligence gathering during the Cold War, as the technological practice of 'remote sensing' and the psychic practice of 'remote viewing' both involved the use of coordinate scanning of distant locations for the purpose of mapping and surveillance. Psychics also depended on satellite reconnaissance photographs to confirm their findings, just as satellites occasionally depended on the information provided by psychics to locate their targets. Many psychic spies began their careers as satellite reconnaissance photograph analysts, as this position appears to have provided them with the necessary training in visualization techniques. Moreover, the practice of remote viewing was not simply inspired by the development of satellite surveillance; rather, it extended the functionality of satellite surveillance systems through increased mobility, improved resolution and accelerated transmission rates. Despite the fact that this practice was never endorsed by the scientific establishment, it nevertheless represented the ultimate realization of the same engineering goals that informed the development of reconnaissance satellites. Both of these research programs were also driven by an urgent sense of paranoia, as researchers argued that it was important for the United States to maintain a leading position in the race for outer space as well as 'the race for inner space.'[5]

These parallels between satellite research and psychical research show how the distinctions between fact and fiction or science and pseudoscience were often unclear, as satellite photographs and psychic visions were both used to construct imaginary representations of the earth that could serve either utopian or imperialist purposes. Indeed, fact and fiction tended to feed each other in an endless loop, as advancements in satellite technology often inspired psychical researchers to develop new techniques, and advancements in psychical research often prefigured innovations in satellite technology.[6] The parallels between these covert military programs are particularly illuminating because they illustrate the strategic objectives of satellite surveillance, which was designed to reframe the entire planet as a global theater of operations. They also demonstrate how the satellization of the planet was thought to penetrate into the interiority of the self by transforming human subjects into mechanized and automated extensions of the military-technological gaze, as psychic spies were understood as being capable of objectively recording and relaying information without conscious mediation. Both of these programs were thus guided by the ideology of Cold War rationality, which sought to automate the process of intelligence gathering and analysis by replacing humans with intelligent machines.

I Remote sensing

In order to understand the parallels between satellite research and psychical research it is first necessary to demonstrate how the history of remote sensing, or the technical measurement of physical properties from a distance, was largely synonymous with the history of aerial reconnaissance. Aerial reconnaissance began in the late eighteenth century, when the French army realized the strategic potential of hot air balloons at the battle of Fleurus (1794), the siege of Mantua (1796–97) and the invasion of Egypt (1798).[7] In the nineteenth century, balloons were employed for the purpose of aerial photography, which led to experiments in photographic topography and aerial photogrammetry. In the 1860s, for example, French army engineer Aimé Laussedat (1819–1907) suspended a camera from an unmanned balloon and used the images to create a bird's-eye map of Paris. The US Army also created a special Balloon Corps, and during the American Civil War this unit reported on enemy troop movements and performed geographical surveys that enabled more accurate mapping of battlefields. In the 1880s a military school of ballooning was established in the United Kingdom as well, and during the Second Boer War (1899–1902) balloons were used to observe enemy troop movements, direct artillery fire and produce more accurate maps of South Africa.[8]

The French army began taking reconnaissance photographs from airplanes during the Agadir Crisis (1911), but the most significant advances in aerial reconnaissance photography occurred during the First World War, including the development of motorized cameras, filters to penetrate haze and dampers to reduce aircraft vibration. The analysis of reconnaissance photographs also became more sophisticated, as overlaps between photographs were used to create stereoscopic images that enabled analysts to identify trenches, ammunition stockpiles, camouflaged batteries and other forms of enemy activity.[9] The Second World War brought further advances in photographic intelligence gathering, as aerial reconnaissance units became capable of rapidly mapping territories using automatic cameras. The development of early rocket photography was similarly employed for the purpose of military mapping.[10]

Following the Second World War, the United States routinely made use of aerial reconnaissance to assess the military strength of the Soviet Union. In 1956 the United States implemented the Genetrix program, which used meteorological balloons that reached altitudes of 15,000–30,000 meters. Due to their low rate of success and the political problems caused by these overflights, President Dwight D. Eisenhower (1890–1969) approved the construction of high-altitude U-2 spy planes, which was known as the Aquatone program. However, these planes could only fly over the edges of countries and were extremely vulnerable, which was made particularly apparent in 1960 when U-2 pilot Francis Gary Powers (1929–77) was shot down over Soviet territory. The United States subsequently agreed not to fly reconnaissance planes over the Soviet Union and other communist nations, which were declared 'denied territories.' Eisenhower then decided to devote military resources to the development of satellites, which were not covered by the 'denied territories' agreement. The CIA was placed in charge of designing them, while the Air Force was given the responsibility of launching them into orbit. The development of satellite technology was thus primarily driven by military interests.[11]

The Air Force initially planned to develop a Manned Orbiting Laboratory (MOL) that would allow astronauts to photograph strategic sites in the Soviet Union. This project would have been a natural extension of earlier forms of aerial reconnaissance, but it was cancelled due to the exorbitant costs of a manned mission. Instead, the CIA focused on the development of automated satellite surveillance systems that could be operated remotely. The CIA's first satellite surveillance system, which was known as the CORONA program, was designed to track the production and location of long-range ballistic missiles in the Soviet Union, and its first satellite, later designated Keyhole-1 (KH-1), carried a single panoramic camera with a resolution of twelve meters. The satellite would spin along its main axis so that it would remain stable and the camera would take photographs only when pointed at the earth. When the entire roll of film was exposed, the canister was ejected using a reentry capsule that could be retrieved by a passing airplane. The first 13 attempts to launch these satellites and retrieve the capsules failed, but the 14th satellite provided more intelligence than all previous U-2 overflights combined.[12]

There was a series of improvements in camera and guidance systems between 1960 and 1972. By 1963 the KH-2 and KH-3 cameras had achieved a resolution of three meters. Beginning with the KH-3 satellite, side thrusters were also used to align the satellite with the earth so that the camera pointed in the right direction at all times. The KH-4 satellite deployed a pair of higher-resolution cameras that rotated in synchronization. By 1967 these cameras had a resolution of 1.8 meters and could be employed simultaneously to produce three-dimensional stereoscopic images.[13] The most significant problem with these satellites was the delay between recording and analysis, as it could be days or even weeks after a photograph was taken before the film capsule could be retrieved, and many of them never were. In order to address this problem the Air Force developed an alternative program in 1961 that was named SAMOS (Satellite and Missile Observation System). These satellites were designed to develop film in orbit and transmit TV scans of the photographs to earth, but they failed to provide any useful intelligence due to the poor quality of the scans.[14]

Subsequent satellite systems introduced a wide range of technological improvements designed to increase mobility, capacity, resolution and range. The immediate successors to SAMOS, the KH-5 Argon satellites (1961–64), were primarily designed for mapping, and they traded lower image resolution for greater coverage. The KH-6 Lanyard satellites, which orbited during the same period, used a higher-resolution camera with a longer focal length lens. These two systems were designed to be used in tandem, as low-resolution, wide-area images from a KH-5 could identify targets for high-resolution reconnaissance by a KH-6.[15] The KH-7 (1963–67) and KH-8 (1966–84) Gambit satellites employed cameras with a 6.4-kilometer wide coverage area and a resolution of 90 centimeters. These satellites could also transmit scans using a 'film read-out' feature, although the resolution was often poor. The KH-9 HEXAGON satellites (1971–86) were able to cover a wider area, and a larger film capacity enabled longer missions. As a result, they were the first satellites to photograph every square foot of the planet.[16] The KH-11 KENNEN satellites (1976–) employed a digital sensor and charge-coupled device (CCD), which enabled the electronic transmission of film-quality images from orbit without the need to retrieve film capsules. This meant that the satellites could remain in orbit for years rather than days, and they were reportedly used to locate hostages during the Iran hostage crisis in 1980 and plan the air attack on Libya in 1986.[17]

At the same time that the CIA was developing satellite surveillance systems, the US military was also working on space surveillance systems to track enemy satellites. The first space surveillance system, Minitrack, predated the launch of Sputnik 1 in 1957, although it was not able to track 'dark' or 'passive' satellites that did not employ standard international frequencies. In addition, the Baker-Nunn telescopic camera systems were used to photograph and identify orbiting satellites, although they were only effective at night in clear weather. The need for an improved satellite tracking system led to the creation in 1960 of the Space Track network, which integrated data from roughly 150 electro-optical, radio and radar sensors, and the Navy's Space Surveillance System (SPASUR), which reflected signals from orbiting objects and then calculated their trajectories using an electronic computer. In 1961 Space Track and SPASUR became part of NORAD's Space Detection and Tracking System (SPADATS), and there were a number of improvements to this system in the 1970s and 1980s, including the Cobra Dane phased array radar in the Aleutians, the Perimeter Acquisition Radar from the cancelled

Safeguard anti-ballistic missile system and the PAVE Phased Array Warning System.[18]

The 1990s was a period of transition, as the US military sought to redirect its space efforts towards the real-time enhancement of military operations using the Global Positioning System (GPS). During the Persian Gulf War (1990–91), for instance, handheld GPS devices allowed US ground troops to navigate the desert areas of southeastern Iraq, and bombers delivered air-launched cruise missiles whose navigation systems employed GPS. General Thomas Moorman (1940–) later wrote that the war 'opened the eyes of senior military leaders' to the military value of space, and Air Force Chief of Staff Merrill A. McPeak (1936–) even described the conflict as the 'first space war.'[19] The tactical use of GPS technology thus demonstrated that the operational value of satellites was not limited to surveillance, as they could simultaneously function as strategic orbital platforms from which military maneuvers could be directed.

While satellite surveillance was clearly designed for the purpose of gathering military intelligence, theorists argue that it also had a significant impact on how the planet was imagined. Unlike Brand, who used satellite images of the earth to promote the idea of global ecological interdependence, or Lovelock, who was inspired by these images to imagine the planet as a single, living organism, Canadian media theorist Marshall McLuhan (1911–80) argued that they actually transformed the world into a 'global theater': 'Since Sputnik and the satellites, the planet is enclosed in a manmade environment that ends "Nature" and turns the globe into a repertory theater to be programmed.'[20] Instead of reinforcing a utopian narrative of global unity and harmony, McLuhan's concept of 'global theater' thus had more in common with French philosopher Jean Baudrillard's (1929–2007) theory of simulations, which was also inspired by the development of satellite technology. In his 1981 book *Simulacres et simulation*, for example, Baudrillard argued that the 'satellization' of the planet generated 'models of simulation' and 'planetary control,' where no one was free – not even the superpowers that possessed the technology:

> By the orbital establishment of a system of control [...] all terrestrial microsystems are satellized and lose their autonomy. All energy, all events are absorbed by this excentric gravitation, everything condenses and implodes on the micro-model of control alone [...]. [E]very principle of meaning is absorbed, every deployment of the real is impossible.[21]

Like McLuhan, therefore, Baudrillard argued that satellite vision replaced the real with a technological simulation, as it converted the surface of the planet into an informational grid. French philosopher Paul Virilio (1932–2007) similarly suggested that the gaze of the satellite reflected 'the aesthetics of the electronic battlefield, the military use of space whose conquest was ultimately the conquest of the image: the electronic image of remote detection, the artificial image produced by satellites as they endlessly sweep over the surface of continents drawing automatic maps.'[22] The practice of remote sensing thus inspired Cold War-era theorists to conceive of satellites not as communications technologies but rather as instruments of vision, knowledge and control, as they transformed every space on the globe into an image and replaced the conquest of territories with the conquest of images.

Following the Cold War, many theorists continued to conceive of satellite surveillance as a technological gaze that exerted power over the entire surface of the planet. For example, American media theorist Mark Poster (1941–2012) described satellites as 'a Superpanopticon, a system of surveillance without walls, windows, towers, or guards.'[23] Some theorists were keenly aware that satellites were first and foremost military technologies, and they emphasized the tactical significance of satellite images. For example, Irish political theorist Gearóid Ó Tuathail (1962–) argued that 'daily electronic mappings of the surface of the globe are produced by spy satellites, AWACS (Advanced Warning Airborne Command System) planes, stealthy aircraft, and new generations of unmanned aerial vehicles,' and this 'organization of the surface of the earth into digitized information [...] has technologized geo-power to an unprecedented degree,' as 'everything that can be seen can be destroyed.'[24] Like Baudrillard and Virilio, therefore, Tuathail concluded that the 'real' was being replaced by 'information,' which transformed every space on the globe into a potential military target.

However, few theorists addressed the significance of the fact that these satellites were unmanned and that human intelligence gathering was being replaced by automatic opto-electronic processes. For instance, the information provided by surveillance satellites was considered to be reliable because it was arbitrary, as satellites acquired information about all kinds of places for no particular reason. The process of analyzing satellite reconnaissance photographs also became increasingly automated, as operators developed computer software to correlate satellite positions and photographs which 'provided almost automatic geo-referencing of the photographs and their image features.'[25] The idea of satellite surveillance as objective was thus dependent on the mechanical nature of the photographic apparatus, which 'embodied a positive ideal of the observer' as 'patient, indefatigable, ever alert.'[26] Practically, it was dependent on the computerization of photogrammetry, which allowed these devices to record, measure and graph the surface of the planet without human intervention or oversight. The mechanization and automation of intelligence gathering and analysis thus promoted the technological simulation of the real as well as the disappearance of the human.

These developments provide a vivid illustration of what historians have called 'Cold War rationality.' It was the product of a particular moment in US history when 'technology had intensified human capacities for destruction, while simultaneously accelerating the pace of decision making beyond humans' ability to reason effectively.' In order to avoid global annihilation and enable rapid problem-solving, scientists sought to develop algorithms and rules that could guide decision-making processes in the absence of human subjects. Advocates of Cold War rationality were thus guided by a belief that machines 'reason better than human minds' and that they would successfully determine the most efficient means of attaining certain prescribed goals.[27] The practice of remote sensing was a natural extension of this concept, as the optimization of satellite surveillance required the elimination of human intervention through the mechanization and automation of intelligence gathering and analysis. The end result of these automated opto-electronic processes was the realization of a global theater of war dominated by intelligent machines, in which no one was free – not even the superpowers responsible for the development and deployment of these systems.

II Remote viewing

Unlike 'remote sensing,' which refers to technical processes, 'remote viewing' refers to the ability to 'access and describe, by means of mental processes, remote geographical locations [...] along with the real-time activities of persons at the target site.'[28] While this practice was derived from older psychic techniques, like clairvoyance and astral projection, the use of this term first emerged in the United States at a particular historical moment when researchers were beginning to promote the potential military applications of psychic phenomena. In their 1970 book *Psychic Discoveries behind the Iron Curtain*, for example, Sheila Ostrander and Lynn Schroeder claimed that Soviet intelligence agencies were conducting a nationwide search for the most talented psychics, whose gifts would be employed against the West, and that they had already established 'twenty or more centers for the study of the paranormal with an annual budget estimated in 1967 at over 12 million rubles ($13 million).'[29] Hence, Ostrander and Schroeder urged the United States to develop a comparable research program, as the race for outer space was equally as urgent as the race for inner space: 'If Westerners had bothered to read Soviet publications in the 1950s, we would have seen that much data on the development of Sputnik was published long before it shot into space and astounded the world. Today we are still not keeping up with material readily available in Soviet publications and scientific papers, particularly in the field of parapsychology.'[30]

Ostrander and Schroeder's book had a tremendous impact on the US intelligence community, as it suggested that psychical research could potentially influence the outcome of the Cold War. In July 1972 the Defense Intelligence Agency (DIA) commissioned a report, which noted that 'the major impetus behind the Soviet drive to harness the possible capabilities of telepathic communication, telekinetics, and bionics are said to come from the Soviet military.'[31] The report also described the potential consequences of the Soviet Union's psychic superiority:

> Soviet efforts in the field of psi research, sooner or later, might enable them to do some of the following: (a) know the contents of top secret U.S. documents, the movements of our troops and ships and the location and nature of our military installations, (b) mold the thoughts of key U.S. military and civilian leaders at a distance, (c) cause the instant death of any U.S. official at a distance, (d) disable, at a distance, U.S. military equipment of all types, including spacecraft.[32]

While the US military was already worried about the possibility of a 'missile gap,' it now faced the additional possibility of a 'psi gap.'[33]

As a result of this report, US intelligence agencies began to monitor the progress of psychical research in the Soviet Union. In April 1972 CIA personnel from the Office of Strategic Intelligence (OSI) met with a member of the Stanford Research Institute (SRI) named Russell Targ (1934–) to gather information on the latest developments in Soviet psychical research. In addition to his work on laser physics, Targ had initiated several studies on psychical research and had even developed an 'ESP-teaching machine,' which he had presented at a NASA conference on speculative technology earlier that year.[34] Targ was eagerly looking for research funding, and he was pleased to discover that many of the other conference participants believed in psychic phenomena, including astronaut Edgar Mitchell, rocket pioneer Wernher von Braun (1912–77), NASA's New-Projects Administrator George Pezdirtz (1933–) and NASA Director James Fletcher (1919–91). Fletcher was particularly concerned that the Soviets were ahead of the Americans, and NASA subsequently offered Targ a contract to start a research program at SRI, which brought him to the attention of the CIA.[35]

In addition to providing research material on Soviet psychical research, Targ also submitted a report on an experiment with an American psychic named Ingo Swann (1933–2013). OSI then contacted the Office of Research and Development (ORD), whose officers visited SRI in August 1972. They explained that there was 'increasing concern in the intelligence community about [...] Soviet parapsychology being funded by the Soviet security services' and that 'they had been on the lookout for a research laboratory outside of academia that could handle a quiet, low-profile classified investigation.' Targ's colleague Harold E. Puthoff (1936–) agreed to organize a demonstration with Swann. The officers were evidently impressed, as their report to the Office of Technical Services (OTS) recommended that SRI's research should be continued and expanded.[36] In October 1972 OTS contracted SRI for an eight-month pilot study, and Kenneth A. Kress was assigned to monitor their progress.[37] Kress later confirmed that this decision was driven by fears of a 'psi gap,' as it 'was known that the Soviet government was supporting the evaluation and development of paranormal phenomena.'[38]

Swann had joined SRI earlier that year to conduct a series of 'beacon' or 'outbounder' experiments, in which he attempted to visualize the locations of remote targets.[39] Puthoff and Targ decided to use the CIA funding to conduct another series of experiments, in which they placed nine 'remote viewers' in isolated rooms while targets were sent to randomly selected locations in the San Francisco Bay Area. The viewers were then asked to describe the target locations, and according to Puthoff and Targ the initial results were encouraging. Swann allegedly described the Palo Alto City Hall, including details like the shape of the windows, the number of trees, the designs on the pavement and the fact that the fountain was not running.[40] Another viewer named Pat Price (1918–75) reportedly described a regional landmark (the Hoover Tower at Stanford University) and a swimming pool complex in Palo Alto.[41] In another experiment viewers were asked to describe Puthoff's location each day at 1:30 p.m. while he was traveling through Central and South America. Twelve descriptions were collected before his return, including one from Targ himself, who described an 'ocean at the end of a runway' and an airport building with a large rectangular overhang, which proved to be correct (Figure 10.1).[42] In each of these experiments, Puthoff and Targ noted the similarities between remote viewing and remote sensing. According to Targ, for example, Price's out-of-body experiences closely resembled aerial reconnaissance:

> When he looked at the various sites he was able to describe them for us, it was as though he came zooming in from thousands of feet in altitude, scanned the bay area until he found the people at the target site, and then described what they were looking at. It was not as though he was reading their minds, because frequently he would describe items at the site that the outbound experimenters hadn't even seen.[43]

Another remote viewer, Paul Smith, similarly noted that 'viewers frequently described additional details beyond what the beacon team actually at the target was able to observe, which nevertheless turned out to be true.' These details often 'could not have been observed by the people at the site, such as closed-off areas or tops of buildings.'[44]

Despite their success, Swann feared that this technique would not appeal to members of the intelligence community, as it could only be used to observe sites where an agent had already been placed, and analysts would most likely consider the agent's firsthand knowledge more valuable than the impressions of a remote viewer. In April 1973 he proposed a new method called 'coordinate remote viewing' (CRV), which employed geographical coordinates to identify target sites. Puthoff and Targ conducted the first CRV experiments later that month. The preliminary results were encouraging, as Swann accurately described a joint French-Soviet meteorological station on Kerguelen Island in the southern Indian Ocean (Figure 10.2).[45] The coordinates had been provided by a contact at OSI, and it was considered a military target because the Soviets were rumored to be using the station to track missiles.[46] The CIA report later confirmed that Swann's description of the installation was 'correct to the limits of KH-4 photography.'[47]

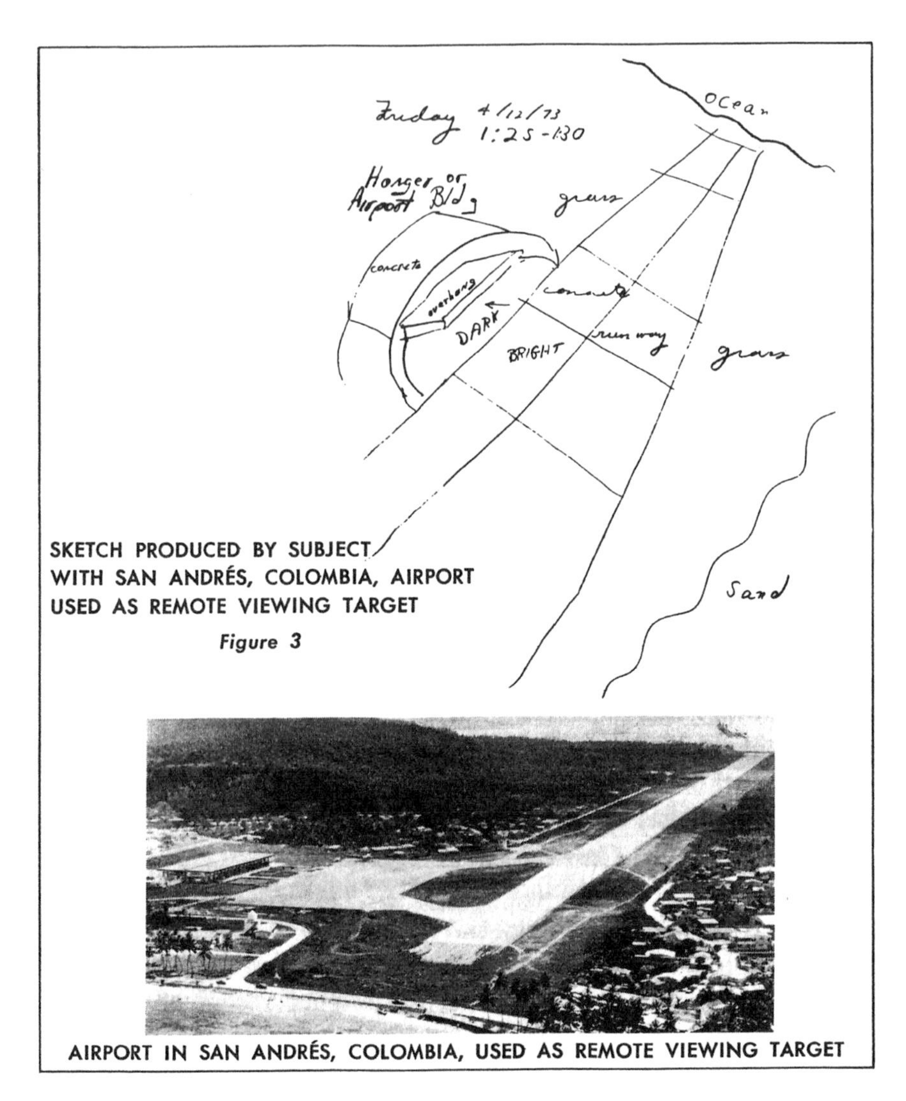

Figure 10.1 Sketch produced by Russell Targ (1934–) with San Andrés, Colombia, airport used as remote viewing target, 1973.

Source: Harold E. Puthoff and Russell Targ, *Mind-Reach: Scientists Look at Psychic Abilities*, Charlottesville: Hampton Roads, 2005, 12. Courtesy of Hampton Roads Publishing/Red Wheel/Weiser.

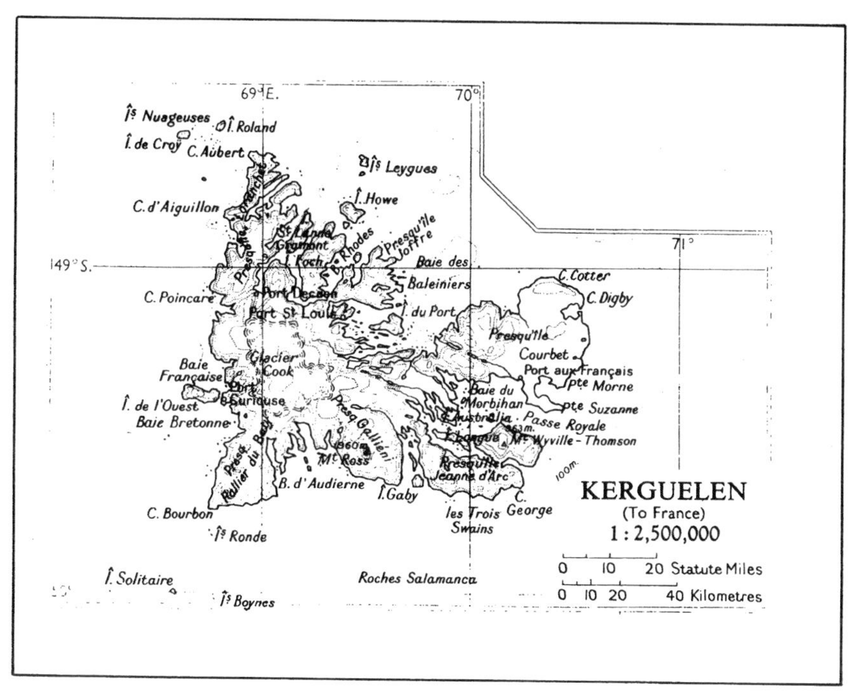

KERGUELEN ISLAND

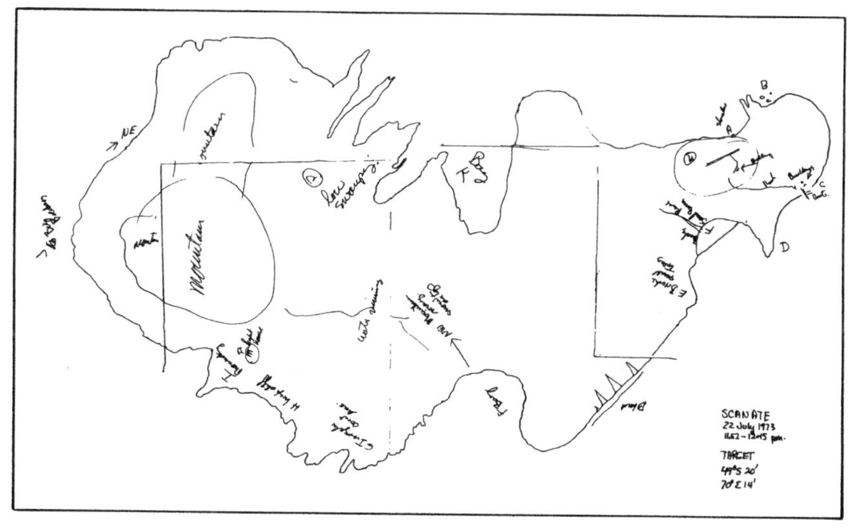

DRAWING BY SWANN OF KERGUELEN ISLAND

Figure 10.2 Sketch produced by Ingo Swann (1933–2013) with Kerguelen Island in the southern Indian Ocean used as remote viewing target, 1973.

Source: Puthoff and Targ, *Mind-Reach*, 32. Courtesy of Hampton Roads Publishing/Red Wheel/Weiser.

In May 1973 Puthoff arranged a demonstration of this new technique, and Christopher 'Kit' Green provided coordinates for a cabin in West Virginia that belonged to one of his colleagues. Instead of seeing a cabin, however, Swann produced a map that featured 'rolling hills, a curved driveway, a cluster of buildings, and an underground bunker' (Figure 10.3). Several days later, the same coordinates were given to Price, who provided more details: 'Price [...] offered to "go inside" the bunker, where he found a file cabinet with names on the drawers. He read off the names and gave us the code name ("Hay Stack") of the facility.'[48] These psychics were effectively working in tandem, like KH-5 and KH-6 satellites, as Swann performed a 'low-resolution' scan to determine the topographical features of the area, while Price performed a 'high-resolution' scan of a particular structure, which provided more detailed information. Furthermore, this demonstration illustrated the perceived advantages of CRV, as it not only enabled virtually unlimited levels of resolution, but it was also capable of penetrating into the interiors of buildings. Green initially believed that the experiment was a failure, but he was intrigued by the similarities between the two descriptions, so he drove to his colleague's cabin and was surprised to discover that a secret National Security Agency (NSA) facility was located a few kilometers away: 'Much of it was underground, and nuclear-hardened, because it would be one of the first targets the Soviets hit if war broke out.'[49] Kress concluded that Swann's 'information concerning the physical layout of the site was accurate' and that Price had provided the actual 'codename of the site' as well as a 'list of project titles associated with current and past activities including one of extreme sensitivity.'[50]

Deputy Director for Intelligence John N. McMahon (1929–) considered CRV to be an 'extremely attractive' approach to intelligence gathering, and he was eager to move from experimentation to application.[51] In 1974 he provided coordinates for a 'Soviet site of great interest' in the Kazakh Soviet Socialist Republic adjacent to the Semipalatinsk nuclear test area, which was known within the intelligence community as URDF-3 (Unidentified Research and Development Facility-3). Price described the layout of the complex, which included a giant gantry crane and several tall, silo-sized cylinders. Inside one of the buildings he also saw a large interior room where a 'sixty foot-diameter metal sphere' was being put together from 'thick metal gores' that resembled a giant orange peel. Satellite reconnaissance photographs confirmed the presence of a gantry crane (Figure 10.4) as well as several gas cylinders that matched Price's sketch, but no sixty-foot sphere could be seen. In their final report, Puthoff and Targ concluded that 'the exceptionally accurate description of the multi-story crane was taken as indicative of probable target acquisition.'[52] Three years later, *Aviation Week* published an article that described the spheres assembled at Semipalatinsk: '[H]uge, extremely thick steel gores were manufactured [...]. These steel segments were parts of a large sphere estimated to be about 18 meters (57.8 feet) in diameter [...]. U.S. officials believe that the spheres are needed to capture and store energy from nuclear driven explosives or pulse power generators.'[53] Targ later described the accuracy of Price's description as a 'miracle,' and he claimed

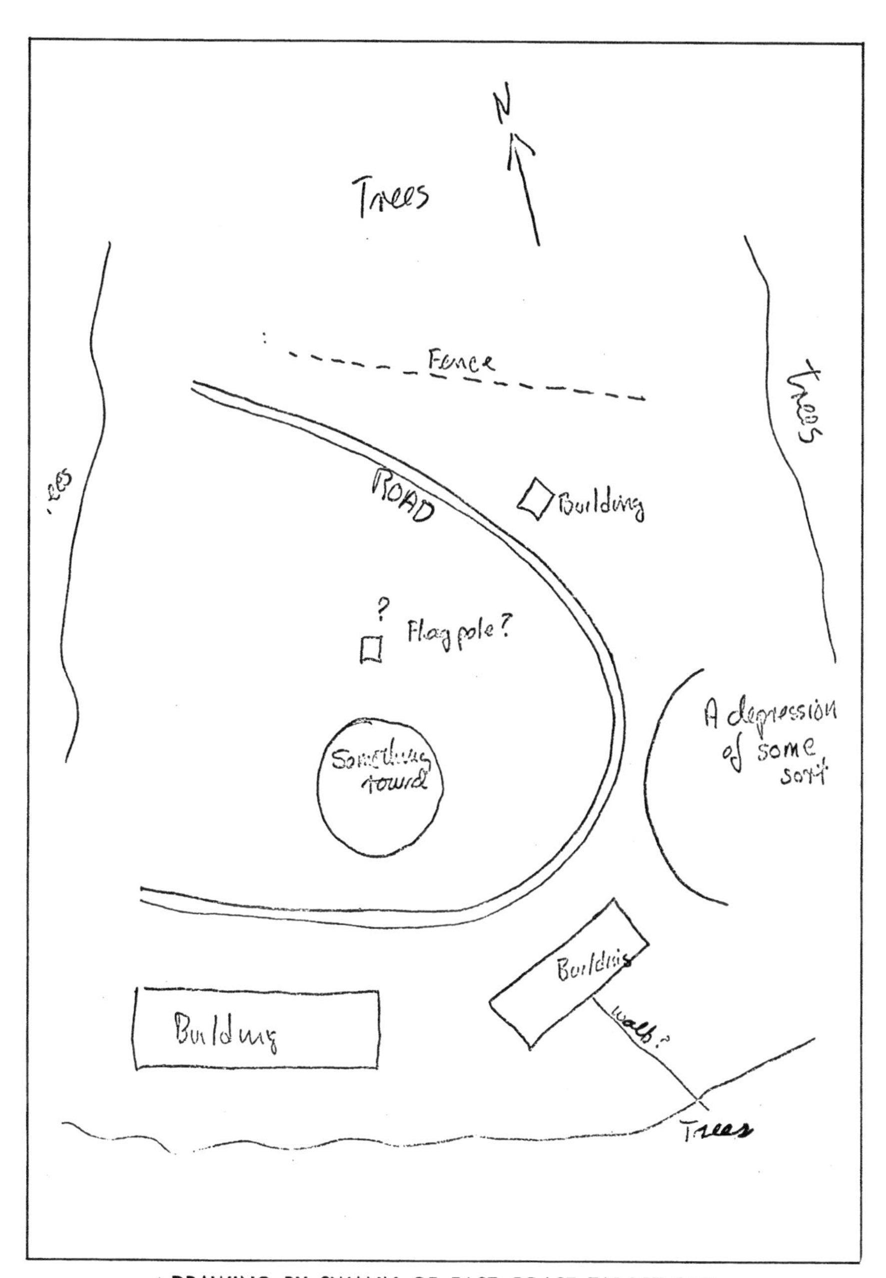

Figure 10.3 Sketch produced by Ingo Swann with National Security Agency facility in West Virginia used as remote viewing target, 1973.
Source: Puthoff and Targ, *Mind-Reach*, 3. Courtesy of Hampton Roads Publishing/Red Wheel/Weiser.

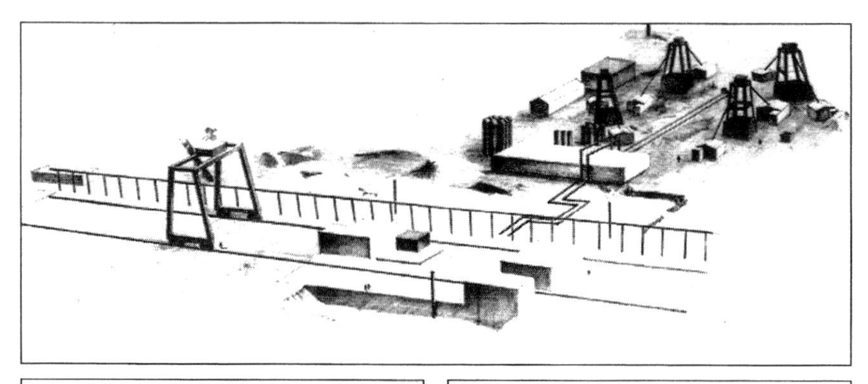

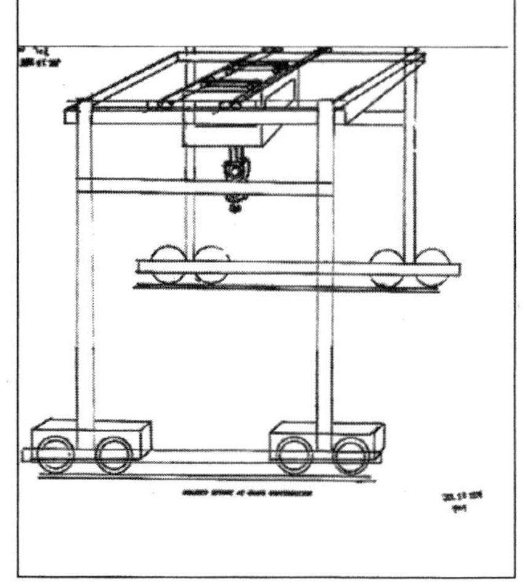

Figure 10.4 Sketch produced by Pat Price (1918–75; right) with Unidentified Research and Development Facility in the Kazakh Soviet Socialist Republic used as remote viewing target and sketches of satellite reconnaissance photographs (above and left) of the same facility, 1974.
Source: Puthoff and Targ, *Mind-Reach*, 3. Courtesy of Hampton Roads Publishing/Red Wheel/Weiser.

that it 'was so outstanding that it alone assured our funding for the next several years.'[54] This episode thus shows not only that remote viewing was verified through the use of satellite surveillance but also that it was understood as a more efficient and cost effective method of gathering intelligence, as it took years for satellites to gather the same information that Price allegedly provided in a matter of minutes.

Despite Price's apparent success, OTS officers remained skeptical as to whether this technique had any operational value. As a result, they organized another demonstration in which Price was given the coordinates of two foreign embassies whose interiors were already known. Price was instructed to view the embassies remotely, locate the code rooms and describe any distinguishing features. According to Kress, Price 'correctly located the code rooms' and 'produced copious data, such as the location of interior doors and colors of marble stairs and fireplaces that were accurate and specific.' The operations officer consequently concluded that 'this technique – whatever it is – offers definite operational possibilities.'[55] Price was then given coordinates for a site in Libya, and he produced a map of the installation as well as a related underwater sabotage training facility several hundred kilometers away. This information was then given to the Libyan Desk, which evaluated the report after obtaining satellite reconnaissance photographs. According to Kress, Price's descriptions were once again accurate, although he died several days later of a heart attack.[56] The fact that these tests could only be verified through the use of satellite reconnaissance photographs illustrates once again how the practice of remote viewing was dependent on satellite surveillance.

In 1976 the Air Force's Foreign Technology Division (FTD) at Wright-Patterson Air Force Base in Dayton, Ohio, agreed to fund SRI on an exploratory basis. FTD's contract allowed Puthoff and Targ to publish several unclassified reports in academic journals, which primarily focused on the physical properties of remote viewing phenomena. According to Puthoff and Targ, the information received by remote viewers could not be transmitted using conventional electromagnetic waves because 'an increase in the distance from a few meters up to 4000 km [...] does not in any apparent way degrade the quality or accuracy of perception.'[57] They concluded that the information must be transmitted using extremely low frequency (ELF) waves, which would explain the low bit rates (0.005–0.1 bits/sec) and the ineffectiveness of electromagnetic shielding. In order to make these calculations, it was necessary to apply information theory to psychic phenomena, as this theory made it possible to quantify psychic performance regardless of the underlying mechanism: 'Observation of the phenomenon implies the existence of an information channel in the information-theoretic sense. Since such channels are amenable to analysis on the basis of communication theory techniques [...] channel characteristics such as bit rate can be determined independent of a well-defined physical channel model.'[58] In order to develop military applications for psychical research, it was thus necessary to evaluate psychic abilities according to the technical standards of communications technologies.

In 1978 Lieutenant Frederick Holmes 'Skip' Atwater proposed that the Army's Intelligence and Security Command (INSCOM) create its own 'in-house' remote viewing team at Fort Meade, Maryland, which became known as the Gondola Wish program.[59] By the end of 1978 government-funded remote viewing experiments were therefore being conducted at both Wright-Patterson Air Force Base and Fort Meade. In March 1979 both groups were tasked with locating a Soviet bomber that had crashed in Zaire. The bomber had been refitted as a reconnaissance platform, and several countries were attempting to retrieve its contents. A viewer named Rosemary Smith reportedly identified the location of the wreckage to within five kilometers, and US forces were able to recover it, which once again demonstrated the superiority of remote viewing over conventional aerial reconnaissance.[60]

Following a congressional investigation, Representative Charles Rose (1939–2012) became an avid supporter of the program. In a 1979 interview Rose not only expressed his belief in remote viewing, but he also revealed that these experiments had been verified by aerial reconnaissance photographs: 'What these people "saw" was confirmed by aerial photography. There is no way it could have been faked.' Consequently, Rose saw remote sensing and remote viewing as complementary methods of intelligence gathering, but the primary advantage of remote viewing was that it provided 'a hell of a cheap radar system.' He also emphasized the importance of remaining competitive with the Soviet Union: '[T]he Russians are very interested in psychic phenomena [...]. They have a national screening program to detect [...] psychic abilities in schoolchildren. The CIA, on the other hand, spends next to nothing in this area [...]. And if the Russians have it and we don't, we are in serious trouble.'[61] To this end, he attempted to secure funding for a 'psychic neuron disruptor,' which was designed to induce temporary insanity by interfering with the connections between neurons in the brain, and he described the effect as 'similar to that of hallucinogenic drugs such as LSD.'[62] He was so convinced of the strategic importance of psychical research that he even urged the US government to develop a 'psychic Manhattan Project' that would restore America's superiority in the field.[63] These claims not only show that psychic or inner space was conceived as a site of military conquest and competition, but they also reveal how attempts to weaponize psychical research were often described as military applications of the US counterculture.

INSCOM subsequently contracted SRI to screen potential viewers. They selected six recruits from a pool of 117 personnel. Two of these recruits – Melvin C. Riley (1946–2020) and Hartleigh Trent – had previously worked as satellite reconnaissance photograph analysts at Fort Meade and the National Photographic Interpretation Center. Their skill at analyzing photographs was thought to be directly related to their psychic abilities. According to historian Jim Schnabel, for example, Riley's exceptional performance as a photograph analyst demonstrated his heightened sense of perception:

> As a photo analyst, he seemed able to see things in a reconnaissance image that no one else could see. One of his commendations came after he had analyzed a photo of an East German industrial area and found what seemed to be a new Soviet artillery piece, [...] which hadn't yet been seen in that particular area [because it] was actually hidden by a canvas canopy. But Riley somehow had a feeling about what it was, and sketched it anyway. Satellite photographs later confirmed the sketch.[64]

In other words, Riley's psychic abilities appeared to explain his skill as a photo analyst, which shows how remote sensing and remote viewing were inextricably connected.

In 1979 the DIA's Scientific and Technical Intelligence Directorate, headed by Jack Vorona, was chosen to oversee all remote viewing efforts. The two groups were then consolidated and the program was redesignated Grill Flame.[65] In November 1979, during the Iranian hostage crisis, this newly consolidated group was tasked with locating the kidnapped diplomats, and a viewer named Joseph McMoneagle (1946–) was reportedly able to 'describe the location where three of the hostages had been taken.'[66] Viewers were also tasked with locating General James L. Dozier (1931–) after he was kidnapped by the Italian Red Brigades in 1981. McMoneagle provided a vivid description of one of these sessions:

> I started the session with an almost perfect image of a coastline, the right-hand side of Italy toward the north. [...] I followed my instinct and began moving away from the coastal city due west along the main road [...]. I suddenly found myself hovering directly over a fairly large town not far from the coast and just south and southeast of a very large mountain range. [...] I produced an overall map depicting the location of the city, seemingly specific enough to say that the city was Padua [...]. I then sketched as best I could the city street map that I had seen [...]. I pointed out the location of the apartment house on the street map, and said that the apartment he was being held in was on the second floor.[67]

McMoneagle thus described the experience of remote viewing as a form of psychic aerial reconnaissance, as he was able to shift from a wide, orbital view of the entire territory to a narrower, more detailed view of a specific city. He also claimed to have provided the geographical location of the city as well as a detailed street map, which was allegedly drawn to scale. He even surpassed the capability of surveillance satellites by identifying the precise address and apartment where the target was located. Although this information was never forwarded to the Italian authorities, he insisted that 'all of my drawings were accurate and would have contributed significantly to finding the building and floor within which he was being held.'[68]

When the Grill Flame program was terminated in 1982, Major General Albert Stubblebine (1930–2017), commander of INSCOM, assumed responsibility for the program, which was redesignated Center Lane. After Stubblebine's retirement in 1984, Vorona arranged for the unit to be transferred to the DIA's Scientific and Technical Intelligence Directorate, where it was redesignated Sun Streak.[69] It was at this time that the unit began to focus increasingly on space surveillance, as declassified government documents reveal that several viewers were tasked with locating and identifying various orbital structures, including the Spacenet 3 satellite, which was launched in 1985, and the Soviet Mir space station, which was launched in 1986.[70] According to these documents, viewers were able not only to determine the position of these structures, but also to describe their shape, color and movement. These experiments thus paralleled the development of space surveillance systems like Space Track and SPASUR, which were similarly designed to track and identify passing satellites.

In 1990 Dale Graff became head of the program, which was redesignated Stargate. When the Persian Gulf War began, there were only four remaining viewers. During Operation Desert Storm they were tasked with locating Scud missile launchers and Iraqi army units, which were torching oil wells in Kuwait.[71] David Morehouse (1954–) described one of these sessions as follows:

> I started moving in large circles, surveying the ground beneath me. [...] The tracks of hundreds of vehicles scarred the sand, almost all going north or northwest. I followed them [...] until I found a spot where [...] I could get close enough to glimpse the source. [...] Everywhere, as far as I could see, blazing torches sprang out of the ground, belching flame and smoke.[72]

Like McMoneagle, Morehouse described remote viewing as a form of psychic aerial reconnaissance, yet in this case the operational value of his intelligence seemed questionable. As he explained: 'Obviously they know about this — who could miss it?' While remote viewing had long been considered faster, cheaper and more effective than remote sensing, it was apparently unable to provide useful real-time tactical information for soldiers who were actively engaged in combat. The inability of remote viewing to provide this information, coupled with the end of the Cold War, eventually led to the cancellation of the program in 1995.[73]

Overall, government support for psychical research in the United States during the 1970s and 1980s was predominantly motivated by fears of a 'psi gap' with the Soviet Union, which forced researchers to emphasize the potential military applications of psychic phenomena. Furthermore, psychical research was only able to receive government support at this time because it seemed to complement satellite research, which was informed by the same strategic objectives. Unlike satellite reconnaissance, however, psychic reconnaissance was eventually cancelled, as the analysis of psychic intelligence could not be automated as efficiently as that of satellite intelligence and psychics were unable to meet the growing demand for real-time tactical information. As a result, the aspects of remote viewing that had made it seem so attractive to the intelligence community during the Cold War – namely its speed, mobility and low cost – no longer seemed relevant.

III The satellization of the self

This chapter has illustrated the striking parallels between the development of satellite research and psychical research in the United States during the Cold War. Although satellites were considered more reliable and were given far more funding than psychics, they were both conceived as mobile platforms that facilitated aerial reconnaissance by mapping 'denied territories' and locating potential targets. In addition, researchers claimed that satellite surveillance and psychic surveillance employed complementary methods of image analysis, as photo analysts and remote viewers were both capable of perceiving details that were not accessible to ordinary observers. Remote sensing and remote viewing can thus be seen as parallel extensions of the military-technological gaze, which transformed the surface of the planet into a global theater of operations while simultaneously monitoring and controlling outer space by tracking the locations and trajectories of orbital structures. The most significant difference between these practices – and the main reason why satellite surveillance was ultimately seen as more reliable and objective – was the fact that remote sensing was a mechanized and automated process, while remote viewing placed human subjects firmly at the center of intelligence gathering and analysis. This difference suggests that remote viewing may have been motivated by a desire to reassert the autonomy and integrity of the individual subject at a time when humans were being increasingly displaced by intelligent machines. Researchers such as Targ and Keith Harary (1953–) argued, for example, that 'when we find ourselves confronted with exploding technology and increasing alienation, psychic functioning can help keep us aware of the difference between human beings and machines.'[74]

While the practice of remote viewing may appear to challenge Cold War rationality by embodying the military-technological gaze in human subjects, it is important to note that researchers justified their requests for funding by emphasizing the objective nature of psychic phenomena. These justifications required the use of information theory to present remote viewing as a fundamentally mechanical and automatic process. Like remote sensing, for instance, remote viewing was described as objective and accurate because the information it provided was allegedly recorded without the conscious mediation of the viewer. Puthoff and Targ even emphasized that viewers should avoid reflecting on what they see, as 'their unanalyzed perceptions are almost always a better guide to the true target than their interpretations of the perceived data.'[75] The results of these experiments were quantitatively analyzed to determine the range, accuracy and channel capacity of viewers, much like an engineer would evaluate the performance of a technical apparatus. Indeed, many viewers were keenly aware that they were being treated like machines, as Paul Smith noted in his autobiography: 'It suddenly dawned on me that, as a remote viewer, I was not a soldier or a federal employee — I was a piece of equipment. My colleagues and I were "turned on" just like any other piece of equipment and "turned off" when not needed.'[76] The claim that remote viewing was an objective and reliable form of intelligence gathering was thus based on the presumably mechanical and automatic nature of psychic phenomena. In other words, it was not only conceived as a psychic technique that would fulfill the needs of the intelligence community, but it also embraced the logic of Cold War rationality, which was based on the fundamental unreliability of human subjects.

Instead of representing a form of resistance against Cold War rationality by reasserting the autonomy and integrity of human subjects, remote viewing can be better understood as a model of technical precision, standardization and optimization that was primarily driven by military interests. Instead of reaffirming the importance of human intelligence gathering and analysis, which was being increasingly displaced by machines, it sought to replace the uncertainty of human agency with the absolute rationality of machinic logic by transforming human subjects into technical automata. Indeed, one might even go so far as to say that the purpose of remote viewing was to demonstrate the degree to which the automatic functioning of the psychic apparatus could replicate and potentially surpass that of the technical apparatus. The development of satellite technology and psychical research thus resulted in the satellization of the planet as well as the satellization of the self, as consciousness itself was effectively colonized, instrumentalized and weaponized. While historians often describe how satellite images and psychic phenomena inspired utopian narratives of global unity and harmony, there was clearly nothing utopian about remote sensing and remote viewing once they eliminated the possibility of human intervention and reduced consciousness to the measurement of channel capacities and bit rates.

Notes

1. See Arthur C. Clarke, 'Extra-Terrestrial Relays: Can Rocket Stations Give World-Wide Radio Coverage?,' *Wireless World* 51.10 (October 1945), 305–8; idem, 'The Social Consequences of Communications Satellites,' in idem, *Voices from the Sky: Previews of the Coming Space Age*, New York: Pyramid Books, 1967, 113–22, here 121; Stewart Brand, *Whole Earth Catalog: Access to Tools*, Menlo Park: Portola Institute, 1968; James Lovelock, 'Gaia: A Model for Planetary and Cellular Dynamics,' in William Irwin Thompson, ed., *Gaia, a Way of Knowing: Political Implications of the New Biology*, Great Barrington: Lindisfarne Press, 1987, 83–97; Jody Berland, 'Mapping Space: Imaging Technologies and the Planetary Body,' in Stanley Aronowitz, Barbara Martinsons and Michael Menser, eds, *Technoscience and Cyberculture*, London: Routledge, 1996, 123–37, here 129; and Lisa Parks, *Cultures in Orbit: Satellites and the Televisual*, Durham, NC: Duke University Press, 2005, 2. All Internet sources were last accessed on 15 July 2020.
2. See 'Drugs Studied to Aid Astronauts,' *Missiles and Rockets* 16.11 (15 March 1965), 33; Timothy Leary, *Exo-Psychology: A Manual on the Use of the Human Nervous System According to the Instructions of the Manufacturers*, Los Angeles: Peace Press, 1977; O. T. Oss and O. N. Oeric [Terence and Dennis McKenna], *Psilocybin: Magic Mushroom Grower's Guide*, Berkeley: And/Or Press, 1976, 8–9; and Walter Breen, 'The Blown Mind on Film,' *Warhoon* 24 (August 1968), 16–24, here 24. On *2001*, see Robert Poole, 'The Myth of Progress: *2001: A Space Odyssey*,' in

Alexander C. T. Geppert, ed., *Limiting Outer Space: Astroculture after Apollo*, London: Palgrave Macmillan, 2018, 103–29 (= *European Astroculture*, vol. 2).

3. See Pierre Teilhard de Chardin, *L'Avenir de l'homme*, Paris: Editions du Seuil, 1959, 214; Leary, *Exo-Psychology*, 36; Edgar D. Mitchell, 'An ESP Test from Apollo 14,' *Journal of Parapsychology* 35.2 (June 1971), 89–107; and Alexander C. T. Geppert, 'European Astrofuturism, Cosmic Provincialism: Historicizing the Space Age,' in idem, ed., *Imagining Outer Space: European Astroculture in the Twentieth Century*, Basingstoke: Palgrave Macmillan, 2012, 3–24, here 8 (= *European Astroculture*, vol. 1).
4. Many of his stories also represent outer space as a site of military conflict. 'The Firing Line,' for example, depicts the use of "chaff" aluminum foil strips to confuse enemy radar, which represents the extraterrestrial application of a military device invented in 1943 by American astronomer Fred Whipple, who was stationed with the US Army in the United Kingdom. See George O. Smith, 'The Firing Line,' *Astounding Science Fiction* 34.4 (December 1944), 69–97.
5. Jack Anderson, 'Psychic Studies Might Help U.S. Explore Soviets,' *Washington Post* (23 April 1984), B14. See also Matthew Connelly, 'Future Shock: The End of the World as They Knew It,' in Niall Ferguson, Charles S. Maier, Erez Manela and Daniel J. Sargent, eds, *The Shock of the Global: The 1970s in Perspective*, Cambridge, MA: Harvard University Press, 2010, 337–50, here 343.
6. While critics like Andrew May insist that 'there is no significant overlap between the "consumers" of pseudoscience and those of real science,' as science is a 'practical discipline' and pseudoscience is a 'branch of the entertainment industry,' such distinctions often obscure the interrelationships between orthodox and unorthodox scientific theories. This is not to say that accounts of psychical research should be considered equally as legitimate as the development of satellite technology; indeed, such a claim would be impossible to sustain, as many of the pseudoscientific texts cited in this chapter have been severely criticized for containing gross inaccuracies and false statements. As this chapter shows, however, science and pseudoscience remain deeply intertwined and interconnected, as they often feed on each other through reciprocal lines of influence; see Andrew May, *Pseudoscience and Science Fiction*, Basel: Springer, 2017, viii.
7. William E. Burrows, *Deep Black: Space Espionage and National Security*, New York: Random House, 1986, 28.
8. See John Noble Wilford, *The Mapmakers: The Story of Great Pioneers in Cartography. From Antiquity to the Space Age*, New York: Vintage Books, 1982, 233; John Sweetman, *Cavalry of the Clouds: Air War over Europe 1914–1918*, Stroud: History Press, 2010, 13.
9. James B. Campbell, 'Origins of Aerial Photographic Interpretation, U.S. Army, 1916 to 1918,' *Photogrammetric Engineering & Remote Sensing* 74.1 (January 2008), 77–93. See also Terrence J. Finnegan, *Shooting the Front: Allied Aerial Reconnaissance and Photographic Interpretation on the Western Front – World War I*, Washington, DC: National Defense Intelligence College, 2006.
10. Jon C. Leachtenauer and Ronald G. Driggers, *Surveillance and Reconnaissance Imaging Systems: Modeling and Performance Prediction*, Boston: Artech House, 2001, 55–6; Ryan Edgington, 'An "All-Seeing Flying Eye": V-2 Rockets and the Promises of Earth Photography,' *History and Technology* 28.3 (September 2012), 363–71.

11. R. Cargill Hall, 'Postwar Strategic Reconnaissance and the Genesis of Corona,' in Dwayne A. Day, John M. Logsdon and Brian Latell, eds, *Eye in the Sky: The Story of the Corona Spy Satellites*, Washington, DC: Smithsonian Institution Press, 1998, 86–118. See also Burrows, *Deep Black*, 78.
12. For declassified records on this project, see http://www.nro.gov/foia/declass/mol.html. Dwayne A. Day, 'The Development and Improvement of the Corona Satellite,' in idem, Logsdon and Latell, *Eye in the Sky*, 48–85, here 59–61. See also the respective contributions by Alexander Geppert and Tilmann Siebeneichner, and by Michael Neufeld, Chapters 1 and 2 in this volume.
13. Frederic C. E. Oder, James C. Fitzpatrick and Paul E. Worthman, *The CORONA Story*, Washington, DC: National Reconnaissance Office, 1987, 109.
14. Day, 'Development and Improvement of the Corona Satellite,' 70–4.
15. Burrows, *Deep Black*, 215.
16. Ibid., 241.
17. Ibid., 249.
18. Paul B. Stares, *The Militarization of Space: U.S. Policy 1945–1984*, Ithaca: Cornell University Press, 1985, 131–3, 206.
19. Thomas S. Moorman Jr., 'The Explosion of Commercial Space and the Implications for National Security,' *Airpower Journal* 13.1 (Spring 1999), 6–20, here 6; Merrill A. McPeak, 'The Air Force's Role in Space,' in idem, *Selected Works 1990–1994*, Montgomery: Air University Press, 1995, 207–13, here 207. On GPS, see Paul Ceruzzi's contribution, Chapter 13 in this volume.
20. Marshall McLuhan and Wilfried Watson, *From Cliché to Archetype*, New York: Viking, 1970, 9–10.
21. Jean Baudrillard, *Simulacres et simulation*, Paris: Galilée, 1981, 61.
22. Paul Virilio, *Guerre et cinéma*, vol. 1: *Logistique de la perception*, Paris: Éditions de l'étoile, 1984, 146.
23. Mark Poster, *Mode of Information: Poststructuralism and Social Context*, Chicago: University of Chicago Press, 1990, 93.
24. Gearóid Ó Tuathail, *Critical Geopolitics: The Politics of Writing Global Space*, Minneapolis: University of Minnesota Press, 1996, 197.
25. John Cloud, 'Imaging the World in a Barrel: CORONA and the Clandestine Convergence of the Earth Sciences,' *Social Studies of Science* 31.2 (April 2001), 231–51.
26. Lorraine Daston and Peter Galison, 'The Image of Objectivity,' *Representations* 40 (Fall 1992), 81–128, here 119.
27. Paul Erickson et al., *How Reason Almost Lost Its Mind: The Strange Career of Cold War Rationality*, Chicago: University of Chicago Press, 2013, 16–17, 4.
28. Harold E. Puthoff, Russell Targ and Edwin C. May, 'Experimental Psi Research: Implications for Physics,' in Robert G. Jahn, ed., *The Role of Consciousness in the Physical World*, Boulder: Westview Press, 1981, 37–86, here 38.
29. Sheila Ostrander and Lynn Schroeder, *Psychic Discoveries behind the Iron Curtain*, Englewood Cliffs: Prentice Hall, 1970, 7.
30. Ibid., 249.
31. John D. LaMothe, *Controlled Offensive Behavior: USSR*, Washington, DC: Defense Intelligence Agency, 1972, 24.

32. Ibid., 39–40.
33. See also P. T. Van Dyke and M. L. Juncosa, *Paranormal Phenomena: Briefing on a Net Assessment Study*, Santa Monica: RAND, 1973. On the 'missile gap,' see Christopher Gainor's contribution, Chapter 3 in this volume.
34. Russell Targ and David Hurt, 'Learning Clairvoyance and Precognition with an ESP Teaching Machine,' *Proceedings of the Parapsychological Association* 8 (September 1971), 9–11. For a critical assessment of this machine, see Martin Gardner, 'Mathematical Games,' *Scientific American* 233.4 (October 1975), 114–18.
35. Russell Targ and Jane Katra, *Miracles of Mind: Exploring Nonlocal Consciousness and Spiritual Healing*, Novato: New World Library, 1999, 37.
36. Harold E. Puthoff, 'CIA-Initiated Remote Viewing Program at Stanford Research Institute,' *Journal of Scientific Exploration* 10.1 (Spring 1996), 63–76, here 65.
37. Jeffrey T. Richelson, *The Wizards of Langley: Inside the CIA's Directorate of Science and Technology*, Boulder: Westview Press, 2001, 178.
38. Kenneth A. Kress, 'Parapsychology in Intelligence: A Personal Review and Conclusions,' *Journal of Scientific Exploration* 13.1 (January 1999), 69–85, here 83.
39. Swann initially referred to this technique as 'remote sensing,' although he later rejected this term because he 'didn't just sense the sites, but experienced mental-image pictures of them in a visualizing kind of way.' By 1971 he was 'referring to the long-distance experiments as remote-viewing ones, since this term seemed the most suitable.' See Ingo Swann, 'On Remote-Viewing, UFOs, and Extraterrestrials,' *Fate* 46.9 (September 1993), 73–82, here 78.
40. Harold E. Puthoff and Russell Targ, *Mind-Reach: Scientists Look at Psychic Abilities*, Charlottesville: Hampton Roads, 2005, 37.
41. Ibid., 52. See also Jim Schnabel, *Remote Viewers: The Secret History of America's Psychic Spies*, New York: Dell, 1997, 152.
42. Ibid., 11–13.
43. 'The Case of E.S.P., 1983–84,' *Horizon*, BBC 2, first aired on 26 September 1983.
44. Paul H. Smith, *Reading the Enemy's Mind: Inside Star Gate. America's Psychic Espionage Program*, New York: Forge, 2005, 74.
45. Puthoff and Targ, *Mind-Reach*, 33.
46. Schnabel, *Remote Viewers*, 120.
47. Richelson, *Wizards of Langley*, 181.
48. Russell Targ, *Limitless Mind: A Guide to Remote Viewing and Transformation of Consciousness*, Novato: New World Library, 2004, 36.
49. Schnabel, *Remote Viewers*, 110.
50. Kress, 'Parapsychology in Intelligence,' 73. See also Puthoff and Targ, *Mind-Reach*, 4; and Smith, *Reading the Enemy's Mind*, 65.
51. Richelson, *Wizards of Langley*, 179.
52. Harold E. Puthoff and Russell Targ, *Perceptual Augmentation Techniques: Final Report (Covering the Period from January 1974 to February 1975)*, Menlo Park: Stanford Research Institute, 1975.

53. Clarence A. Robinson Jr., 'Soviets Push for Beam Weapon,' *Aviation Week and Space Technology* 106.18 (2 May 1977), 16–22.
54. Russell Targ, 'Remote Viewing at Stanford Research Institute in the 1970s: A Memoir,' *Journal of Scientific Exploration* 10.1 (Spring 1996), 77–88, here 87 and 77.
55. Kress, 'Parapsychology in Intelligence,' 78.
56. Ibid., 79.
57. Harold E. Puthoff and Russell Targ, 'A Perceptual Channel for Information Transfer over Kilometer Distances: Historical Perspective and Recent Research,' *Proceedings of the IEEE* 64.3 (March 1976), 329–54, here 330.
58. Ibid., 350.
59. Schnabel, *Remote Viewers*, 13–20.
60. See Puthoff, 'CIA-Initiated Remote Viewing Program,' 75; Smith, *Reading the Enemy's Mind*, 97–8; and Schnabel, *Remote Viewers*, 215–19.
61. William K. Stuckey, 'Psi on Capitol Hill,' *Omni* 1.10 (July 1979), 24.
62. Ronald M. McRae, *Mind Wars: The True Story of Government Research into the Military Potential of Psychic Weapons*, New York: St. Martin's Press, 1984, 48.
63. Ibid., 47.
64. Schnabel, *Remote Viewers*, 9–10.
65. Ibid., 23–5.
66. James Adams, 'Day of the Pentagon Mindbenders: Psychic Warriors Were Trained to Enact "Thought Theft" on Kremlin,' *Sunday Times* (3 December 1995), 21.
67. McMoneagle, *Stargate Chronicles*, 116–17.
68. Ibid., 120. See also Schnabel, *Remote Viewers*, 285.
69. Ibid., 280, 319.
70. For declassified reports on these sessions, see http://www.remoteviewed.com.
71. Schnabel, *Remote Viewers*, 380.
72. David Morehouse, *Psychic Warrior: Inside the CIA's Stargate Program*, New York: St. Martin's Press, 1996, 167–8.
73. Supporters of the program, like former intelligence officer and practicing psychic W. Adam Mandelbaum, still insisted that 'remote viewing could be operationally useful in areas where the presence or undetected insertion of an agent was impracticable or outright impossible.' According to Mandelbaum, therefore, remote viewing was superior to satellite surveillance systems because it offered increased mobility, heightened resolution and faster transmission rates. See W. Adam Mandelbaum, *The Psychic Battlefield: A History of the Military-Occult Complex*, New York: St. Martin's Press, 2000, 135.
74. Russell Targ and Keith Harary, *The Mind Race: Understanding and Using Psychic Abilities*, New York: Villard, 1984, 245.
75. Puthoff and Targ, 'Perceptual Channel,' 346.
76. Smith, *Reading the Enemy's Mind*, 227.

PART IV

Mounting Combat Infrastructures

CHAPTER 11

Architectures of Command: The Dual-Use Legacy of Mission Control Centers

Regina Peldszus

As central nodes of the large-scale, distributed infrastructure of space operations, mission control centers multiplied and flourished during the burgeoning space programs of the United States, the Soviet Union and Europe during the Cold War. At this time, undertaking activity in space served a political agenda as much as an operational one: Mission control – the practice of deploying, operating and safeguarding an asset in the remote and hostile natural environment of earth orbit and beyond – also meant the projection of power in a strategically important domain. As visible sites of operations, control centers fused human agency and technology on different scales. On the one hand, they tangibly and often dramatically set the scene for the 'human-in-the-loop' in complex, high-risk missions. On the other, in transcending their mission priorities they also manifested the overarching programmatic values of their operating organizations as establishments of international prestige, sophistication and capability.[1]

Recent works on the history of space systems, predominantly in the United States, have sought to understand significant aspects of control centers from the perspectives of human-machine interaction, collaborative practices, systems design and evolution.[2] In charting the transformation of control centers in general, or describing minutely the practices located in specific locales, these in-depth examinations complement the autobiographical accounts by individual flight directors or program managers who shaped operations in the

Regina Peldszus (✉)
DLR Space Administration, Bonn, Germany
e-mail: regina.peldszus@dlr.de

Alexander C. T. Geppert et al. (eds), *Militarizing Outer Space*
European Astroculture, vol. 3
https://doi.org/10.1057/978-1-349-95851-1_11

advent of spaceflight, as well as the body of applied research on operations in mission control.[3] In adding to these contributions, this chapter addresses another intrinsic yet underexamined aspect of technology governance and policy of operations centers, which otherwise feeds an abundant field of discourse in space policy: the dimension of dual use.

As a notion and attribute, 'dual use' denotes the capability of systems, infrastructures or practices to serve both civilian and military uses in a given domain. For space technology specifically, the dual aspect has come to represent a characteristic 'indivisible essence' through an initial military investment and 'continuous backing and association' of military and civilian actors in the development and deployment of space systems.[4] In practice, this entails that a distinction of military and civilian space systems is highly contingent upon context and can rarely be made on the basis of a technology per se. Payloads for military and civilian systems would, for instance, be distinguished primarily in use rather than design. Even the identity of the operating organization of a system, its user and programmatic purpose cannot always serve as a reliable indicator of civilian systems. This rings true to the fundamental dictum of technology being neither neutral nor unambiguous by default.[5]

Yet the notion of ambiguity that policy-makers, practitioners and scholars have attempted to explore and structure by employing the concept of dual use also emerged as a feature of the concept itself. During the Cold War, the intrinsic potential of a technology to be of dual use was seen primarily as an arms-control issue in view of export control and proliferation of defense technology. As a central but somewhat elusive notion at the intersection of important strategic sectors, dual use was prone to being co-opted as a policy instrument to foster defense integration. In the post-Cold War period, the validity of 'dual use' was questioned, both as a framework to think and as a practical instrument to guide policy formulation. Subsequently, observations and theory on the plasticity of the notion of dual use were quickly superseded by a pragmatic understanding of the concept in terms of industrial concerns. The convergence of government actors with defense contractors and commercial actors gained such momentum in the 1990s that commentators remarked that the 'borderlines between civilian and military high technology goods [...] have become meaningless and technical parameters that qualified equipment as being military [...] are now useless,' and that the relationship between defense and aerospace had been rendered increasingly indistinct.[6] The military heritage of space technology was put in the service of civilian purposes – civilianized – through experts and organizations that developed, integrated, tested, validated and refined the emerging operational systems. The related artifacts, codified and tacit knowledge, management principles, production and research facilities, labor skills and expertise could be jointly developed or transferred between actors of both domains. Their exploitation for greater military-civilian integration helped maintain industrial bases for defense technology, in light of a post-Cold War slowdown in military spending.[7]

Control room elements were re-infused with the expertise, norms and technologies of defense through an emerging sector that bundled and expanded civilian and military portfolios. This development was fostered by government policies: They reconciled the tension between export control concerns and greater market integration, aligned civilian goals with military technology, and promoted the cooperation between the military, commercial and science sector. Academics were encouraged to found companies or corporations to build computing facilities for academia or the military; in the process, close ties were forged, for instance, in defense computing development.[8]

In the course of this process, the idea and interpretation of dual use of space technology shifted from a problem to a desirable potential of technology. This dynamic applied, and continues to apply, not only to space technology in general, but seamlessly to control systems and control centers in general. With their blueprint in guidance and control technology of ballistic missile weapon systems, the basic elements of control centers were inherited from defense. Integrated into the civilian space programs from the preparation of the first crewed spaceflights in 1959 to the post-Cold War multilateral operations emerging in the mid-1990s, mission control ingested many of the structural settings, operational practices and organizational features of its military origins. As a consequence, the 'control room' as a generic site offers rich material to retrace both temporally and structurally not only what constituted its dual elements in the past and current operational environment: It also invites us to observe how the interpretation of the concept of dual use in space technology itself evolved, as it was debated and employed in increasingly nuanced and affirmative terms, particularly in post-Cold War Europe.

I The structural setting of control rooms and its military kernel of 'war rooms'

Notwithstanding subtle differences in the terminologies of different space programs structurally, nationally and historically, a control room essentially forms the core element of an operations or control center. As a larger facility run by a space agency, commercial operator or military branch, the operations center may aggregate several control rooms for different missions or space programs. The control center itself is part of the so-called ground segment of a space mission, which also includes ground stations for telemetry, tracking and command, launch sites and facilities such as data centers and payload operations. The control center's personnel are responsible for overall mission planning and execution, primarily focusing on the behavior and integrity of the spacecraft or a fleet of satellites, and the success of the mission.

In its basic configuration, a control room consists of one or more terminals that allow an operator or controller to command a spacecraft and receive telemetry via an antenna link. Scaled up to the iconic setting of a large room nestled within a suite of support rooms, it accommodates the group of experts that make up the mission control team. On consoles arranged around or towards a set of central situational displays, they collaboratively handle routine, special or contingency operations, such as the launch and early orbit phase (LEOP) of a mission, critical maneuvers or recovery activities after serious anomalies. The flight control team interacts with a spacecraft by receiving data from onboard sensors to monitor the spacecraft's state. By conveying telecommands – or verbal instructions to guide human operators in crewed spacecraft through their onboard activities – they affect a desired mode or behavior of the spacecraft, in order to complete a procedure in fulfillment of an overall schedule, the mission timeline.[9]

Beyond control centers conceived for the military programs of the main space powers after the Second World War, the origins of civilian space mission operations emerged from the same military capabilities. These included missile technology, radar tracking, computing and related developments, as well as design approaches such as operations research or systems engineering.[10] As a precursor of launch control, the first control terminals for missile applications consisted of a 'radio mounted on a Jeep,' or the firing panels for anti-aircraft guns in the 1930s and 1940s.[11] Early facilities also evolved as part of the A4/V-2 guided ballistic missile development in Peenemünde, Germany, during the Second World War. They were composed of sheltered sites located at a suitable distance from a launch pad to allow for remote firing and observation. The related infrastructure included wiring systems for issuing verbal commands to human operators at different workstations, ignition lines and measuring points. The test stand or launch pad was positioned in the vicinity of a bunker in which controllers would monitor gauges, meters, alarm lamps and television footage from the launch rig. They also observed the launch pad directly through periscopes and remotely controlled the actuation of commands through levers or switches according to a predetermined plan. These control facilities – at that point still as part of validation and test stands, and not flight operations as such – also involved the first fully electronic analog computers for trajectory calculation in the context of ballistic missile guidance.[12]

As they evolved in scope after the Second World War, the fundamental patterns of space mission control rooms were derived from this setup of test ranges for intercontinental ballistic missiles (ICBM) and informed by control settings in industrial and commercial test ranges. This included aviation navigation systems and communications networks, air traffic control, processing or manufacturing plants, experimental science facilities and power systems. Their control rooms hosted vast boards and monolithic consoles that allowed the remote manipulation of large-scale installations through electric switches.[13] Subsequently, the control center typology of the growing space programs diversified into distinct launch control and mission control centers, with the latter further distinguishing for spacecraft control and payload operations, or crewed versus uncrewed missions.

While on the Soviet side the first crewed and uncrewed space missions did not yet link to a dedicated control facility with specialized staff, the physical and operational development of control rooms and their teams began in the United States with NASA's first and second human spaceflight programs. Mercury (1958–63) and Gemini (1961–66) aimed at operating crewed orbital spacecraft in preparation for more complex missions. The operational infrastructure of Mercury mission control was designed by the Army Corps of Engineers and supported for pre-launch, launch, communications and tracking activities by the Army and Air Force. It was an inconspicuous building close to one of the several launch pads at Cape Canaveral in Florida, which at the time resembled an 'oil field, with towering structures, dirt and asphalt roads.'[14]

These 'minimal, even primitive' facilities included a control center and the so-called blockhouse, a reinforced structure in closer proximity to the launch pad that housed the team observing and controlling the rocket. The first control room for the Mercury program, set up in 1959, housed a military Univac computer and 14 flight controllers. They huddled together facing a large wall-mounted map across which a 'toy-like spacecraft model, suspended by wires, moved [...] to trace the [ground track of an] orbit.'[15] The engineers staffing the control centers developed operational concepts on the fly, experimenting as prudently as possible in such a high-risk setting. They adapted what they could from existing aviation and missile operations, and drew on their own professional experiences during and after the war. As rudimentary as it was, the co-located setups from Mercury and its follow-on program Gemini already included key elements of what was to become the iconic setup of a control room: operators in functional arrangements, a large screen displaying key information for the entire team and the background second-tier seating separated by a glass pane (Figure 11.1).

These elements were expanded and refined significantly to accommodate demanding activities beyond launch and orbiting, including longer and more sophisticated operational timelines for the lunar missions of the Apollo program (1961–72). Infrastructure was distributed across sites in a network of control centers, with a launch pad at Cape Canaveral and mission control proper having moved to Johnson Space Center in Houston, Texas, by 1964. In contrast to its experimental predecessors, the new Apollo control facility was clad in a white oyster shell and glass façade on a landscaped campus with clipped lawns and carefully tended shrubs. The sleek, monolithic character of the facility betrayed the immense complexity of its internal workings. Inside, several floors housed a string of control rooms, including support and briefing rooms, and computer processing facilities that significantly expanded the capabilities of the earlier Mercury control center. The control rooms were staffed round the clock by a cadre of shift-working young operations engineers, many of whom were former US Air Force (USAF) engineers or pilots. As the newly banded Flight Control Team, they had adopted a civilian

Figure 11.1 The flight control area in Mercury Mission Control during a pre-launch simulation at Cape Canaveral, Florida, USA, 1962. The setup shows the characteristic rows of consoles oriented towards a central display, here photographed from the viewing gallery. *Source*: Courtesy of NASA.

uniform of white shirts and black ties. Each console included two monochrome television screens, on which the operator could manually bring up information from 70 different input channels. This system was fed by an elaborate array of screens culminating in a phalanx of display screens towards which all operators were oriented.[16]

Even for those contemporaries directly interacting with flight control personnel, many of whom also had a military background, the resulting setting came across as completely novel. One Apollo astronaut in training – of the first civilian intake, in fact – likened the 'antiseptic atmosphere, futuristic skylights and arrangement of consoles' surrounding enormous screens in mission control to the 'war room' in Stanley Kubrick's 1964 film *Dr. Strangelove*.[17] The design of Kubrick's fictional war room was arguably not based on government facilities, unlike some of the other weapon systems that were portrayed rather realistically in the same film. Nevertheless, it synthesized and amplified key elements of command and control by assembling the tools that afforded situation awareness and collaborative decision-making. And indeed, the 'war room' notion constituted the basic blueprint of control rooms: The conceptual roots of NASA's mission control related back to the combat centers and situation rooms of the Second World War, and the virtual command spaces and computing rooms of the Strategic Air Command. Although NASA was explicitly set up as a civilian administration, it drew on the same base of industrial contractors as the agencies of the defense realm. For instance, having equipped the command and control network of the US Air Force's Semi-Automatic Ground Environment (SAGE), IBM was well positioned to furnish space mission control facilities with real-time computing capabilities.[18]

The fundamental layout and operational paradigm consolidated for Apollo would remain an enduring concept. It propagated into the setup of Europe's own control centers, as NASA and its contractors such as Boeing and Lockheed interacted with the agencies and industry involved in Europe's nascent space programs. Initially, this learning process occurred by way of formal exchange in bilateral working groups with NASA and through official month-long fact-finding trips to sites of operations. Technical and operational expertise was also gained from participating in Apollo operations directly, or through European controllers training at NASA facilities. In 1969 a generation of Spanish engineers, for instance, infused the European market with NASA expertise after having been trained to support the Apollo network of ground tracking stations on the Canary Islands and near Madrid.[19] Later, expertise migrated informally through the return of European staff who had spent their postdoctoral years or significant portions of their professional lives at institutions such as the Jet Propulsion Laboratory (JPL), which itself had transformed from handling military systems to operating interplanetary probes when it was incorporated into NASA during the latter's foundation in 1958.[20]

In 1967 the precursor of the European Space Agency (ESA), the European Space Research Organisation (ESRO), controlled its first satellite mission and the emerging ground station tracking network from a modest control facility located at the European Space Technology Centre (ESTEC) in the Dutch coastal town of Noordwijk. Its 'operations area' featured maps and displays of the tracking stations and their status. Divided by glass partitions, the displays were also visible from two side rooms that housed additional equipment. As missions grew in nature and duration, ESRO inaugurated a dedicated operations facility, the European Space Operations Centre (ESOC), in Darmstadt, Germany, in 1968. It essentially consisted of an operational control room flanked by an IBM-equipped, real-time computing facility, a communications center and a briefing room (Figure 11.2).[21]

Earlier design requirement studies conducted by ESRO and ESA had already identified key operations functions for a control center. These included facilities to perform information visualization and analysis, diagnostics, problem-solving, command transmission, verification of command execution and communication among the flight control team. Similar to NASA's Apollo program, ESA's operations involved a flight director responsible for

Figure 11.2 The main control room at the Operations Centre of the European Space Research Organisation (ESRO) in Darmstadt, Germany, 1968–72. The wood-paneled operations rooms would become the first of a whole set of dedicated control rooms and support suites nestled around it as the center's operations expanded.
Source: Courtesy of European Space Agency.

final decision-making. He was supported by various operations engineers who monitored and coordinated procedures with launch site and tracking stations, by project representatives from the agency, by scientists and program managers. Satellite controllers situated in the control rooms would send commands to the spacecraft. In the early days dedicated operators were also responsible for updating information displays.[22]

Over the course of the 1980s, ESOC in Darmstadt emerged as a prominent operations facility that regularly controlled science, commercial and earth observations satellites. The newly refurbished Operations Control

Centre building included a suite of control rooms, a computing complex and centralized control room for the ground station network. The operations center's heart, the main control room, featured the capability to support launch and early orbit operations of up to two spacecraft in parallel. The colored lamps and cumbersome strip-chart recorders of the past had been replaced with wall-to-wall high definition color monitors framed by dimmable lights on a black ceiling that appeared to evoke the space environment and new alpha-numeric displays on the consoles. A simulation control room in the basement housed personnel tasked with extensively exercising the entire control center systems and staff before a launch.[23]

During and after the setup of ESOC, Europe's national space agencies built equivalent facilities in order to operate their first research satellites. In 1968, after commissioning a ground tracking station, the German Aerospace Center's precursor Deutsche Versuchsanstalt für Luftfahrt (DVL) replicated the NASA setup for the control of the first German research satellite, AZUR, and follow-up projects in cooperation with France.[24] The German Control Center (GCC) in Oberpfaffenhofen near Munich eventually complemented the expertise from satellite manufacturer and defense contractor Messerschmitt-Bölkow-Blohm (MBB) with insights gathered during a six-week business trip of GCC managers to NASA's Multi-Satellite Control Center at Greenbelt, Maryland, in January 1969. A decade later, for the joint Spacelab mission with NASA, DLR built the first control center for crewed operations outside the United States and the Soviet Union, the German Space Operations Center (GSOC) in Oberpfaffenhofen. Meanwhile, the French space agency Centre National d'Etudes Spatiales (CNES), headquartered in Paris, set up an operations center in Toulouse in 1968, controlling satellites from 1970 onwards. CNES developed sophisticated control systems that emphasized automated routine control, which later allowed different sites to be used as standalone or interfacing command and control centers, in view of cost-reduction and multi-satellite operation.[25]

The European, German, French and Italian space agencies then spawned other control centers by providing the conceptual blueprints, control systems, personnel and services to other commercial and multilateral governmental satellite operators such as Eutelsat, Inmarsat and Eumetsat in the rapidly growing communications and meteorological sectors established in the course of the late 1970s and mid-1980s. Towards the end of the Cold War, the dual-use nature of components such as microprocessors – and the ensuing complexities of export control – resulted in the desire to independently develop new generations of technology in Europe.[26] Yet a lively dialogue with international partners was upheld also in light of collaborative operations; inter-organizational fact-finding continued within the European and international sector. This information cross-pollinated back into ESA through surveys of reference sites for their own control centers from commercial and experimental operations as well as for telecommunication satellites and scientific missions.[27] Thus, as scaled-down derivatives of Apollo operations, the physical composition of control rooms in Europe remained relatively static between intervals of occasional upgrade, rearrangement or facility expansion. Other elements, however, were much more dynamic: Operational practices moved both geographically between the global and nascent space players throughout the 1970s and 1980s, and effortlessly crossed the conceptual boundaries of civilian and military application.

II The operational practice of command and control

Beyond the physical layout of the control rooms, a control center's physical structure was laced with the informational architecture of software, the mission control system. The operational practice of 'commanding' was permeated with terminologies and concepts shared with, or originating from, defense and computing.[28] As typical products of the Cold War, the emerging approaches of operations research and systems engineering were rooted in the design development of ICBMs and nuclear weapons in the Manhattan Project. They went into the development of the Apollo spacecraft and were reflected in its operations philosophy. The Associate Administrator of NASA's Office of Manned Spaceflight, George E. Mueller (1918–2015), who had acquired and refined the methods of technology management while working at Bell Labs, promoted systems engineering to expedite the development of the Apollo program and firmly lodged its culture of precision, planning and refinement in his organization.[29]

Systems engineering as a discipline sought to optimize and harmonize the 'ensemble of subsystems and components – machines, communication networks, humans [...] – all related by channeled flows of information, mass and energy.'[30] In this context, and taking cue from practices and standards evolving in parallel to the nuclear and military regulatory authorities, operators and their equipment were considered not discrete actors, but elements of one workstation. Infrastructures such as SAGE for air surveillance, which processed radar data from different sites into an integrated airspace image, were regarded not simply as technical, but socio-technical systems. They were composed of highly heterogeneous but dovetailing elements of various types of human operators, air bases, guided missiles, radio communication and interceptors, air surveillance rooms and operations centers.[31] This fusion of hardware, software and the 'fundamental human unit' of the operations team at NASA's Apollo control center constituted a joint operational system that integrated all other elements into an effective process.[32] Operators were organized into a seating arrangement where each console represented a particular aspect of the mission or subsystems of the spacecraft. Each individual linked back to dozens of experts in adjacent support rooms and auxiliary facilities. Their activities were bundled within the central authority of the flight director who would oversee real-time information sharing among all participants and a high-level timeline of activities, taking responsibility for final decision-making.

Systems thinking also informed the functional layout of the control room to afford collaborative work in what was predominantly an information environment. Key activities included the monitoring and exchange of data and information, determination of the overall situation and subsequent decision-making. Control, here, denoted the 'deliberate guidance or manipulation [...] to achieve a prescribed value of a variable,' that is, the state or mode of the spacecraft or its subsystems and the ongoing processes within it, for instance, by triggering the execution of specified events such as maneuvers, or a response to anomalies. Control systems linked to a ground station with an antenna for transmitting commands, receiving downlinked telemetry, archiving and displaying this data, and alerting the operator in case predefined parameter limits of sensors onboard the spacecraft were reached.[33]

In their physically manifested hierarchy, the diverse tasks within a flight control team were unified by a single common goal: increasing the likelihood of mission success. To this end, the flight control team interacted with a digital command and control system. Despite an increase of automation in control operation particularly since the 1980s, the information architecture of a generic space operations environment has endured to the present day due to fundamental technical requirements and concepts of control. Then and now, the mission control system was mediated by displays on individual consoles that grouped telemetry and commanding display with a voice loop to communicate across larger flight control teams. To execute an action, the operator would send instructions encoded in frames as part of a message. These commands could be given in real-time or were executed according to a pre-scheduled plan onboard the spacecraft for routine control in the form of single commands, command sequences or – for more complex commanding – entire flight procedures that resulted in a state change in a subsystem or payload, the mode of the entire spacecraft or the initiation of a maneuver.[34]

At ESOC in Darmstadt, such a 'centralized command, control and display capability' was integrated in the main control room and in the various dedicated control rooms. In the advent of operating multiple missions sequentially or simultaneously in the mid-1970s, the reusable Multi-Satellite Support System (MSSS) was developed to support different types of missions. For manual commanding, that is, releasing instructions to the spacecraft not by automated procedures but direct input into the control system, the control software interface and its later iterations contained a critical set of buttons labeled 'arm' and 'go.' The operator would type or select a command – or arm it – in the so-called command stack, a pre-assembled list, and then release it at a specific point in time. This button combination translated into a qualification of an intention. As a coupled practice, it presented a safety mechanism and epitomized a fundamental concept in command and control – whether this was in the context of sending a simple instruction to a comparably tiny scientific probe millions of kilometers out in deep space, or launching an intercontinental missile from an underground silo. In fact, in nuclear defense, similar practices were a rite performed by two operators: Upon receiving an order, they had access to the keys unlocking the launch mechanism and were only able to execute a command jointly.[35]

This four-eye principle, whether implemented by two operators at the same time or sequentially through separate double-checks, still constitutes a fundamental rule in routine and special operations today. But also the spread of quality, reliability and safety practices from Air Force and missile defense complemented operations and mission management in order to render sensitive systems less vulnerable to external disturbances or internal mishaps. Potential failure rates were analyzed using techniques that were otherwise used mainly in military safety, such as fault tree analysis that emerged in the early 1960s around the Minuteman ICBM. Security mechanisms prevented unauthorized access to the facility and governed access to the various control of the system within the operations staff, including limiting access to commanding functions.[36]

The first in a series of generic mission control systems, the MSSS underwent six major implementations. In its twenty years of operation it formed the base for other control systems in Europe, including the commercial sector.[37] Aside from operational practices due to growing transatlantic partnerships, critical elements such as mission control systems developed in Europe like the MSSS percolated through European agencies and industry. Software had a greater plasticity than hardware components of operations. As mobile, malleable and somewhat modular artifacts, control systems could proliferate between actors, be modified and further developed. The successor generations of control systems at ESOC, such as the Spacecraft Control and Operations System (SCOS) and SCOS-2000 introduced in the 1980s and 1990s, were used and adapted widely in the European space sector. The software and the expertise of those who created it was disseminated to operating organizations through workshops, training and through contracts that encouraged suppliers to develop and utilize the control infrastructure 'within and beyond ESA' in tight cooperation between ESA and the European software industry.[38] A growing ecosystem of contractors took part in this process, whereby the dual-use dimension came full circle: Some started out as small groups providing control personnel or software for civilian centers, and post-1989 expanded their portfolios towards broader aerospace and defense electronics. It became evident that the technology and expertise for space operations was applicable across sectors, and that serving different domains made business cases more appealing and robust. ESA itself was acutely attuned to this potential and became instrumental in fostering collaboration across the private sector, by encouraging private multinational consortia to work together.[39]

As part of this process, European operations developed a more distinct profile. Technical elements may have remained continuous in their evolution;

yet organizational practices that conveyed space operations to the public were much more shaped by influences such as cultures of multilateral cooperation and science in Europe.[40] This interaction with external state-actors occurred on a spectrum of opacity and transparency, and was embedded in the wider frame of security, both in geopolitical and facility access terms. The priorities that the work domain of the control room had furthered – mission success, mission safety and assurance – also expanded towards the realm of security. Such a dimension had been deeply ingrained in space operations from the outset, but newly surfaced as an issue before coming to prominence in Europe after the Cold War.

III Organizational architecture: opacity and transparency in view of security

Mission control not only communicated with the spacecraft and its network of support nodes. On a higher level, beyond agency staff, they also played a part in informing external mission participants such as scientists, and educating political stakeholders and the public. Operations had to be visible to internal personnel, both in military and civilian contexts. Already in the Apollo program transparency became a pervasive feature that was enhanced or limited by architectural and organizational design choices. One way to promote transparency quite literally utilized viewing galleries or glass panes. In addition to a separate briefing room for operational needs, the mission control center building 30 at NASA Johnson Space Center, for instance, featured an observation auditorium on the first floor. Enclosed by glass, high-level interest groups involved in, but not critical to, actual operational processes could follow operations. Virtually all high-profile mission control rooms, regardless of their origin, included large glass panes separating their console areas from a briefing area, from other support teams and exclusive audiences.

Other approaches employed mediation. The Apollo missions were highly televised, facilitated by a public relations officer on an official console position inside the control room, thereby reassuring the Western public of US capabilities.[41] The Soviet ground segment itself had been quite opaque to even Warsaw Pact audiences and closely aligned with the military during the Cold War. The 1975 Apollo-Soyuz Test Project with NASA, however, was to occur 'under the glare of the global press and televised widely.'[42] The original Soviet control center in Yevpatoria in Crimea had been constructed using scavenged technology from battleships and submarines, and was inaugurated in 1965. It controlled missions of the Vostok, Soyuz and later Almaz and Salyut programs. Initially, it was operated by the engineers who had designed and developed the space systems themselves, until dedicated control teams were trained to staff the center in 1967.[43] By the early 1970s the control center was so small, aging and spartan that a visiting US delegation was certain that their Russian partners had shown them a fabricated facility, instead of what they believed must in fact be a secret underground bunker complex located elsewhere. Hence, the need for a presentable yet secure mission control center was met in 1973, with a facility in Kaliningrad, for the operation of the Russian Salyut and MIR stations. It featured a grand cinema-style hall with a main screen and consoles. Flight controllers were wary of political officials observing them closely while working on critical tasks on the consoles. The architectural layout thus reconciled the demands of an undisturbed work setting and external observation through a selective access system that led visitors including foreign delegations, dignitaries and press through a separate entrance to a mezzanine with views onto the main screen.[44]

Another element that facilitated both internal decision-making and external engagement was the centerpiece of the control room environment, the large display of a general situational picture (Figure 11.3). A 1975 report by the Royal Aircraft Establishment in the United Kingdom poignantly summarized this 'classical' composition of a satellite control center, which prioritized information flow, hierarchies of decision-making and featured omnipresent ashtrays for the spacecraft controllers: Besides consoles for operators and their actual operations-critical tasks, '[a]ll that are now needed are a few large graphic displays to decorate the walls and a VIP viewing gallery at the back [...] [that] will be quite dramatic in its operation and [...] look well on the television.'[45] Almost as if they were concerned about appearing uneventful due to the lack of an astronaut's presence in the voice loop that had lent drama and vividness to operations, organizations operating uncrewed missions sought to amplify some of those by-now classical elements, such as the main screen.

The idea of integrating public engagement purposes in the functional requirements for control rooms was owed to the need and desire for transparency. It was both necessary to satisfy or inform external parties of certain operational and technical capabilities, but also to make an essentially remote process control setting attractive to the taxpaying audience. Primarily the large central display helped different experts in larger teams to maintain an impression of other operational aspects than their own, for instance, different subsystems of their flight controller colleagues. Yet it also served the situational awareness and overview for senior decision-makers, such as flight directors and managers, as well as the observers on the viewing galleries. Ironically, in fiction, providing such transparency through a 'big board,' was not without conflict between the defense and security officials on the one hand, and public policy-makers and diplomats on the other. In the hypothetical exchange in *Dr. Strangelove*, the tension between information disclosure and concealment sets the scene for the concerned general of the nuclear missile forces challenging the president's order of admitting the Russian ambassador entrance to the war room: 'Are you aware of what a serious breach of security that would be? [...] He'd see everything! He'd see the *big board*!' – 'That is

Figure 11.3 ESRO-2 control room at the European Space Operations Centre (ESOC), Darmstadt, Germany, May 1968. The ESRO-2B spacecraft was the first mission to be controlled by the newly inaugurated operations center.
Source: Courtesy of European Space Agency.

precisely the idea, General, that is precisely the idea.'[46] Information sharing was thus a double-edged sword: On the one hand, it was necessary to keep all those involved or potentially affected in the loop about an unfolding incident, including adversaries in the spirit of averting a global incident; on the other, one risked giving away too much about indigenous systems and strategies.

Other affordances of transparency were temporal. This was epitomized by a first open day in October 1973 at ESA's Space Operations Centre in Darmstadt, when the control center hosted over 10,000 local visitors who were

> leaving a trail of waste and destruction in their wake. Distracted secretaries arriving the following Monday found cupboards, which had been inadvertently left ajar, deplete of everything bearing the ESOC insignia; programmers, who had been foolish enough to leave their programs to run overnight, dejectly [*sic*] collected the remnants together (the visitors – not satisfied with the standard PR material – had selected punch cards at random as souvenirs).[47]

The account recalled in an informal corporate history of ESOC highlights how reaching out to the public meant occasionally inviting them in – especially in the case of civilians. Open days and guided tours, in addition to displays during the public days of aerospace industry shows, became a staple of the public relations strategies of space agencies. As opposed to their commercial or military colleagues, whose mission was to safeguard assets such as communications satellites for maximum availability and reliability, national or intergovernmental agencies with civilian science programs also had to educate, inform and even enchant the public and their political representatives, as they were gatekeepers to public funding. Allowing access to control centers conveyed insights into an operational domain that was naturally intractable and for most observers rather abstract. It exposed its glamour and banality at the same time: That bespoke high technology of a satellite tens of thousands of kilometers away was paired with procedures clipped into off-the-shelf ring-binders from the regional office supply, and international teams of highly qualified experts convened in a set of unremarkable buildings and temporary office containers at an urban periphery. Nevertheless, a cannibalization of the innards of a control center such as that during the ESOC open day would be unthinkable in a military command facility or indeed civilian sites today. In the context of mechanisms ensuring the integrity of classified information and technology through access control, the composition of visitor groups are usually constrained to professional and political delegations.

The tension between transparency and obscurity during the Cold War was also played out in the choices of location of infrastructure. Military or joint civilian-military operators had to physically protect and conceal their facilities in order to keep them out of the public, or an adversary's, eye and reach, particularly since mission control facilities represented a key vulnerability of an entire space systems architecture.[48] Command centers were drilled into mountain ranges. Philco-Ford, the contractor who had supplied the setup surrounding the computers for mission control in Houston, built a nearly identical system deep inside Cheyenne Mountain in Colorado for the North American Air Defense Command (NORAD).[49] Aside from the essence of dual use in this context – the literal replication and deployment of the same technology for two purposes – a hidden underground location rendered a technology inconspicuous. With a similar aim, sites such as the Soviet Baikonur launch complex operated by the armed forces were deliberately misnamed, disguised or their exact locations undisclosed.

Civilian programs needed to consider such security aspects less prominently. But their facilities, such as launch sites, were also often located in what – for European actors – represented geographically inaccessible locations. The European launch site in Kourou was constructed from 1964 onwards on a then malaria-infested coastal strip bordering the Amazon rainforest in French Guiana (Figure 11.4).[50] Next to launch sites and operations centers, other infrastructural elements were also often distributed in relatively remote regions. Due to requirements of tracking spacecraft launched into a certain orbit, Europe's network of ground stations for real-time commanding, ESA Tracking Stations (ESTRACK, since 1964), was located in sites including higher latitudes, such as Fairbanks, Alaska, and Svalbard, Norway. Their local control rooms were much less elaborate than the facilities at ESOC in Darmstadt – then still called the European Space Data Acquisition Centre (ESDAC). The ground stations were staffed by controllers and scientists who coped with long polar winters, interrupted only by fortnightly dispatches of magnetic tapes containing telemetry data to mainland Europe.

Figure 11.4 The control room at the launch facility of the European Launcher Development Organisation (ELDO) equatorial base at the Guiana Space Centre in Kourou, French Guiana, in the early 1970s. The facility was later significantly expanded into the Jupiter flight control center of the Centre Spatial Guyanais (CSG).
Source: Courtesy of European Space Agency.

Since tracking technology hailed from systems such as radar technology developed for missile ranges, early warning systems or surveillance and reconnaissance, their deployment in certain strategic locations sometimes raised suspicion.[51] ESTRACK, for instance, included one ground station on the Svalbard archipelago operated by the Norwegian Technical Science Research Council for ESA's precursor organization ESRO. The ground station's location in the High Arctic, essential for tracking polar orbits vital for earth observation, posed a geopolitical and logistical challenge. Due to its strategic location, the Svalbard Treaty from 1920 prescribed the archipelago to be a demilitarized zone, not to be used for what was termed 'warlike' purposes. At the time, this predominantly concerned the establishment of naval bases and fortifications, but did allow the installation of research and logistics facilities. When in 1964, however, ESRO obtained permission from the Norwegian government to build a telemetry station for satellite missions of purely civilian research, this was met with protest: The Soviets alleged that the humble building, with its two plastic radome-covered antennas constituted, in fact, a secret espionage station to serve the surveillance satellites of NATO. After fresh Norwegian affirmations as to the peaceful scientific nature of the site in compliance with the Svalbard Treaty failed to convince their interlocutors, the incident was in the end resolved by adopting an 'open door policy' granting Soviet and other scientists the right to visit the station whenever they wished.[52]

Ultimately, this transparency measure ensured security at an otherwise volatile setting. After the ESRO telemetry station ceased operations in 1974, it paved the way for a successful and sustained presence of telemetry stations serving ESA and other actors in Svalbard today.[53] Generally, building a facility and having others discover or visit it – like the European ground station, or the aforementioned Soviet control center – also always entailed the managing of expectations, assumptions and suspicions about a site's real purpose, character or even whereabouts.

From 1985 the Americans and Europeans, and from 1993 also the Russians, cooperated more closely for the International Space Station (ISS) program, together with their Canadian and Japanese partners. With a first resident crew onboard from 2000, the act of sharing information on command and control capabilities was both mindful of security concerns, but also deferred to the expertise of those most knowledgeable or invested in the respective sub-systems of the orbital platform. For operations of the different ISS segments, this included mission control centers in Houston and Moscow, as well as dedicated payload control centers for the European Columbus module at GSOC near Munich, and later for the Automated Transfer Vehicle (ATV) supply vehicle in Toulouse, France.[54] The complexity of missions, their international crew composition and sheer scale required global coverage of sensors and command structures. It thus brought about an expansion from single control centers to a globally distributed network, and by default, a greater degree of transparency.

Transparency was a prerequisite for, and in turn promoted, cooperation. The early fact-finding trips in the late 1960s of European operators to the advanced space actors in the United States, and their less frequent but insightful exchanges with their Soviet counterparts, evolved to the status of transparency and confidence-building measures (TCBMs) post-1989. This was a pivotal time: Europe's postwar science-oriented and collaborative culture of operations was idiosyncratic compared to the more unilateral, military-rooted engineering cultures in the United States and Soviet Union. Yet it began to frame and re-order its space activities and increasingly started to develop the dual or defense dimension of space technology, applications and infrastructure. While 'new space' actors across the Atlantic today lend an air of transparency to some of their operations – from glass-enclosed control rooms to generic furniture and live-streamed transmissions to the public – and institutional European operators increasingly employ sophisticated public engagement strategies for some of their programs and missions, Europe's space community has also been sensitized to the emerging needs of commercial and military actors, in view of an increasingly comprehensive interpretation of security.

IV Control rooms as sites of securitization

Throughout their evolution operational centers of civilian programs have been representing sites of international prestige, of technical mastery and engagement with external actors. Meanwhile, the concept of the control center navigated the blurring fault-line of dual use. The tangible makeup of the prototypical space mission control room manifested a structure of command that evolved from decision-making environments in the military. Its operational practices and terms of commanding through a human-machine interface had a legacy in the command and control paradigm of ballistic missile defense (BMD). And the organizational strategy of fostering a culture of transparency or opacity of operational facilities was initially driven by the need to showcase, in a civilian context, capabilities potentially to be employed for defense purposes and deterrence. Beyond their real-time operational functions, certain design features and organizational practices provided transparency for space agency and control center staff, fueled the imagination of both the public and the international security community, promoted trust and reassurance or demonstrated power towards partners and adversaries alike. Control centers thus functioned as a kind of tangible subtext. As a political tool, this transparency was rooted in the defense and security realm as much as in the civilian, and is today considered part of the repertoire of space security practice.[55]

As civilian control centers evolved, their heritage of defense elements was dynamic and fed by the formal collaboration and informal exchange of expertise between the different institutions and their personnel. Civilian control centers would initially wholly absorb military concepts and then shed some of the layers of this legacy by developing their own practices of control that fused transparency with greater informality. Despite the formal enfranchisement and organizational distinction from their defense pedigree, however, the deeply ingrained conceptual core of defense legacy would never be dismantled.

Instead, it was cautiously cultivated. While previously outer space had primarily been framed as a medium for civilian scientific and commercial applications, in mid-1990s Europe the fusion of civilian and military undertakings and the expanding concept of 'security' were considered more prominently. Today, control rooms for satellite fleets of programs such as earth observation and navigation are civilian-operated, but often form an integral part of systems-of-systems that are also understood as critical infrastructure and are positioned in the context of security.[56] Ensuring the safety and security of assets in space is paramount for the reliable provision of routine services and applications that are central to the functioning of societies and industries. At the same time, they lend a degree of autonomy for the operating state actor – including supranational entities such as the European Union. In light of this, the raison d'être of European flagship programs such as Copernicus or Galileo are conceived, or explicitly re-framed during their life cycle, to also fulfill military or security needs.[57]

In the early 1990s the immediate post-Cold War environment foreshadowed an initially latent, then explicit security-imperative of civilian space systems. This was increasingly shaped by diverse partnerships of civilian, military, security, science, governmental and commercial actors, as well as the cross-pollination of their respective cultures.[58] Against a backdrop of the 'securitization' of space and an increasingly widely framed concept of both space security and security per se beyond the traditional remit of defense, European actors have since been cultivating an ambiguity towards the dual dimension of space applications.[59] Control rooms are the epitome of this ambiguity. Their utilization for programs in a civilian context and the related organizational efforts to make operations transparent, even inclusive, allowed control rooms in Europe a partial conceptual emancipation from their military legacy.

However, this did not usher in a period of divestment from the engrained fabric of dual use. Instead, the notion of dual use as a strategy for both industrial and security policy retains its relevance today, particularly in Europe. Previously, dual use was regarded a byproduct, an intrinsic feature that presented export control headaches. Today it is promoted as unique potential, synonymous with synergies that not only allow, but also demand the exploitation of multiple applications, when other industrial sectors compete for public funds. Policy-makers in Europe thus simultaneously invoke the dual nature of space and defense technology as driver of the consolidating aerospace and defense industry, and in response actively seek to promote and develop dual programs. This reasoning is fostered further through the consideration of space in European defense policies and the growing links between traditional and emerging space actors. This is somewhat particular in Europe, where military space forces are still comparably less integrated into the modus operandi of civilian programs than in the United States, China or Russia. Expedited through the increased emphasis and integration of security aspects and, to an extent, military requirements in civilian programs and nascent policy considerations on a European level, this development is likely to continue and even accelerate.[60]

Thus the doubly civilian and military character of the generic control room that had remained a key feature throughout the Cold War assumes renewed importance today. Many of the early operations sites are still active. Their heritage is alive not only in organizational histories, but also coded into their future: Whether missions have been operating for a couple of decades, or whether control rooms are gutted to host new operations, the elementary concept of the control or operations center is as enduring as its internal settings are malleable. Various configurations can accommodate needs on the spectrum of security for purely civilian, purely military, explicitly joint military-civilian control centers or those commercially run on behalf of military or civilian customers that populate the booming ecosystem of space operations in its broadest sense today. Stripped bare of obvious cultural insignia of their operating organization – and without wading into the fine-grained depth of component-level measures such as encryption for classified networks – a distinct categorization of control centers on any but the highest few levels of abstraction remains elusive. Most ground-based space infrastructure today is relatively immobile once deployed and likely to be operated over a long time as part of continually upgraded legacy systems. Yet some of their key components are highly dynamic, as human expertise, software and tacit practice migrate between state and non-state actors through technology transfer, personnel transition, procurement policies, cross-institutional collaboration and corporate reorganization of the industrial sector of prime contractors building space systems. This inherent fluidity invites the conclusion that plasticity and ambiguity between the civilian and the military will continue to be a salient feature in the future, as the space sector attracts an increasingly heterogeneous community of actors.

As it bundles distinct strands of human agency, the concept of the 'control room' serves as a platform containing the subtle but symptomatic markers of a possible securitization of space in Europe, and reflects how space has become an increasingly complex operational domain today. Orbital regimes such as polar or geosynchronous orbits have been consolidated as sites for critical infrastructure. If it ever was, space is no longer the locale of optimistic projection, a backdrop for a grander narrative of the human condition or an aspirational destination. Instead of an end, it has truly become a means: Access to various orbits that societies and economies rely on for real-time positioning, observation and communication services, a vehicle for security policy, a driver of industrial development, a global commons like the high seas and Antarctica – and for a growing community, a warfighting domain. The provenance and aggregation of major elements of mission control, which links us to the otherwise inaccessible and hostile domain of space, ceases to be attributable indiscriminately to a distinct realm on the dual spectrum. In light of this, it is particularly the European development in the global context of space operations which ought to be viewed progressively through the lens of securitization even more so than that of militarization.

Notes

1. On programmatic values in complex socio-technical systems, see the work of Kim J. Vicente; for example, idem, *Human-Tech: Ethical and Scientific Foundations*, Oxford: Oxford University Press, 2010; and idem, *Cognitive Work Analysis: Towards Safe, Productive, and Healthy Computer-Based Work*, Mahwah: Erlbaum, 1999. All Internet sources were last accessed on 15 July 2020.
2. For a de- and re-construction of the intricacies of Apollo control technology, see David A. Mindell, *Digital Apollo: Human and Machine in Spaceflight*, 2nd edn, Cambridge, MA: MIT Press, 2011, while William J. Clancey has performed detailed ethnographic research of contemporary science operations at NASA for *Working on Mars: Voyages of Scientific Discovery with the Mars Exploration Rovers*, Cambridge, MA: MIT Press, 2012. For detailed discussions of the coming-in-to-being of US combat and mission control centers, see Layne Karafantis, *Under Control: Constructing the Nerve Centers of the Cold War*, PhD thesis, Baltimore: Johns Hopkins University, 2016; in comparison to Europe, see Michael Peter Johnson, *Mission Control: Inventing the Groundwork of Spaceflight*, Gainesville: University Press of Florida, 2015. In *Spacesuit: Fashioning Apollo*, Cambridge, MA: MIT Press, 2011, Nicholas de Monchaux dedicates a chapter to the physical and information architecture of the Apollo control room, including hardware supplied by defense contractors.
3. For detailed accounts of the operational environment of mission control, see Christopher C. Kraft, *Flight: My Life in Mission Control*, Waterville: Thorndike Press, 2001; Gene Kranz, *Failure Is Not an Option: Mission Control from Mercury to Apollo 13 and beyond*, New York: Simon & Schuster, 2000; Rick Houston and Heflin J. Milt, *Go Flight! The Unsung Heroes of Mission Control, 1965–1992*, Lincoln: University of Nebraska Press, 2015; and Loyd S. Swenson Jr., James M. Grimwood and Charles C. Alexander, *This New Ocean: A History of Project Mercury*, Washington, DC: NASA, 1989.
4. Bernard Lovell, *The Origins and International Economics of Space Exploration*, Edinburgh: Edinburgh University Press, 1973, here 20 and 33.
5. See, for example, John A. Alic 'The Dual Use of Technology: Concepts and Policies,' *Technology in Society* 16.2 (April–June 1994), 155–72, here 157. For a comparative framework of characteristic differentiations of civilian and military systems, see Stephen E. Doyle, *Civil Space Systems: Implications for International Security*, Geneva: United Nations Institute for Disarmament Research, 1994, 89; Melvin C. Kranzberg, 'Technology and History: "Kranzberg's Laws",' *Technology and Culture* 27.3 (July 1986), 544–60; and James R. Hansen, *Technology and*

the History of Aeronautics: An Essay, Washington, DC: US Centennial of Flight Commission, 2004.

6. The evolution of dual use, its ambiguity and potential post-1989 'uselessness' are explored in Roger Handberg, *Seeking New World Vistas: The Militarization of Space*, Westport: Praeger, here 53–5; Patrick A. Salin, 'Privatization and Militarization in the Space Business Environment,' *Space Policy* 17.1 (February 2001) 19–26, here 19 and 22. See also Alexander Geppert and Tilmann Siebeneichner's introduction, Chapter 1 in this volume.
7. Jordi Molas-Gallart, 'Which Way to Go? Defence Technology and the Diversity of "Dual-use" Technology Transfer,' *Research Policy* 26.3 (February 1997), 367–85; Haico te Kulve and Wim A. Smit, 'Civilian-military Co-operation Strategies in Developing New Technologies,' *Research Policy* 32.6 (June 2003), 955–70, here 955–6. Note that these observations predominantly refer to the NATO context, specifically Western Europe, although the conceptual trend is also applicable to states of the former Warsaw Pact.
8. Philip Kraft, *Programmers and Managers: The Routinization of Computer Programming in the United States*, New York: Springer, 1977, 35. For the military-science-corporate convergence in space, see also James C. Moltz, *The Politics of Space Security: Strategic Restraint and the Pursuit of National Interests*, Stanford: Stanford University Press, 2008, 317, 325; Max M. Mutschler, *Arms Control in Space: Exploring Conditions for Preventive Arms Control*, Basingstoke: Palgrave Macmillan, 2013, 2 and 194; and Alasdair W. M. McLean, *Western European Space Policy*, Aldershot: Dartmouth, 1992, 19.
9. Tommaso Sgobba and Firooz A. Allahdadi, 'On-Orbit Mission Control,' in eidem et al., eds, *Safety Design for Space Operations*, Amsterdam: Elsevier, 2013, 371–410, here 371.
10. See, for instance, John Krige, *American Hegemony and the Postwar Reconstruction of Science in Europe*, Cambridge, MA: MIT Press, 2006, 228–37. For the guided missile as key dual-use technology, see, for example, Michael J. Neufeld, *The Rocket and the Reich: Peenemünde and the Coming of the Ballistic Missile Era*, New York: Free Press, 1995, here 270.
11. 'Fuel-Cell Flight,' *Time Magazine* 86.17 (27 August 1965), 46–54, here 54; Kranz, *Failure*, 27. For control as a concept of fire control, see Paul E. Ceruzzi, *Computing: A Concise History*, Cambridge, MA: MIT Press, 2012, 39.
12. Neufeld, *Rocket*, 106. For anecdotal descriptions, see also Walter Dornberger, *Peenemünde: Die Geschichte der V-Waffe* [1952], Berlin: Ullstein, 2013, 11–26. For JPL's history as ballistic missile range, see Michael Peter Johnson, *'This Is Ground Control': The Invention of Mission Control Centers in the United States and Europe*, PhD thesis, Auburn: Auburn University, 2012, 60. For practicalities of a missile control setting in the full-scale program for nuclear weapons systems during the Cold War, see Eric Schlosser, *Command and Control: Nuclear Weapons, the Damascus Accident, and the Illusion of Safety*, New York: Penguin, 2013. On ICBMs, see also the contribution by Christopher Gainor, Chapter 3 in this volume.
13. Thomas P. Hughes, *Human-Built World: How to Think About Technology and Culture*, Chicago: University of Chicago Press, 2004, 53; Donald F. Lamberti and C. F. Wyke, *Study of Requirements of Operational Support Tools for Multisatellite Control Center: Final Report*, Paris: ESA, 1993, here 13; Elliott Automation

Group, 'In Aviation: Farnborough 1962' [advertisement], *Flight International* (30 August 1962), 26; 'Altitude Plant: Rolls-Royce's Privately-Financed High-Altitude Establishment,' *Flight* (3 October 1958), 551; William G. Osmun and Evan Herbert, 'Air Traffic Control,' in Robert Colborn, ed., *Modern Science and Technology*, Princeton: Van Nostrand, 1965, 306–22. For the command paradigm in process control settings that 'evolved from military command and control systems' at the time, see, for example, 'Display-command Operator/Machine Interface,' in Douglas M. Considine, *Encyclopedia of Instrumentation and Control*, New York: McGraw Hill, 1971, 202–11.

14. Viktor Blagov and Wladimir Samsonov, 'Raumfahrtkontrollzentrum Koroljow: Standleitung zur Internationalen Raumstation ISS,' in Philipp Meuser, ed., *Architektur für die russische Raumfahrt: Vom Konstruktivismus zur Kosmonautik: Pläne, Projekte und Bauten*, Berlin: DOM, 2013, 232–45; Kranz, *Failure*, 19–20.
15. Ibid., 23. For detailed accounts of both shaping and working in the built environment of mission control, see also Kraft, *Flight*; Houston and Milt, *Go Flight!*; and Swenson, Grimwood and Alexander, *New Ocean*.
16. For a detailed architectural history of site, building and interiors architectural, see Ray Loree, *MCC Development History*, Washington, DC: NASA, 1990. See also Patricia Slovinac, *Mission Control Center/Building 30: Historical Documentation*, Houston: NASA, 2010; Archaeological Consultants, *Assessment Report: Inventory of Historic Artifacts from the Mission Control Center, Cape Canaveral Air Force Station, Florida*, Houston: NASA, 2009.
17. Brian O'Leary, *The Making of an Ex-Astronaut*, Boston: Houghton Mifflin, 1970, 217; *Dr. Strangelove or: How I Learned to Stop Worrying and Love the Bomb*, directed by Stanley Kubrick, UK/USA 1964 (Columbia Pictures). David Hayles, 'Is This the Best Film Set Ever Designed? On Dr Strangelove's War Room,' *New Statesman* (5 November 2014), available at http://www.newstatesman.com/culture/2014/11/best-film-set-ever-designed-dr-strangelove-s-war-room, notes that although designer Ken Adam points out he 'certainly didn't base the war room on (government facilities),' he had indeed previously been exposed to military infrastructure and equipment while serving as fighter pilot in the Royal Air Force during the Second World War.
18. In his history of the Apollo space suit, Nicholas de Monchaux identifies control rooms as an additional information layer of the suit system, exposing an inextricable operational link; see idem, *Spacesuit*, 280–2. Incidentally, Kubrick also drew on the same corporations and research institutions to enhance the representation and extrapolation of the aerospace systems depicted in *Dr. Strangelove* and *2001: A Space Odyssey*. See, for example, Peter Krämer, *Dr. Strangelove or: How I Learned to Stop Worrying and Love the Bomb*, London: British Film Institute, 2014; and Regina Peldszus, 'Speculative Systems: Kubrick's Interaction with the Aerospace Industry during the Production of *2001*,' in Tatjana Ljujic, Peter Krämer and Richard Daniels, eds, *Stanley Kubrick: New Perspectives*, London: Black Dog, 2015, 234–52.
19. See *European Experts Operate Apollo Tracking Network*, 16 July 2009, available at http://m.esa.int/Our_Activities/Operations/Estrack/European_experts_operate_Apollo_tracking_network; 'European Apollo Ground Station Pioneer' 15 July 2009, available at http://www.esa.int/Our_Activities/Operations/European_Apollo_ground_station_pioneer.

20. See ESOC's Director of Operations, the Austrian Kurt Heftman, 'New Chief at Esoc,' *Flight International* (3 September 1983), 622.
21. Geoff C. Tootill, *Capabilities of the ESRO Control Centre and Tracking Network*, Noordwijk: European Space Technology Centre, 1967, 2, 8; David E. B. Wilkins, 'Mission Control: ESA's New Control Centre,' *Satellite Technology International Space Report* 1.2 (May 1985), 9–13.
22. H. Marin and R. Fraysse, *Définition provisoire et partielle de l'organisation et de l'équipement du Centre de Contrôle sur the plan opérationnel*, Noordwijk: ESTEC, 1965; David E. B. Wilkins, 'The Operations Control Centre,' *Journal of the British Interplanetary Society* 40.6 (June 1987), 243–56.
23. 'Mission Control,' *Flight International* (21–28 December 1985), 36.
24. Hubertus Wanke, '"We Have a Mission!" Das Raumfahrt-Kontrollzentrum in Oberpfaffenhofen,' *DLR Nachrichten: Special Issue 35 Jahre Raumfahrt-Kontrollzentrum* 105 (August 2003), 64–71.
25. Jean-Paul Abadie, 'A New Line of Products Designed for Mission Control Centers for Low Earth Orbit Satellites,' *Acta Astronautica* 40.2–8 (January–April 1997), 211–21, here 221.
26. For instance, key parts for onboard and control systems in space operations were subject to the International Traffic in Arms Regulation (ITAR); see Jens Eickhoff, *Onboard Computers, Onboard Software and Satellite Operations: An Introduction*, Heidelberg: Springer, 2012, 45.
27. Reference sites included control centers of Italsat and Olympus in Fucino; Intelsat in Washington, DC and Fucino; the GSOC in Oberpfaffenhofen; Eutelsat in Paris; and NASA's Goddard Space Operations Center in Greenbelt, Maryland. See Lamberti and Wyke, *Study of Requirements*, 1993.
28. For Konrad Zuse's commanding chart in the operation of computers and the 'command and time table' utilized in missile launch operations, see Ceruzzi, *Computing*; and Dornberger, *V-Waffe*. For terminology, see Defense Technology Integration Center, *DTIC Mil Doctrine: Department of Defence Dictionary of Military and Associated Terms*, Washington, DC: Office of the Secretary of Defence, 2010; and Barry Smith, Kristo Miettinen and William Mandrick, 'The Ontology of Command and Control (C2),' *Proceedings of the 14th ICCRTS, International Command and Control Research and Technology Symposium, 15–17 June 2009, Washington, DC*, Washington, DC: Department of Defence, 2009.
29. On Mueller's systems engineering background, see Arthur L. Slotkin, *Doing the Impossible: George E. Mueller and the Management of NASA's Human Spaceflight Program*, New York: Springer, 2012; for the significance of systems engineering in Apollo leaders, see Mindell, *Digital Apollo*, 37; and for a comprehensive outline, Stephen B. Johnson, *The Secret of Apollo: Systems Management in American and European Space Programs*, Baltimore: Johns Hopkins University Press, 2002.
30. Richard C. Booton Jr. and Simon Ramo, 'The Development of Systems Engineering,' *IEEE Transactions on Aerospace and Electronic Systems* 20.4 (July 1984), 306–10, here 306–7.
31. James Martin, *Telecommunications and the Computer*, Englewood Cliffs: Prentice-Hall, 1976, 74; Rebecca Slayton, *Arguments That Count: Physics, Computing, and Missile Defense, 1949–2012*, Cambridge, MA: MIT Press, 2013, 37–8.
32. Joost Jongert and A. M. Bos, *Advanced Workstations for Meteosat Command and Control*, vol. 1, Utrecht: BSO/Aerospace, 1987; Richard Holdaway, 'Ground

Stations,' in Peter W. Fortescue, John P. W. Stark and Graham G. Swinerd, eds, *Spacecraft Systems Engineering*, New York: John Wiley, 2003, 477–500, here 494–5.

33. Considine, *Encyclopedia*, 153; John T. Garner and Malcolm Jones, *Satellite Operations: Systems Approach to Design and Control*, Chichester: Horwood, 1990.
34. Layne Karafantis, 'Nasa's Control Centers: Design and History,' *Engineering & Technology Magazine* 8.12 (16 December 2013), available at https://eandt.theiet.org/content/articles/2013/12/nasas-control-centers-design-and-history/; Monchaux, *Spacesuit*; Nigel Head, 'SCOS II: ESA's New Generation of Control Systems,' *Acta Astronautica* 35.1 (January 1995), 515–24, here 517; Eickhoff, *Onboard Computers*, 226; and James E. Tomayko, *Computers in Spaceflight: The NASA Experience*, Washington, DC: NASA, 1988.
35. See Schlosser, *Command and Control*, 27–9, for the launch procedure of the Titan missile system; David Velupillai, 'Europe's Equatorial Launch Site,' *Flight International* (17 February 1979), 467–71.
36. Harold R. Booher, 'Introduction: Human Systems Integration,' in idem, ed., *Handbook of Human Systems Integration*, New York: John Wiley, 2003, 1–30, here 19; see also Richard W. Pew and Anne S. Mavor, *Human-System Integration in the System Development Process*, Washington, DC: National Academies Press, 2008.
37. David E. B. Wilkins, 'Mission Control: ESA's New Control Centre,' *Satellite Technology International Space Report* 1.2 (May 1985), 9–13, here 13; Mario Merri et al., 'Happy Families: Cutting the Cost of ESA Mission Ground Software,' *ESA Bulletin* 130 (May 2007), 62–9.
38. Andrea Baldi et al., 'The Evolution of ESA's Spacecraft Control Systems,' *ESA Bulletin* 89 (February 1997), available at http://www.esa.int/esapub/bulletin/bullet89/baldi89.htm; Michael Jones and Nestor Peccia, 'Making European Industry More Competitive,' *ESA Bulletin* 126 (May 2006), 55–61, here 57–8; Merri et al., 'Mission Ground Software.'
39. See, for instance, Tim Furniss, 'Logical Control,' *Flight International* 14.20 (November 1990), 28–30; and Nayef Al-Rodhan, *Meta-Geopolitics of Outer Space: An Analysis of Space Power, Security and Governance*, Basingstoke: Palgrave Macmillan, 2012, 18.
40. Stacia E. Zabusky, *Launching Europe: An Ethnography of European Cooperation in Space Science*, Princeton: Princeton University Press, 1995.
41. Slovinac, *Mission Control Center*, 13; *Failure Is Not an Option*, directed by Rushmore DeNooyer and Kirk Wolfinger, USA 2003 (History Channel). See in this context Lorenz Engell, 'Das Mondprogramm: Wie das Fernsehen das größte Ereignis aller Zeiten erzeugte und wieder auflöste, um zu seiner Geschichte zu finden,' in Friedrich Lenger and Ansgar Nünning, eds, *Medienereignisse der Moderne*, Darmstadt: Wissenschaftliche Buchgesellschaft, 2008, 150–71; and Michael Allen, *Live from the Moon: Film, Television and the Space Race*, London: I. B. Tauris, 2009.
42. Brian Harvey, *Race into Space: The Soviet Space Programme*, Chichester: Horwood, 1988, 219; see also Pat Norris, *Spies in the Sky: Surveillance Satellites in War and Peace*, Chichester: Springer, 2008.
43. Dominic Phelan, 'Sir Bernard Lovell and the Soviets,' *Spaceflight* 56.9 (September 2014), 336–8.

44. ESA, *Informationen zur Astrolab Mission*, Noordwijk: Erasmus User Centre, 2006; Blagov and Samsonov, *Raumfahrtkontrollzentrum*.
45. Donald D. Hardy, 'Satellite Operations and Control – Proliferate or Automate?,' *Proceedings of the Australian Astronautics Convention, Perth, Australia, 27–30 August 1975*, Perth: Astronautical Society of Western Australia, 1977, 65–79, here 65.
46. Kubrick, *Dr. Strangelove*; the 'big board' refers to the large wall display of higher level strategic information.
47. Madeleine Schäfer, *How to Survive in Space: A Light-Hearted Chronicle of ESOC*, vol. 1: *1963–1986*, Darmstadt: ESA, 1997, 53–4.
48. For the contemporary notion leveraging the ground segment to compromise the integrity of a space asset, see, for example, Dean Cheng, 'Chinese Concepts of Space Security,' in Kai-Uwe Schrogl et al., eds, *Handbook of Space Security*, vol. 1: *Policies, Applications, Programs*, New York: Springer, 2015, 431–51, here 447.
49. Monchaux, *Spacesuit*, 277, 292. See also Slayton, *Arguments*, 65; Joseph A. Carretto Jr., *Military Man in Space: Essential to National Strategy*, Washington, DC: National Defense University, 1995; and Layne Karafantis, 'NORAD's Combat Operations Center: A Distinctively Cold War Environment,' *Information & Culture: A Journal of History* 52.2 (May/June 2017), 139–62.
50. In his partially auto-ethnographic reflection on the contemporary setting of Kourou after the Cold War, Peter Redfield has examined the launch site as a 'placeless space'; see his 'Beneath a Modern Sky: Space Technology and Its Place on the Ground,' *Science, Technology and Human Values* 21.3 (Summer 1996), 251–74, here 252–3 and 264.
51. For the related topic of tracking in missile defense, respectively reconnaissance, in the Arctic, see also the outline of the network developed for the Atlantic Missile Range in Stanley J. Macko, *Satellite Tracking*, New York: John F. Rider, 1962.
52. The incident is recounted in the context of international security and the Svalbard Treaty by Willy Østreng, *Politics in High Latitudes: The Svalbard Archipelago*, London: Hurst, 1978, 57–8; and, from a national industrial and research policy perspective, in Olav Wicken, 'Cold War in Space Research: Ionospheric Research and Military Communication in Norwegian Politics,' in John Peter Collett, ed., *Making Sense of Space: History of Norwegian Space Activities*, Oslo: Aschehoug, 1995, 41–73.
53. The interpretation of 'warlike purposes' in Article IX of the Svalbard treaty from 1920, much as the notion of 'peaceful use' of space in the Outer Space Treaty (OST) from 1967, has evolved to accommodate the needs of dual use and security aspects of infrastructure and missions. On the Outer Space Treaty, see Luca Follis, 'The Province and Heritage of Humankind: Space Law's Imaginary of Outer Space, 1967–79,' in Alexander C. T. Geppert, ed., *Limiting Outer Space: Astroculture after Apollo*, London: Palgrave Macmillan, 2018, 183–205 (= *European Astroculture*, vol. 2).
54. Michael K. Fawcett, 'From Centralised to Distributed: The Evolution of Space Station Command and Control,' *Acta Astronautica* 38.4–8 (February–April 1996), 637–45, here 643–4.
55. Scott Pace, 'Security in Space,' *Space Policy* 33.2 (August 2015), 51–5, here 54; Jana Robinson and Michael Romancov, 'The European Union and Space: Opportunities and Risks,' *Non-Proliferation Papers* 37 (January 2014), here 2; on

TCBM specifically, see Jana Robinson, 'Transparency and Confidence-Building Measures for Space Security,' *Space Policy* 27.1 (February 2011), 27–37; and for transparency and cooperation in the disarmament context, see Michael C. Mineiro, *Space Technology Export Controls and International Cooperation in Outer Space*, New York: Springer, 2012, 212.

56. See Markus Hesse and Markus Hornung, 'Space as Critical Infrastructure,' in Schrogl et al., *Handbook of Space Security*, 187–201, here 196. Infrastructure as a term itself found its way into the language of military logistics during the 1950s through NATO and was itself inherently understood as dual use; see Dirk van Laak, 'Infra-Strukturgeschichte,' *Geschichte und Gesellschaft* 27.3 (July–September 2001), 367–93, here 370–3. On 'security' as a concept, see also Daniel Brandau's contribution, Chapter 7 in this volume.
57. See, for instance, Paul Stephenson, 'Talking Space: The European Commission's Changing Frames in Defining Galileo,' *Space Policy* 28.2 (May 2012), 86–93; Péricles Gasparini Alves, ed., *Evolving Trends in the Dual Use of Satellites*, Geneva: United Nations Institute for Disarmament Research, 1996; and, above all, Paul Ceruzzi's contribution, Chapter 13 in this volume.
58. Theresa Hitchens and Tomas Valasek, 'The Security Dimension of European Collective Efforts in Space: Military Spending and Armaments 2005,' in Alyson J. K. Bailes, ed., *SIPRI Yearbook 2006: Armaments, Disarmament and International Security*, Stockholm: Stockholm International Peace Research Institute/Oxford University Press, 2006, 565–77, here 547.
59. The notion of ambiguity with which European actors such as ESA, the European Commission or the European Defense Agency frame the intersection of the civilian and military dimension in space security is addressed, for instance, by Frank Slijper, 'The EU Should Freeze Its Military Ambitions in Space,' *Space Policy* 25.2 (May 2009), 70–4, here 73; Columba Peoples, 'The Growing "Securitization" of Outer Space,' *Space Policy* 26.4 (November 2010), 205–8; Michael Sheehan, 'Defining Space Security,' in Schrogl et al., *Handbook of Space Security*, 7–21; and Cesar Jaramillo, 'The Multifaceted Nature of Space Security Challenges,' *Space Policy* 33.2 (August 2015), 63–6.
60. For an extensive discussion, see Nunzia Paradiso, *The EU Dual Approach to Security and Space: Twenty Years of European Policy Making*, 2013, Vienna: European Space Policy Institute, 2013. For the change in emphasis, see, for instance, Bertrand de Montluc, 'SSA: Where Does Europe Stand Now?,' *Space Policy* 28.3 (August 2012), 199–201, here 200; and Salin, 'Privatization,' 20–1. See also Gérard Brachet and Bernard Deloffre, 'Space for Defence: A European Vision,' *Space Policy* 22.2 (May 2006), 92–9, here 98; 'First Steps Into Space,' *European Defence Matters* 1 (May 2012), 16–19; and Max M. Mutschler and Christoph Venet, 'The European Union as an Emerging Actor in Space Security?,' *Space Policy* 28.2 (May 2012), 118–24.

CHAPTER 12

Space Spies in the Open: Military Space Stations and Heroic Cosmonauts in the Post-Apollo Period, 1971–77

Cathleen Lewis

In the 1970s the Soviet space program turned from its Cold War origins and developed its own, new and more internally defined course that maintained the many mythologies that had originated in the 1960s.[1] While the Apollo 11 moon landing marked a watershed in the Space Race, from the Soviet perspective all that changed was that there was no longer a head-to-head public competition between the two sides. The competition, however, had not ended. The United States had truncated planned Apollo lunar missions and turned its focus solely on developing a reusable orbiter for the 1970s. The Soviet Union found itself with human spaceflight hardware built and prototyped for a moon program and a plan to place humans in low-earth orbit for weeks at a time. They also had orbital hardware that had been overlooked due to internal infighting among design groups.[2] What resulted was a human spaceflight program split between one devoted to a series of human-tended military space stations and a very similar civilian one. Both had been cobbled together from a previous era's hardware that folded a civilian program and a military-curated program together.

The result of the two space powers quickly changing directions of their human spaceflight programs after the Moon Race was that public perceptions

Cathleen Lewis (✉)
Smithsonian National Air and Space Museum, Washington, DC, USA
e-mail: LewisCS@si.edu

Alexander C. T. Geppert et al. (eds), *Militarizing Outer Space*
European Astroculture, vol. 3
https://doi.org/10.1057/978-1-349-95851-1_12

turned on their heads. Americans became quickly bored with the Apollo spectaculars as demonstrated by their television viewing.[3] In the Soviet Union, there were much more concrete justifications for a loss of interest – Soviet leadership in the program had faded away.[4] Without a clear, definable objective and a dwindling list of remaining firsts to be accomplished, the public could no longer measure the relative position of the competitors in this new era of the Space Race. The United States closed out the Apollo program through a rump space station program called Skylab by repurposing Apollo hardware and then took a long hiatus in human spaceflight, leaving an impression of the Americans turning away from human spaceflight. In contrast, the Soviet Union progressed headlong into a venture in sustained human presence in space that also upheld their public presence in space. They appended a new military program to the space station. This program included a shielded project that seemed to fulfill what had been an early 1960s US Air Force aspiration of establishing a military space station. The Americans had discarded the duplicative effort of running both military and civilian human spaceflight programs once it became clear that robotic reconnaissance was the better alternative. The Soviet Union took up the dual missions. The new space station program had a public face of continued flights of increasing duration, using two separate hardware systems that had been designed in the previous decade. And the Soviet Union continued the much-denied program to send humans to the moon. They also took on a fourth mission of cooperating with the United States in human spaceflight rescue and safety at the same time culminating in the Apollo-Soyuz Test Project (ASTP) in July 1975. From the outside, the Soviet Union seemed to be ascending in space as the United States dismantled the Apollo program. Internally, the Soviet Union was quadrupling its human spaceflight programs in spite of the fact that funding and resources had declined in the mid-1960s.[5]

Examining how the Soviet Union managed these multiple purposed programs, especially the introduction of a shadow, military program, touches on many issues of how the Soviet space effort transitioned from the heyday of human spaceflight in the 1960s that captivated the world's imagination to an age of stagnation in the 1980s. Soviet spaceflight in the 1970s operated in a political environment that was increasingly disconnected from domestic civil culture in the arts, literature and human rights that would ultimately undercut communist authority for the next decade. In the 1960s the space program had celebrated a hope for a bold new future for the country. By the 1970s the human spaceflight program represented yet another bureaucratic tool of the Brezhnev regime that fixated on the present and offered few if any promises for the future. Key to unraveling this mystery is an understanding of the climate and assumptions under which the Soviet space program and especially that of human spaceflight operated during the 1970s. Planetary exploration had found another institutional home in the Lavochkin design bureau in 1965.[6] Although linked by technology and scientific doctrine both programs were distinct in their politics during the first decade of the Space Age. The second and more difficult comparison is to overall Soviet military doctrine of the time to assess if the military human space program did operate and change according to overall Soviet military doctrine or if it remained a battleground for bureaucratic infighting. This suggests that in the case of human-tended military operations in space, the Soviet military lacked the influence to override other, civilian concerns outright, but sought a rare compromise.

The US and USSR space programs diverged at this point, not only because the Americans had won the Moon Race, but because the era of thrilling firsts was now over.[7] The military origins of spaceflight were clear on both sides of the Cold War. Each side used the technology that emerged from the Second World War in remarkably similar ways for both military, strategic and propaganda purposes. The rockets, navigation systems and even the first men to fly in space had their origins in the military. The military benefits of spaceflight were evident: Weather satellites and navigation systems, as well as those of communications and earth observation, were space-based infrastructure that had become necessities first to national defense.[8] Where this argument falls apart is the fixation on human spaceflight. The origins of the hardware and technology of human spaceflight were military; however, there was little or no military justification for the first decade of human activities in space. The United States had made this clear with the creation of NASA, a civilian agency, in 1958. On the other hand, the Soviet Union relied on a more precarious balance between military and civilian activities absent an administrative demarcation. The late 1960s and early 1970s were a period of rapid change in Soviet military policy. These changes addressed the complexities of the political and technological nature of the Cold War, the transition from the immediate postwar era of Khrushchev to that of Brezhnev, and the increasingly fragile and strained Soviet economic and foreign political conditions.[9]

American motivations for continuing human spaceflight during the post-Apollo period cannot be applied to Soviet planning. The Soviet Union did not face the sharp edge of cost accounting that was routine in the United States and became more commonplace in the mid-1980s in Russia. As had been true since 1957, outer space had both a public and military face. Human spaceflight retained value in public relations that could offset monetary costs. Space stations maintained the illusion of an emerging Soviet utopia in space that rested on the heroic legends of the previous decade. For the first time in Soviet history, the state offered the public a taste of the future.[10] The military aspect of the program satisfied lingering need to demonstrate the military competition against the Cold War foe. Where Soviet Premier Nikita Khrushchev (1894–1971) had relied frequently on rhetorical and symbolic competition, his successor Leonid Brezhnev (1906–82) returned to direct military confrontation after taking over in 1964. Twentieth-century historians of the region have generally accepted the notion of the militarization of Russian and Soviet society.[11] The tradition of maintaining a large standing army, external threats notwithstanding, is well documented under Tsarist and

Soviet rule. The infusion of military culture into the population was a tried and true way to maintain legitimacy and served as a reinforcing strength to party authority.[12] Khrushchev had recognized the need for substitute assurances for a society that was facing its first adult generation that had no personal memory of war or civil war. While he had sought to placate the generation with promises of a future in space, easy harvests and full communism within their lifetimes, Brezhnev turned towards time-honored references to military traditions without completely abandoning the communist hopes for the future.

I Routinization of human spaceflight

For both the United States and the Soviet Union, human spaceflight captured the collective imagination during the 1960s in large part due to the rapid succession of first-time accomplishments. For a public that had long dreamed of spaceflight, each day seemed to break new barriers. The process of making human spaceflight routine also included the challenge of managing the high expectations of the 1960s. Routine should not become boring. The Soviets had an added challenge in managing expectations, because this transition coincided with the political task of rolling back expectations from the ambitious projections of Nikita Khrushchev. During his eight years in power, Khrushchev had made outrageous promises for the immediate Soviet future. He used spaceflight as a prophecy of things to come.[13] Unpeeling the links between the accomplishments of the 1960s from the promises of accomplishing socialism within the next decade was a difficult task. Brezhnev's overarching policy of removing evidence of Khrushchev's management of the Soviet Union notwithstanding, the Soviets had had no intention of abandoning human spaceflight in the late 1960s and early 1970s. Cuts in the program had eliminated plans for an all-woman crew on a Voskhod spacecraft or any subsequent Voskhod flights after Aleksey Leonov (1934–2019) and Pavel Belayev's (1925–70) Voskhod 2 flight in 1965 during which Leonov became the first human to make a spacewalk. Male cosmonauts maintained their training for a series of planned, but unfunded, missions that sustained administrative momentum. The design and testing for an operational Soyuz spacecraft had been the primary delay to launching humans into space from 1967 until 1969. However, in contrast to the N-1 launch vehicle, the Soviet lunar rocket, the Soyuz problem was soluble despite the limited resources that the program had at its disposal. The flights of Soyuz 3–9 demonstrated that the spacecraft that had killed Cosmonaut Vladimir Komarov (1927–67) had been made reliable for human flight. Yet a ferry craft without a mission to the moon had to have another destination. The continued N-1 moon rocket failures up to 1974 placed the moon beyond the Soviet's grasp. An orbiting space station in low-earth orbit seemed to offer an achievable accomplishment, one that could give the impression of a completely new direction with only a modest investment. At least two design bureaus recognized this next step. Each prepared their own design of a space station that would rely on the existing heavy-lift launch vehicle, the Proton. Originally proposed as a heavy-lift intercontinental ballistic missile (ICBM), the Proton had been the reliable vehicle to send probes to explore the solar system.[14]

A Proton launch vehicle sent the Salyut 1 space station into orbit on 19 April 1971. The official yet secret Soviet designation of the station was DOS-1, DOS being the acronym in Russian for 'Long-Term Orbital Station.' With only one docking port, the 15.8-meter long and 4.15-meter diameter station had a 90-cubic meter habitable volume that is about the size of a large studio apartment. Four days after the station entered orbit and once preliminary systems turned on, Soyuz 10 launched to dock with the space station with a three-man crew that included Cosmonauts Vladimir Shatalov (commander, 1927–), Aleksey Yeliseyev (flight engineer, 1934–) and Nikolay Rukavishnikov (test engineer, 1932–2002). Shatalov had flown onboard two previous Soyuz missions, Soyuz 4 and 8.[15] Yeliseyev had spacewalked from Soyuz 5 to 4 and was the flight engineer onboard Soyuz 8. For Rukavishnikov, this was the first flight in space. The new docking adaptor would allow internal access to the space station from the Soyuz. Unfortunately, the Soyuz failed to achieve a hard dock and the mission was abandoned. The space station would not receive any occupants at the time.[16]

Six weeks later, on 6 June 1971, Soyuz 11 launched with a new three-man crew. Georgy Dobrovolsky (commander, 1928–71), Vladislav Volkov (flight engineer, 1935–71) and Viktor Patsayev (test engineer, 1933–71). Soyuz 11 successfully docked with Salyut 1, the cosmonauts entered the station, and for 22 days Soviet media updated the public with their activities onboard Salyut 1, including live television broadcasts, earth observations and photography.[17] When a fire broke out on the eleventh day of their mission, mission control allowed them to continue with their flight plan. The Soyuz 11 crew broke the 18-day orbital mission record of Soyuz 9, undocked from the space station and returned to earth on 29 June 1971. Recovery crews arrived at the landing site and opened the landing capsule, but they discovered that all three cosmonauts were dead, two firmly strapped into their seats. The third had made initial movements in attempt to close a valve in his last seconds of consciousness. A breathing ventilation valve had opened prematurely during descent instead of automatically adjusting cabin pressure at an altitude of 168 kilometers, and the gradual and complete loss of pressure was fatal. Previous Soyuz crews had worn space suits only as a necessity for spacewalks to gain external access to another spacecraft and provided a life support backup. In order to squeeze three people into a Soyuz spacecraft, cosmonauts had foregone space suits in their capsule. This plan was not beyond the Soviets' experience of risk. In 1964, in order to preempt the US Gemini program, they had launched a crew of three in a modified single-passenger Vostok spacecraft without the backup support of space suits. The Voskhod crew had returned safely without pressurization incidents. Thus came the normalization of the risk of travelling into space without space suits. Good luck had made a bad practice acceptable. During the years 1967–71, the only space suits that the Soviet cosmonauts used were for extravehicular activities. The luck did not hold. The Salyut 1 space station mission had been a success; it was the landing that cost the Soyuz 11 crew their lives.[18]

II Camouflaging the military Salyuts with civilian programs

Plans for an Orbiting Piloted Station (OPS), code-named Almaz (Russian for diamond) had been approved in 1968 about the same time that the DOS space stations were designed.[19] The DOS stations arose as a direct response to the Soyuz legacy that the design bureau of influential rocket engineer Sergey Korolyov (1907–66) faced in the late 1960s.[20] The original target launch date for the Almaz space stations had been the hundredth anniversary of Lenin's birth in April 1970. Vladimir Chelomey (1914–84), Korolyov's professional rival, had designed the station as a complete system with its own non-Soyuz cosmonaut transport system, the Transport Supply Spacecraft (TKS).[21] Chelomey's goal had never been to propose an explicitly military station, but to offer a successful human spaceflight alternative to Korolyov's 1960s plans that had dominated the Soviet space program under Khrushchev. Chelomey's designs found institutional patronage in the Ministry of Defense under Ustinov.[22] The reality of building a completely new system and having it ready for launch caused delays in the schedule, forcing Chelomey and the Ministry to defer their first launch to the Korolyov Design Bureau's DOS station first. Another casualty of schedule was the TKS ferry vehicle that would deliver cosmonauts to Salyut. Both programs had to rely on the Soyuz ferry that the Korolyov team had designed for DOS. The Almaz system would have to adopt Soyuz navigation and docking systems.[23]

The Almaz system demonstrated how the layer of militarization was applied to the human spaceflight program. The US Air Force Manned Orbiting Laboratory (MOL) project, on the other hand, had begun as a large-scale, human-tended space station program that was fully independent of NASA's Gemini and Apollo programs. As it progressed, ambitions contracted to US Air Force pilot-astronaut crews to inhabit a "heavy" Gemini program. The Almaz program began with the concept of consolidated crew training, with sub-groups designated for specialized missions.[24] The crews were to be shared, as were the launch facilities and, ultimately, the transport vehicles. In this way, the Soviets could have avoided having to operate dual overt military and civilian programs at the same time. Once the reality of having to operate two separate programs set in by the beginning of 1970, the economics of consolidation were lost. The diaries and memories of Nikolay Kamanin (1909–82) and Boris Chertok (1912–2011) are full of notes on the logistics of sending crews to the appropriate locations for training. Published documents show the end resolutions of these discussions over priority and mission of the hardware that was currently in orbit.[25] That is how the Soviet Union came to orbit the world's only series of military space stations.

Chelomey's Almaz stations were military only because the Soviet Union designated them as such.[26] Chelomey had designed a station as an alternative to his rival Korolyov; it was only when he found patronage in the Ministry of Defense that the Almaz became a military station. The cost of this patronage was the public fame of his project, the complete deployment of his system and the burden of a piecemeal military program.[27] This was not the first time that military and civilian programs shared staff and hardware. From the very beginning of the human spaceflight program, the Ministry of Defense had played a role in directing another, related program. During the early 1960s Korolyov's design bureau OKB-1 was obliged to develop and keep to schedule the simultaneous development of the manned Vostok capsule programs and that of the unmanned photoreconnaissance capsule, Zenit.[28] The hardware buses of both programs were identical. It was the payload that determined which was military or civilian.

The first Almaz (OPS-1) finally launched on 4 April 1973. TASS formally announced it as the second in the sequence of the Salyut program. Although it launched and initially orbited successfully, within two days the unmanned Salyut 2 began losing pressure and its flight control system failed. Analysts attributed the cause of the failure to shrapnel from the discarded and exploded Proton rocket upper stage that pierced the station. On 11 April 1973, seven days after launch, an unexplainable accident caused both solar panels to be torn loose from the space station, cutting off all power. Salyut 2/OPS-1 re-entered on 28 May 1973. Another Proton launch vehicle sent the Salyut 3/OPS into orbit on 25 June 1974. This second attempt at an Almaz launch maintained operations in orbit successfully. Despite the Soviet's best efforts to maintain that Salyuts 1, 2 and 3 were part of the same program, Western space observers almost immediately noticed a difference in the electronic signals between the DOS and Almaz stations. Astute and experienced listeners to Soviet space signals detected an additional, special, encrypted channel broadcasting from these stations that the DOS stations did not have. The precise purposes of that extra signal remained secret for decades, even though the significance was obvious to avid Western space watchers.[29]

Salyut 3 was the first station to test a wide variety of installed reconnaissance sensors, including both radio and photographic equipment, and have a return canister for film for analysis. The photoreconnaissance capability of the station proved a marked improvement over previous generations of images from space. The observation deck on which the photographic camera was mounted had a limited ability to turn and track a visual target. The size and position inside a large pressurized module limited the platform's agility, though. That facility allowed the station crew to determine the bearing and direction of ships at sea that travelled under the flight paths of the station. Once the film had been exposed, cosmonauts returned film cassettes to earth inside heavily insulated ablative canisters. It was known by its Russian acronym for Information Return Capsule, KSI (Figure 12.1).[30]

Launched on 22 June 1976, the final military Salyut, Salyut 5 (OPS-3) was the third in the Almaz series. This was the most successful of the Almaz series by boarding crews in two out of three attempts. Salyut 5 re-entered on 8 August 1977. After the final inhabitation of Salyut 5, the Soviet Union recognized the greater potential of more sophisticated automated spy satellites, much as the United States had in 1969, to allow the abandonment of human-tended reconnaissance. Even though Soviet automated spy satellites had a much shorter lifetime than American ones, they cost less and were more versatile than Almaz stations.[31] The final two Salyuts were DOS/civilian space stations – Salyuts 6 and 7 in 1977 and 1982, respectively. The scientific focus for the last two Salyut stations shifted towards civilian research and international prestige. The Soyuz transport ship had a limited service life that long-duration space missions had surpassed. The solution to this hardware problem was to replace the craft docked with the station at a regular interval. The three-person craft required a two-person crew for this ferry mission. The Soviets quickly saw the advantage of filling the third seat with a passenger that would relieve some of the monotony of spaceflight for the long-duration crew. The Interkosmos program already existed to encourage scientific collaboration among Warsaw Pact nations in 1967.[32]

The first pilots were chosen from among the Interkosmos member nations, beginning with Czechoslovak pilot Vladimír Remek (1948–) in March 1978 who thus became the first European in space. The program grew over the next decade to include a widening circle of Soviet allies, including the second and third space travelers from predominantly Muslim countries.[33] This marked a turning point for the Salyut program.

Overall, the Almaz program was not entirely successful. The attempts to match the Korolyov-designed Soyuz transport craft to the Chelomey-constructed Almaz program caused a series of docking malfunctions that the civilian program experienced infrequently. The significance of the reconnaissance bounty of the Almaz series remains a mystery. The film return capsules could not return the level of stereoscope images that the American CORONA program had routinely returned since its inception in the 1960s.[34] But the crew did have the facilities from which to process and transmit radio signals containing image data directly from the station. As the United States turned to automated spy satellites in the early 1960s and then to digital imaging through the Hexagon program in the late 1960s, the relative expense of human-tended reconnaissance quickly outstripped its benefits. From the perspective of the Soviet population, the Almaz program had an enormous drawback. As a secret program, it provided little cause for national celebration. Successful launches and landings made the news announcements, but

Figure 12.1 Almaz film return capsule (Information Return Capsule, KSI) on display at the Smithsonian National Air and Space Museum in Washington, DC. This container brought back exposed film from the Almaz space station Salyut 5's Agat system (1976–77).
Source: Photograph by Mark Avino. Courtesy of Smithsonian National Air and Space Museum.

there was limited or no press coverage of the day-to-day work of the cosmonauts. Their lives in space offered little drama since they were frequently occupied with the business of exercising, eating or maintaining space station equipment. This was not the only time that a military space program in the Soviet Union had operated under the camouflage of a civilian program. That had been standard practice throughout the history of the Soviet program. However, this was the first time that military and civilian programs were interleaved under a single designation.

III The spy cosmonauts

The Soviet space program had never accumulated the wealth and robustness of the American program. The United States had the luxury of operating separate civilian and military enterprises, a situation that continued to exist even after the mid-1960s fall-off of funding and support for the Apollo program. In contrast, hiding military operations among civilian programs in space was nothing new, and the distinction between military and civilian cosmonauts in the Soviet Union was blurred. Nonetheless, its use and witness as a political and propaganda tool declined in the 1970s and never regained the acclaim of the 1960s. The absence of three individuals, Khrushchev, Korolyov and Gagarin, no matter how key, does not explain the withering away of a civil infrastructure that had raised spontaneous crowds to celebrate the Soviet Union's space accomplishments.

The US Air Force Manned Orbiting Laboratory program established its own retinue of Air Force pilot-astronauts in the early 1960s. Selection, training and all forms of administration were entirely separate from that of NASA. Although they came from similar applicant pools as the later generations of Apollo astronauts, these were military astronauts, not Apollo explorers. And yet their skills were transferable. Once the MOL project faced cancellation, those pilot-astronauts who had remained transferred to the civilian NASA program in 1969. All those making the age cutoff eventually flew either in the Apollo program or onboard the US Space Shuttle. Thus, for a few years, the Americans maintained two administratively independent, yet interchangeable astronaut corps.[35] The composition of the Soviet spy satellite cosmonauts did not maintain the interchangeability and independence that the United States had. The Soviets could not afford a separate and dedicated cosmonaut crew for the Almaz program, and drew from the selections of military pilots and engineer cosmonauts who had been training to fly in space since the late 1960s with a few additional Air Force Pilots added to the mix. And even though, unlike the United States, the Soviets did launch missions dedicated to this program, it was not a particularly successful program. Only three of five attempts to inhabit all three Almaz Salyuts were successful. Salyut 2 was a complete failure. Soyuz 14 was the only successful mission to Salyut 3. Soyuz 21 and 24 docked with Salyut 5.

Salyut 5 raised the success rate between the Soyuz ferry craft and the Almaz stations to a two out of three. The crew of Soyuz 21 included Boris Volynov (1934–) and Vitaly Zholobov (1937–). Volynov was a Soyuz veteran, while his flight engineer flew his first mission to this station. And despite an emergency evacuation at the end of the mission, theirs was considered successful. Soyuz 23 did not make a complete dock with the station, and Zudov and Rozhdestvensky returned to earth without staying aboard their station. And finally, the crew of Soyuz 24, Soyuz veteran Viktor Gorbatko (1934–2017) and first timer Yury Glazkov (1939–2008) accomplished the most complete and final military Salyut mission.

There were, however, two changes in Soviet domestic and foreign policy, both seemingly unrelated and inconsequential, that might contribute to an understanding of this sudden flipping of script from public to some more muted, routine and secret military cosmonauts. Both are legacies of the Brezhnev era. They included the spread of state-controlled television broadcasting throughout the eleven time zones of the Soviet Union and development of the Brezhnev doctrine in Soviet foreign policy. The former would become a new tool in spreading the word of Soviet accomplishments in space. The latter would take on a more subtle and secret role in placing the military Salyuts into their proper context within the Soviet domestic culture in the 1970s.

IV Soviet television empire and human spaceflight

The role that the American free press played in shaping US culture and politics in the 1950s and 1960s is well known. Comedians, drama writers and journalists reporting on the Civil Rights movement in the American South and Vietnam were invited guests in US households every night. The Soviet population adopted television at the end of the 1960s when sets penetrated the majority of Russian households. Up until that point, the majority of the population had relied on radio broadcasts and newspapers for their knowledge of current events. The highly centralized television office demanded programming that would respect the 11 time zones and over 400 ethnicities through a single authoritative message.[36] Prior to 1970 the administration of Soviet television programming had been haphazard, reflecting official Soviet inattention to a media that lacked the political importance of film and the national appeal of radio. Individual local producers, largely in Moscow and Leningrad, presented shows that appealed to their own audiences while not offending political operatives. The majority of Soviet citizens learned about Yury Gagarin's Vostok flight in 1961 via radio broadcasts.[37] During the ensuing decade, television audiences had the potential to surpass movie audiences. The overhaul of Soviet media created a State Committee for Television and Radio and placed former diplomat Sergey Lapin (1912–90) at the head. This elevated programming to the Politburo level and set into place centralized broadcasting and policies. Lapin immediately fired large swaths of staff, centralized broadcasts across all times zones and, perhaps most famously, instituted a nation-wide news broadcast, *Vremia*, whose signal was sent into homes throughout the Soviet Union at 9 p.m. Moscow time.[38]

The impact of television reorganization on the public perceptions of spaceflight was twofold. First, the nightly, nation-wide broadcasts of the news provided a handy venue for frequent reports on cosmonaut activities to fill up the remaining minutes of the nightly news that was not taken up by national and international politics as well as economic issues.[39] The fastidious collection of film footage of cosmonauts in training and in space had a ready market receptive to any new, non-overtly propagandistic footage. Second, the staid evening programming that tried to promote national contentment tended to dwell on period costume dramas, detective series and grim spy serials, avoiding provocative programming. Coverage of human spaceflight added drama and excitement to the mix without provoking ideas of discontent. There did remain a lingering fear of resurrecting memories of Khrushchev's 'harebrained schemes' and extreme risk-taking of the 1960s.[40] Whereas space had been the center of public cultural events in the 1960s, Brezhnev television relegated it to the status of an alternative to the nightly grain reports on the news or prerecorded educational science specials. Space was no longer a public spectacle, but transformed into a reassuring coverage of routine Soviet technological achievements (Figure 12.2).

Press coverage of human spaceflight in the Almaz program did not differ at all from that of DOS Salyut programs. Broadcasting images of men in spacecraft served to perpetuate the legend of the Soviet Union continuing its leading role on the road to the cosmos and supported the ruse that Salyut was a single program. The programming softly echoed the Khrushchevian theme of cosmonauts leading armies of civilian workers into space. These men were portrayed as being at the forefront of establishing a permanent Soviet outpost among the stars. Clues to other activities were guarded and only revealed to the public a generation later, often without sufficient explanation even then. The important message was that the cosmonauts were in orbit, not what they were doing there. Unlike the previous decade, there were no breathless announcements of record-breaking accomplishments generated from Salyut missions. In fact, the acknowledgment of the cosmonaut accomplishments onboard the Almaz Salyuts was done secretly and years after the program had terminated.

Figure 12.2 The first space station Salyut 1 as seen by the Soyuz 11 crew in 1971. This image was broadcast to the public through print and electronic media and was adapted throughout the world as a symbol of Soviet progress in human spaceflight. This version of the image was taken from a presentation from a NASA official.
Source: Courtesy of Smithsonian National Air and Space Museum.

V The medal for Distinction in Guarding the State Border of the Soviet Union

Soviet secrecy and public media colluded to make the impression of a single, unified space station program. This fact begs the question of how it was known that there were discrete programs going on. There are three pieces of evidence that have come to light since the Soviet Union collapsed. The first was an award that went to some, but not all Almaz cosmonauts; the second and third were two pieces of hardware that were unique to Almaz missions.

The use of the medal for Distinction in Guarding the State Border (*Za Otlichie v okhrane gosudarstvennoi granitsy SSSR*, DGSB) must have been part of a distinct effort to appear to integrate the Almaz program into Soviet military foreign policy. The association of this medal with the Almaz program cosmonauts only came to light long after their missions. The use of medals and awards to place cosmonauts within the hierarchy of the Soviet State was commonplace, however. There has been a long association between cosmonauts and national awards. The Central Executive Committee of the Soviet Union established the Hero of the Soviet Union award on 5 May 1934. The award was intended to honor feats in service to the Soviet state and society, and to create a new legacy of legitimacy and duty within the rapidly evolving Stalinist state. Modeled on the imperial awards and medals of the previous century, sufficient time had passed and cultural change occurred to impress the public that this was by no means a tsarist medal.[41] The first recipients of the award were the seven pilots who participated in the successful aerial search and rescue of the crew and civilian passengers of the steamship Cheliuskin, which sank in Arctic waters, crushed by ice fields, on 13 February 1934 while attempting to navigate the Northern Maritime Route from Murmansk to Vladivostok.[42]

Recipients of the Hero of the Soviet Union award were not only men. Later in that decade pilot Valentina Grizodubova (1909–93) became the first female recipient on 2 November 1938. She earned the award for her international women's record for a straight-line distance flight.[43] During the Second World War, Zoya Kosmodemyanskaya (1923–41), a famous Soviet partisan, was the first woman to receive the award during wartime on 16 February 1942, albeit posthumously. By the time of the Space Race, the award was the obvious choice for honoring returning cosmonauts. Starting with Yury Gagarin, each cosmonaut of the 1960s and throughout the history of the Soviet Union received the honor, with a maximum of twice in a lifetime. Cosmonauts display two medals that commend their multiple spaceflights on their military uniforms and civilian suits.[44]

The Almaz program brought with it a new honor to Soviet cosmonauts, although one that was shrouded in secrecy until after the program ended. In addition to receiving a Hero of the Soviet Union medal and being dubbed Pilot-cosmonaut of the Soviet Union, some Almaz cosmonauts were given another reward. The DGSB was established on 13 July 1950, by Decree of the Presidium of the Supreme Soviet of the Soviet Union.[45] This was a lower rank than the highest (Hero of the Soviet Union, Hero of Socialist Labor and Heroine Mother). This military award was devised to honor extreme levels of service on the part of soldiers and the original intended recipients were border troops. The medal honored military exploits and special services displayed in the protection of the state borders of the Soviet Union. The list of justifications for the award included range of seven skills and characteristics in border protection.[46] This medal would probably have remained in a low level on anonymity for rank and file border soldiers if not for one recipient: Erich Mielke (1907–2000), East German Minister of State Security who headed the Ministerium für Staatssicherheit from 1957 until the fall of the Berlin Wall in 1989. The Soviet government awarded him the medal for Distinction in Guarding the State Border of the Soviet Union in January 1970, seventeen years before awarding him the more distinguished Hero of the Soviet Union.[47] This inaugural award made it clear that the border defenses were far more political than physical to merit receipt of the award.

Almaz cosmonauts received the DGSB according to a different protocol than did those more famous and traditional awardees. Cosmonauts first garnered their expected decorations for participating in spaceflight immediately after flight. That included the Hero of the Soviet Union award recognizing up to the first two flights in space. Then in 1977 some, but not all, received the border service with distinction award. Both the selection and timing of this award is intriguing. In subsequent years of that decade, six of the ten cosmonauts who launched to the military Almaz stations received the same honor. They were (in order of their missions): Artyukhin, Sarafanov, Volynov, Zholobov, Glazkov and Gorbatko. They represented only crews to Salyuts 3 and 5, included the commander of one failed docking, and the flight engineer, but not the commander, of what had been deemed a successful mission.[48]

The decision as to whom to give the award and not is almost as perplexing as unraveling the political or military contribution that a cosmonaut in low-earth orbit made to border defense. Interestingly, Pavel Popovich (1930–2009), the most senior cosmonaut among the Almaz crews, was not awarded the DGSB, receiving the Hero of the Soviet Union for the second time after this mission and nothing more. An explanation for this oversight is elusive. It might be due to the fact that Popovich alone was one of the original twenty cosmonaut trainees accepted into the space program in 1959/60 and that he had already received the highest award as a recipient of the Hero of the Soviet Union medal. It is even more difficult to understand the reason that the commander of the failed-to-dock mission of Soyuz 15, Gennadi Sarafanov (1942–2005), was a recipient, but his crewmate Lev Dyomin (1926–98) was not. The Soyuz 23 mission to the third Almaz station also did not dock carrying Vyacheslav Zudov (1942–) and Valery Rozhdestvensky (1939–2011), and there was no mention of either of them. The Soyuz 24 crew of Salyut 5, Gorbatko and Glazkov, did receive the honor. The secretive selection process could possibly hint at a sophisticated measure of each cosmonaut's tactical contribution to border security.

The timing of the announcement of these awards is another difficult aspect to decipher. All cosmonaut awards were made after Soviet abandonment of the Chelomey station in 1977. However, there were no formal citations published in the Soviet military press during 1977. It was only much later in the first published compendium of all Soviet and Russian cosmonauts in 2001 that the award is listed among National Honors among some cosmonaut entries.[49] And even in this case the entries were made in idiosyncratic ways. For some, the full title 'For Distinction in the Defense of the Border' is spelled out completely. For others, the listing merely states that they were awarded an 'honor for the defense of the borders,' omitting whether or not this was for distinguished service, which might indicate a lower award.[50] There are no hints as to whether this was a publishing truncation of the name of the award, or if it was an indication that there were in fact two award levels. In trying to decipher this puzzle, it is clear that all the award winners were among the later selections of Soviet Air Force nominees to the cosmonaut corps and did not include the engineers and physicians from Korolyov's design bureau and the Institute for Biomedical Problems. As a result, crews were split, one receiving the award and another not. Popovich (commander) did not receive the award, but Yury Artyukhin did for their Soyuz 14 mission to Salyut 3. And yet Sarafanov received the award, but Dyomin (flight engineer) did not for their failed Soyuz 15 mission to the same station later in 1974. No member of the Soyuz missions to Salyut 4, failed or successful, received the awards at all. All four crew members of the successful Soyuz 21 and 24 missions to Salyut 5 were credited with the honor. Despite the fact that there was no public announcement of the award, there seems to be no evidence of hiding the awards once they were made. One can only assume that cosmonauts proudly wore the red bordered green bar among their other military awards without fanfare.

This symbolic connection between the foreign policy missions and human spaceflight extended the military balance of power out into low-earth orbit. While Mielke took documentable actions to protect the Soviet homeland no matter how repulsive his actions might have been, the steps that the cosmonauts took lacked any political or physical defense of the border. The ex-post facto acknowledgment of their work seemed closer aligned to 1970s Soviet patterns of establishing heroes and legends. This 1970s mythmaking, like that of Stalin a generation before, allowed for no ambiguity. As historian Elana Gomel has noted, 'the Soviet New Man, on the other hand, marches along the one-way road of historical progress toward the revelation of his own glorious self.'[51] The "peaceful" Salyut cosmonauts were continuing the progress of the 1960s, without directly challenging their Cold War enemy, even though the underlying challenge of demonstrating continuous progress was steady. In contrast to would-have-been adversaries, these cosmonauts were designated Cold War warriors after their program had finished, almost as an afterthought, drawing no celebrations.

While ambiguities about the crew training, missions on the stations and their military importance remain today, there are two pieces of equipment that were known to be onboard the Almaz ships that defined the military missions. These pieces of equipment played no role in the continuing mythmaking around Soviet cosmonauts, but they were instrumental in answering the Defense Ministry's demand for a space-based challenge to the United States. The first matched American aspirations to the use of space for reconnaissance; the second answered Soviet anxieties about the imagined coming battlefield in space. It seems that the awarding of the DGSB served most as a consoling acknowledgment of a past and not entirely successful program.

VI The Almaz camera

The first piece of physical evidence of separate military and civilian space stations was the existence of a specialized spy camera onboard the Almaz stations. Electronic signals coming from the stations hinted at its existence from the first, but its material reality was acknowledged only in 1993, and images of it have become public only in recent years. As mentioned before, the external clue that gave away the Soviet secret of the dual identities of the Salyut stations was the radio channel link. The Kettering Group had learned to distinguish between Almaz and DOS stations from their radio signals, the Almaz having a dedicated link for reconnaissance information.[52] Almaz had a shortwave telemetry transmitter, called 'support' telemetry by the Russians. Just like DOS, Almaz had a VHF (very high frequency) transmitter for the main telemetry system. However, the command uplink and command verification downlink systems were completely different. DOS used the command system on Soyuz, similar to the one used on the International Space Station (Zvezda), while Almaz probably employed a system common to military spacecraft and the *Funktsional'no-gruzovoi blok*, Functional Cargo Block (FGB), known as Zarya (sunrise), the first module of the International Space Station (ISS), launched in 1998.[53]

This dedicated link was for the Agat-1 reconnaissance camera and its radio download link. The redundant system provided two paths for reconnaissance data to reach earth. A payload from Almaz could be returned via the Information Return Capsule (KSI), a modified ablated warhead that would return film to earth. Cosmonauts had another option of returning photographed information through a radio link to the ground. A limited amount of film from the reconnaissance cameras would be developed onboard, scanned and transmitted to the ground – all within 30 minutes. This versatility of the Agat camera system had been a selling point to the Ministry of Defense. The Almaz promised to be a substantial improvement over the performance of the Zenit automated spy satellite system.[54] The Zenit program had begun almost as early as the Space Race itself with a first launch in 1961. Zenit used a Vostok capsule bus with the camera and film mounted inside. The entire landing sphere had to be recovered including film and camera in order to retrieve the images. While the Agat system was more versatile in theory, the program was not as robust either in deployment or longevity. The Soviet Union continued to launch Zenit-style spy satellites until 1994, numbering over 500 in a 33-year period.[55] The Agat system was deployed onboard human-tended stations only twice, in Salyut 3 and Salyut 5. The technical success of the Agat system notwithstanding, the system could not live up to the potential of providing timely reconnaissance and surveillance photos to meet national security requirements. There had been a two-year gap between camera deployments and only three out of five planned crews successfully inhabited the stations. Throughout this time, the Soviet Union had been steadily refining its use of the Zenit spy satellite program, deploying both high- and low-altitude satellites that relayed a variety of signal to earth. The Agat system was a static electronic and film system that, despite its limited targeting capability, could not rival Zenit's versatility in orbital choices.[56]

VII The weapon

International treaties sought to discourage the use of weapons in space, but they could not eliminate the attempts to do so. Rumors of the role of the Almaz stations as a weapon preceded public disclosure for decades. Author Nicholas Johnson mentioned stories of a Gatling gun onboard one of the stations in 1987.[57] The key to interpreting the weapon is to withdraw from the general discussion of militarization of space in order to understand its specific role as a space weapon. The distinction between the concepts of militarization and weaponization of spaceflight is a historical one. Space exploration is de facto based on the development of military technology. The actual deployment of weapons in space has been discouraged even while specific concepts have been tested. Atmospheric weapons in space are subject to Newton's third law of physics.[58]

As though to fully demonstrate a turn towards the weaponization of space, the Almaz station was equipped with a machine gun. Suitable for defending against or attacking another aircraft in flight, this machine gun had no clear contemporary target. They did not test this hardware until all operations were complete and the station was about to be deorbited. As far as personal accounts tell, the machine gun was only deployed on Salyut 3. On 24 January 1975 ground crews ordered the tests of what was then reported to be an onboard 23 mm Nudelman aircraft cannon. During the 1970s and through the end of the twentieth century, analysts assumed that the weapon that was deployed onto the military Salyut had been adapted directly from an aircraft cannon. Soviet lore proclaimed that this gun had been mounted on the exterior of the station as a defensive measure. Having inflated and vague target ranges from 500 to 3,000 meters, the gun was clearly not a refined anti-satellite weapon, but one that could ward off a direct physical boarding or capturing attack against the station and was adequate as a demonstration of principle of having a cannon onboard the station.[59]

Although the machine gun existed as no more than rumors and conjecture for decades, its detailed history and description have become public in the past ten years. A photo of the purported gun was published only in 2015.[60] It was a 23-millimeter cannon, as reported at the time, but one designed by Aron Rikhter as a powerful aircraft weapon for the Tupolev TU-22 Blinder supersonic bomber. The adaptation of an aircraft gun to space revealed two major design challenges that could not be overcome. First, in order for the cosmonauts to fire using an optical sight in their cockpit, they would have had to turn the entire 20-ton station to point the cannon toward its target. This was not a practicable operation. The second design obstacle was that the R23M could not overcome basic Newtonian physics. The recoil of the cannon even against a 20-ton station was sufficient that during its only post-inhabitation, pre-deorbit firing, on 24 January 1975, the Salyut-3 ground controllers had to initiate station jet thrusters simultaneously with the firing to counteract the recoil.[61] The Salyut cannon was a weapon that could not be targeted in real time and could not be used while there was a crew inside the station to defend. It could provide no practical, tactical support in space and its shrouded identity prevented it from being of any use for deterrence. And given its location onboard a low-earth orbiting station, other strategic satellites in far higher orbits were beyond its range. Nonetheless, the rumor of its existence prevailed for decades before the Ministry of Defense unveiled it, thus fueling the internal mythology of the defensive military space station deterring capture and boarding. In all likelihood, the cannon was no more than a one-off test of a concept. Unlike the photoreconnaissance equipment that had already been used successfully onboard Salyut 3, no cannon was present on the next Almaz station, Salyut 5.

VIII Conclusion

The de-Khrushchevization of the Soviet state in the 1970s was marked by deliberate efforts to remove all traces that had been embarrassing and challenging from Khrushchev's de-Stalinization.[62] Nikita Khrushchev's confrontational stances, 'harebrained schemes' and negligence of rational planning were easy targets for ridicule. Yet the 1970s were in no way a complete reversal of Khrushchev's contradictions. The decade did not usher in a new age of rationalism, nor was there complete de-Khrushchevization of Soviet idealism. Under Brezhnev, the Soviet state took a small step back from the obviously hypocritical stances that Khrushchev had taken in the effort to absolve himself from Stalin's legacy. But the Soviet leadership remained facing a troubled country still recovering from the Second World War with an increasingly disaffected youth who had no memory of the war, with no personal memories of the Stalin terror. Brezhnev faced the very real challenges of ruling in a world in the midst of Cold War and domestic and nearby challenges to Soviet rule. Human spaceflight was one of the symbolic tools that Khrushchev had left to combat these challenges.

This official evisceration of the legacy of Khrushchev left the country with three characteristics of a Brezhnev-era Soviet culture. The first was the doctrine of the continuing Cold War with the United States that could never be lost. The second was the militarization of civilian culture including human spaceflight without any appreciable change in the programs and outcome. The third change that manifested itself in the new, Brezhnev culture was the shift inward to a more defensively stanced public culture. The result of these changes in Soviet astroculture from the Khrushchev to Brezhnev regimes was an invitation to a broad collection of ideas about the purposes of human spaceflight. The early 1970s were the first time that no single vision of human spaceflight dominated completely. Soviet efforts in the Moon Race continued through the testing of the N-1 launch vehicle. A civilian program cobbled together leftover hardware to maintain the profile that had been established during the 1960s. International cooperation became a spaceflight objective, be it with the United States or with Warsaw Pact allies. And the Ministry of Defense adopted a hardware routine within which to test and demonstrate a program originally conceived to challenge US military space stations. While Khrushchev had been accused of providing trite and often facile answers to the public demand for a more relaxed and robust postwar culture, the Brezhnev response relied on nostalgia for the period that they sought to erase from memory. The public wanted a time and place that was safe and hopeful.[63] Khrushchev had built that illusion with spaceflight.

The major mission distinction between the Almaz and DOS program came from the hardware added to the former orbiting station. The photoreconnaissance missions inspecting potential hot spots during the Cold War and avoiding orbital capture did not promise a comfortable and safe future for the civilian population. The stations never had a sustainable mission for the military. They served to reassure the internal members of the defense industry of three things: First, even though the United States had gone to the moon, Soviet national prestige in space remained intact. The Cold War remained a continuous battle between the two nations that continued beyond the Moon Race. Second, space policy was no longer directed towards a goal of stunt missions and gaining a propaganda advantage over the United States, but it did have a well-defined national security role that was too secret to be public. In effect, Brezhnev had co-opted Khrushchev's promotion of the domestication of the science and technology revolution back to the military.[64] And the third, underlying and possibly unintended message of the combined Salyut program to the Soviet public was one of confidence in the reliability of Soviet hardware. By conflating the military and civilian programs through dual use of the Soyuz hardware and by calling them both Salyut, Soviet planners offered the public an illusion of hardware continuity. This was similar to the continuity and comfort that resulted from the anonymous 'Kosmos' designation that disguised failures, covert and test missions. The irony was that this fake continuity of the Salyuts initially undermined public perceptions of reliability. Western reporters who were anticipating the 1975 Apollo-Soyuz Test Project and the average Soviet citizen could assess the high failure rate among early Soviet space station docking attempts. The combined Almaz/DOS programs made both audiences nervous.[65]

After a total of three failed attempts to inhabit and carry out missions to the Almaz military stations, the Soviet program reverted to the civilian DOS program during the late 1970s with Salyuts 6 and 7. This act did not only relinquish human spaceflight to the civilian realm, but it had other, far-reaching consequences. Surrender allowed for the use of Soviet space stations as diplomatic stages, welcoming foreign Warsaw Pact and subsequently other Allied pilots, including Westerners onboard as guest cosmonauts.

The removal of the military label also opened the door for returning Soviet women to space in 1982 after almost two decades of unexplained absence.[66] In the mid-1980s the launch of the modular Mir space station indicated that the Soviet Union continued to explore the possibility of utilizing more robust and varied hardware in its space station program. Soviet and Russian demonstration of their capabilities to maintain a human presence in space over the 15-year life-space of Mir laid the groundwork for Russia to join the United States, European Space Agency, Canada and Japan to build the International Space Station.

The results of the 1970s brought about brief changes in cosmonaut culture that were not without positive legacies. The political and military climate in the twenty-first century has laid clear three opportunities that could not have been imagined in the 1970s. The salvaged hardware pieces that remained after the Almaz program had been preserved, recycled and reconfigured and became components of the International Space Station. The base block of the ISS, the Zvezda module, is of OPS-5/Almaz heritage, having drawn on the more vigorous space station design from the 1960s (Figures 12.3 and 12.4).[67]

Figure 12.3 Russians are working on the aft portion of the United States-funded, Russian-built Functional Cargo Block (FGB) also known as Zarya (Russian for sunrise). Built at Khrunichev, the FGB began pre-launch testing shortly after this photo was taken. Launched by a Russian Proton rocket from the Baikonur Cosmodrome on 20 November 1998, Zarya was the first element of the International Space Station (ISS).
Source: Courtesy of NASA.

Figure 12.4 This image of the International Space Station (ISS) was taken when Space Shuttle Atlantis (STS-106 mission) approached the ISS for docking in September 2000. At the top is the Russian Progress supply ship that is linked with the Russian-built Service Module or Zvezda, based on Almaz hardware. The Zvezda is docked with the Russian-built Functional Cargo Block (FGB) or Zarya.
Source: Courtesy of NASA.

The Zarya or FGB module (the second Russian module) of ISS is based on the TKS spacecraft originally designed as the transport ship for the Almaz. The TKS design itself was launched and landed but never piloted. During the 1980s it docked with the Mir space station modules once as the anonymous Kosmos 1443 (Figure 12.5). The TKS transport ship is now the basis of the proposed next generation human transport spacecraft from Russia. And finally, and most unlikely, as parts of the US Manned Orbiting Laboratory project have been declassified allowing for the first time a direct comparison between the Cold War powers' military space station hardware, the Almaz program has gained favorable comparison for its technical capability.[68] As it

Figure 12.5 The Soviet Transport Supply Spacecraft (TKS) on display at the Smithsonian Institution National Air and Space Museum in Washington, DC. The ill-fated Chelomey-designed ferry vehicle for the Almaz finally underwent orbital testing when it docked with the civilian Salyut 7 space station in 1983 and 1985.
Source: Photograph by Eric Long. Courtesy of Smithsonian National Air and Space Museum.

turns out, if stretched finances and resources had not being curtailing factors and both sides had been free to complete programs that fulfilled their respective military's ideals of human-tended military stations, the Soviet Union might not have abandoned that secret war of the heroic military space pilots standing guard several hundred kilometers above all borders. The irony is that the hardware of the 1970s endured, but not the attempt to remake Soviet astroculture to reflect military politics or to stretch Soviet fiscal capabilities beyond their very limited resources.

Notes

1. Slava Gerovitch, *Soviet Space Mythologies: Public Images, Private Memories and the Making of a Cultural Identity*, Pittsburgh: University of Pittsburgh Press, 2015, 1–26. All Internet sources were last accessed on 15 July 2020.
2. A highly detailed account of the rivalry between Sergey Korolyov and Vladimir Chelomey is featured in Asif A. Siddiqi, *Challenge to Apollo: The Soviet Union and the Space Race, 1945–1974*, Washington, DC: NASA, 2000, 234, 591–2, 843.

3. Andrew Chaikin states that for many Americans, the Apollo program ended with the moon landing; see idem, 'Live from the Moon: The Societal Impact of Apollo,' in Steven J. Dick and Roger D. Launius, eds, *Societal Impact of Spaceflight*, Washington, DC: NASA, 2007, 53–66, here 58. On the post-Apollo period, see also the contributions to the second volume in this trilogy on European astroculture, Alexander C. T. Geppert, ed., *Limiting Outer Space: Astroculture after Apollo*, London: Palgrave Macmillan, 2018 (=*European Astroculture*, vol. 2).
4. One of Korolyov's deputies, Boris Chertok, tells a poignant story in his memoirs about his encounter with a taxi driver a month after Apollo 11 landed on the moon. The driver blamed a lack of leadership in the Soviet program following Korolyov's death in January 1966; see Boris E. Chertok, *Rakety i liudi: lunnaia gonka*, Moscow: Mashinostroenie, 1999, 13 (Eng. *Rockets and People*, vol. 4: *The Moon Race*, Washington, DC: NASA, 2012).
5. Discussions of issues that emerged while attempting to balance multiple programs can be found in the memoirs and diaries of the time, and the outcome of these negotiations are reflected in the collection of documents that RKK Energiia published after the collapse of the Soviet Union. See Chertok, *Rakety i liudi: lunnaia gonka*, 208. The 20 January 1970 document that established the DOS Salyut system and the Soyuz transportation system can be found in Vasilli Mishin, 'Prikaz nachal'nika predpriiatiia No. 8 ot 16/20 ianvaria 1970 g,' in Yury P. Semenov, ed., *Raketno-kosmicheskaia korporatsiia 'energiia' imeni S. P. Koroleva, 1946–1996*, Korolev: RKK 'Energiia,' 1996, here 264–6.
6. Boris Chertok outlined the conditions and the relief that the Korolyov design bureau felt about transferring the planetary missions from the already overworked Korolyov design bureau, OKB-1 to Lavochkin under designer Georgy Babakin; see Boris E. Chertok, *Rakety i liudi: Goriachie Dni Kholodnoi Voiny*, Moscow: Mashinostroenie, 1997, 311–19 (Eng. *Rockets and People*, vol. 3: *The Hot Days of the Cold War*, Washington, DC: NASA, 2012).
7. Paul Stares, 'U.S. and Soviet Military Space Programs: A Comparative Assessment,' *Daedalus* 114.2 (Spring 1985), 127–49.
8. Ibid., 127–9.
9. Siddiqi, *Challenge to Apollo*, 781. Stuart J. Kaufman offers two models through which to explain changes in Soviet military doctrine at the time that focused on defense of the homeland: belief that overwhelming attack was the best balance against escalation; and avoidance of nuclear war. See idem, 'Organizational Politics and Change in Soviet Military Power,' *World Politics* 46.3 (April 1994), 355–82, here 369–70.
10. Matthias Schwartz, 'A Dream Come True: Close Encounters with Outer Space in Soviet Popular Scientific Journals of the 1950s and 1960s,' in Eva Maurer et al., eds, *Soviet Space Culture: Cosmic Enthusiasm in Socialist Societies*, Basingstoke: Palgrave Macmillan, 2011, 232–50, here 245.
11. Dimitri K. Simes, 'The Military and Militarism in Soviet Society,' *International Security* 6.3 (Winter 1981/82), 123–43, here 124–6.
12. Ibid., 131–5.
13. In his biography of Nikita Khrushchev, William Taubman refers to Politburo member Anastas Mikoyan's memoirs, which stated that Khrushchev did not like

statistics, considered the year 1980 to be far off into the future and wanted most to impress people; see William Taubman, *Khrushchev: The Man and His Era*, New York: Norton, 2003, 512.

14. Anatoli Zak, *Russia in Space: The Past Explained, the Future Explored*, Burlington: Apogee, 2013, 4; Siddiqi, *Challenge to Apollo*, 440.
15. Precise numbering of Soyuz missions, or any series of Soviet space projects, is difficult, as the anonymous designation 'Kosmos' was used for test missions and failures for all launches without impact on sequence numbers.
16. Theodore Shabad, 'Soyuz Orbit Shifted; Link to Lab Awaited: Soyuz Orbit Shift Hints at Lab Link-up,' *New York Times* (24 April 1971), 1 and 58. Since mission intentions were not announced in advance, the objective of the Soyuz 10 mission was never published in the Soviet press. It was Western analysts who made the conjecture that the Soyuz 10 had failed to dock. This conclusion was later confirmed in post-Soviet published sources and memoires.
17. American attention focused sharply on Soviet activities in space as plans for a joint US-USSR mission began in the early 1970s. See Theodore Shabad, '3 Soviet Spacemen Board Workshop After Docking,' *New York Times* (8 June 1971), 1; John Noble Wilford, 'The Soviets Are Building a "City",' ibid. (13 June 1971), E6.
18. Bernard Gwertzman, 'Drop in Pressure Hinted in Deaths of 3 Astronauts: Top Soviet Scientist, at Red Square Funeral, Reports an Unexpected Occurrence,' *New York Times* (3 July 1971), 1.
19. The names of the Russian stations were very similar. The civilian stations built by Korolyov's design bureau, OKB-1, were known as DOS stations, an acronym derived from the Russian for Long-Term Orbital Station (*Dolgovechno Orbytal'naia Stantsiia*). The Almaz stations received the code name of the Russian word for diamond. See Chertok, *Rakety i liudi: lunnaia gonka*, 208; Nikolay P. Kamanin, *Skrytyi kosmos*, vol. 2, Moscow: Izdatel'stvo 'RTSoft,' 2013, 37; and Semenov, *Raketno-kosmicheskaia korporatsiia*, 267.
20. Ibid., 218.
21. Sergey Khrushchev (1935–2020), son of Nikita and an engineer by training, has provided a close to firsthand account of the feud between Sergey Korolyov and Vladimir Chelomey and their competition for scarce resources and patronage within the Soviet rocket community. See Sergei N. Khrushchev, *Nikita Khrushchev and the Creation of a Superpower*, University Park: Penn State University Press, 2001, 746–52.
22. Chertok, *Rakety i liudi: lunnaia gonka*, 210.
23. The decision to abandon, or at least put on hold, the TKS transport system placed an added burden on the cosmonauts and their infrastructure. This required training teams from DOS, Almaz, the lunar program and ultimately the Apollo-Soyuz Test Project to carry out separate and independent training exercises at the same facility at the Cosmonaut Space Flight Training Center. See Rex Hall, David J. Shayler and Bert Vis, *Russia's Cosmonauts: Inside the Yuri Gagarin Training Center*, Chichester: Praxis, 2005, 37; Kamanin, *Skrytyi kosmos*, 373.
24. Ibid., 157–8.
25. Boris E. Chertok et al., 'Pervie orbital'nye stantsiia "saliut",' in Semenov, *Raketno-kosmicheskaia*, here 264–8.

26. Asif A. Siddiqi, 'The Almaz Space Station Complex: A History, 1964–1992, Part 1,' *Journal of the British Interplanetary Society* 54.11/12 (November/December 2001), 389–416, here 390–2.
27. Chelomey had designed the station as a challenge to Korolyov's civilian DOS project. See Siddiqi, *Challenge to Apollo*, 596–7; and Nikolay P. Kamanin, *Skrytyi kosmos, kniga chetvertaia, 1969–1978 gg*, Moscow: Novosti kosmonavtiki, 2001, 127–8.
28. Peter A. Gorin, 'Zenit: The Soviet Response to CORONA,' in Dwayne A. Day, John M. Logsdon and Brian Latell, eds, *Eye in the Sky: The Story of the Corona Spy Satellites*, Washington, DC: Smithsonian Institution Press, 1998, 157–70, here 160–1; Boris E. Chertok, *Rakety i liudi: Fili, Podlipki, Tiuratam*, 2-e izdanie, Moscow: Mashinostroenie, 1999, 319 (Eng. *Rockets and People*, vol. 2: *Creating a Rocket Industry*, Washington, DC: NASA, 2006).
29. Sven Grahn, 'Salyut-1, Its Origin, Flights to It and Radio Tracking Thereof,' http://www.svengrahn.pp.se/trackind/salyut1/salyut1.html.
30. Sotheby's, ed., *Sotheby's Russian Space History*, Sale 6516, 12/11/93 auction catalogue, New York: Sotheby's, 1993, n.p., no. 94.
31. Stares, 'U.S. and Soviet Military Space Programs,' 136.
32. Michael Sheehan, *The International Politics of Space*, London: Routledge, 2007, 59–61; see also his contribution to the present volume, Chapter 4.
33. See Colin Burgess and Bert Vis, *Interkosmos: The Eastern Bloc's Early Space Program*, London: Springer, 2015. Saudi Prince and military pilot Sultan bin Salman bin Abdulaziz Al Saud (1956–) flew onboard the American Space Shuttle *Discovery* in June 1985. The Soviets sent Syrian pilot Muhammed Ahmed Faris to Mir two years later. Afghan pilot Abdul Ahad Mohmand made the same trip in the following year. For a discussion of how Islam has been practiced in space, including the experiences of these men, see Cathleen Lewis, 'Muslims in Space: Observing Religious Rites in a New Environment,' *Astropolitics* 11.1/2 (June 2013), 108–15.
34. Day et al., *Eye in the Sky*. See also Michael Neufeld's contribution, Chapter 2 in this volume.
35. See 'Astrospies: An Elite Corps of Secret U.S. Astronauts is Trained to Gather Intelligence on the Soviets During the Cold War,' *Frontline*, PBS Television (12 February 2008), http://www.pbs.org/wgbh/nova/military/astrospies.html; and J. P. McConnell, 'Memorandum to AFCCS (General Schriever): Selection of Aerospace Research Pilots for Assignment to the MOL Program,' 15 March 1965, National Reconnaissance Officer, Declassified Records, Index, Declassified manned Orbiting Laboratory (MOL) records, http://www.nro.gov/foia/declass/mol/85.pdf.
36. Kristin Roth-Ey, 'Finding a Home for Television in the USSR, 1950–1970,' *Slavic Review* 66.2 (Summer 2007), 278–306, here 281.
37. In an informal newspaper survey of USSR citizens on the fortieth anniversary of Gagarin's flight, respondents most frequently answered that they heard the news of Gagarin's flight on the radio; see A. Komarova, A. Kobzarev and K. Petrov, 'Pervym v Kosmos Poletel Gagarin,' *Novosti Pskova* (Pskov) (12 April 2001).
38. Roth-Ey, 'Finding a Home for Television,' 306.
39. Ellen Mickiewicz, *Split Signals: Television and Politics in the Soviet Union*, Oxford: Oxford University Press, 1988, 106.

40. One of the more recent scholarly biographies of Nikita Khrushchev details the motivations and execution of his removal from power; see Taubman, *Khrushchev*, 578–619.
41. Paul D. McDaniel, Paul J. Schmitt and Paul D. McDaniel Jr., *The Comprehensive Guide to Soviet Orders and Medals*, Arlington: Historical Research, 1997.
42. The seven pilots were Lyapidevsky (certificate number one), Levanevsky, Molokov, Slepanyov, Kamanin, Doronin and Vodopianov. Ronald E. G. Davies, *The Chelyuskin Adventure*, McLean: Paladwr, 2005, tells the English and Russian bilingual adventure of the crew of the Chelyuskin. Although the recounting of the mission and plight are accurate, the book does not challenge the political context that led to the mission.
43. Two historians have devoted their time to uncovering the long-forgotten history of women aviators in prewar and wartime Soviet Union; see Reina Pennington, *Wings, Women, and War: Soviet Airwomen in World War II Combat*, Lawrence: University of Kansas Press, 1997; Kazimiera Janina Cottam, *Women in War and Resistance: Selected Biographies of Soviet Women Soldiers*, Nepean: New Military, 1998, 5–7.
44. This numerical restriction for Hero of the Soviet Union awarded to cosmonauts avoided confusion between their relative contributions to the Soviet state with the contributions made by wartime heroes. The three three-time awardees were all soldiers from the Second World War. The only four-time winners were Leonid Brezhnev and Marshall Georgy Zhukov.
45. The decree creating the medal was originally enacted in 1950 and was revised twice, once in 1977 and again in 1980; see Union of Soviet Socialist Republics, 'Decree of the Presidium of the Supreme Soviet of the USSR of July 13, 1950 (in Russian),' Legal Library of the USSR, signed 13 July 1950, http://www.libussr.ru/doc_ussr/ussr_4786.htm.
46. Ibid. The full criteria were: bravery and selflessness displayed during combat operations aimed at the arrest of violators of the state border of the Soviet Union; leadership of border protection units while ensuring the inviolability of the borders of the Soviet Union; vigilance and proactive actions which resulted in the arrest of violators of the state border of the Soviet Union; skillful organization of border service units and exemplary work to strengthen the borders of the Soviet Union; excellent performance of military duties associated with the protection of the state borders of the Soviet Union; and assistance to border protection forces in their combat assignments aimed at the protection of the state borders of the Soviet Union.
47. David Binder, 'Erich Mielke, Powerful Head of Stasi, East Germany's Vast Spy Network, Dies at 92,' *New York Times* (26 May 2000), C19.
48. The mission assignments of each were: Artyukhin: Salyut 3, Soyuz 14; Sarafanov: Salyut 3, Soyuz 15 (failed to dock); Volynov: Salyut 5, Soyuz 21; Zholobov: Salyut 5, Soyuz 21; Glazkov: Salyut 5, Soyuz 24; and Gorbatko: Salyut 5, Soyuz 24.
49. The first complete Russian guide to all cosmonauts nominated to the program that at least began training was published after the collapse of the Soviet Union; see Yury M. Baturin, ed., *Sovetskie i rossiiskie kosmonavty XX vek: spravochik*, Moscow: Informatsionno-izdatel'skii dom 'Novosti kosmonavtiki,' 2001.
50. See ibid., 52, 62, 64 for Volynov, Glazkov and Gorbatko; and 19, 161 and 80 for Artyurkhin, Sarafanov and Zholobov.

51. Elana Gomel, 'Gods like Men: Soviet Science Fiction and the Utopian Self,' *Science Fiction Studies* 31.3 (November 2004), 358–77, here 362.
52. Peter A. Gorin, 'Zenit: The First Soviet Photo-Reconnaissance Satellite,' *Journal of the British Interplanetary Society* 50.11 (November 1997), 441–8, here 441. The Kettering Group was a group of informal space watchers that grew out of the leadership of science master Geoff Perry at Kettering boys' grammar school. Perry had sought to make science exciting for his charges by having them track the new phenomenon of artificial satellites and map their orbits. The group began its work during the orbits of Sputnik 2 and Laika, and ended with Perry's death in 2000.
53. Sven Grahn, 'The Almaz Space Station Program,' http://www.svengrahn.pp.se/histind/Almprog/almprog.htm.
54. Chertok, *Rakety i liudi: Fili, Podlipki, Tiuratam*, 210.
55. Gorin, 'Zenit,' 441–6.
56. Idem, 'Zenit: CORONA's Soviet Counterpart,' in Robert A. McDonald, ed., *Corona between the Sun and the Earth: The First NRO Reconnaissance Eye in Space*, Bethesda: ASPRS, 1997, 84–107.
57. Nicholas L. Johnson, *Soviet Military Strategy in Space*, London: Jane's, 1987, 76–9.
58. On the difference between militarization and weaponization, see the contributions by Alexander Geppert and Tilmann Siebeneichner, and by Michael Neufeld in this volume, Chapters 1 and 2, respectively.
59. The existence of the cannon had been rumored throughout the closing decades of the twentieth century, but it was only a series of Russian media reports that confirmed it. The process took over seven years. The first report was in the Russian version of *Popular Mechanics*; see Mikhail Zherdev, 'Artilleriia na orbite: boevye orbital'nye stantsii,' *Popmech* (23 November 2008), available at http://www.popmech.ru/science/8406-artilleriya-na-orbite-boevye-orbitalnye-stantsii. The topic then became a matter of discussion on military hardware chat sites as individuals expanded the understanding of the original report; see entries from Judgesuhov on the Livejournal discussion board that built on the original Russian print report: 'Artiulleriia na orbite: Boevyi orbital'nye stantsii,' 21 November 2015, http://judgesuhov.livejournal.com/118410.html. About the same time, the Russian military-affiliated television channel Zvezda TV show *Voennaia priemka* (Military Acceptance) broadcast an image of the cannon from inside a closed, military design bureau museum, finally confirming the identity of the cannon in October 2015.
60. The first English account of the cannon was published in the American version of *Popular Mechanics* in 2015; see Anatoli Zak, 'Here Is the Soviet Union's Secret Space Cannon,' *Popular Mechanics Online* (16 November 2015), http://www.popularmechanics.com/military/weapons/a18187/here-is-the-soviet-unions-secret-space-cannon.
61. Ibid.
62. Taubman, *Khrushchev*, 578–619.
63. Svetlana Boym vividly describes this desire for returning to the safe place from childhood; see idem, 'Kosmos: Remembrances of the Future,' in Adam Bartos and Svetlana Boym, eds, *Kosmos: A Portrait of the Russian Space Age*, New York: Princeton Architectural Press, 2001, 82–99.

64. Susan E. Reid, 'The Khrushchev Kitchen: Domesticating the Scientific-Technological Revolution,' *Journal of Contemporary History* 40.2 (April 2005), 289–316, here 290.
65. Christopher S. Wren, 'Two Soviet Astronauts in Good Health. First Night Landing Hailed in Moscow: Salyut Still Operating No Explanations,' *New York Times* (30 August 1974), 8.
66. Svetlana Savitskaya, the daughter of former Soviet Marshall of the Air Forces, Yevgeny Savitsky, was an internationally recognized aerobatic pilot when she launched onboard the Soyuz T-7 spacecraft to the DOS space station Salyut 7 on 19 August 1982 for a two-week stay. She returned to space just under two years later onboard Soyuz T-12 to become the first woman to perform a spacewalk. Although never formally documented, it is widely assumed that the timing of her two missions were to preempt the United States and Sally Ride's and Kathryn Sullivan's respective missions.
67. See Anatoli Zak, 'Almaz,' *EO Port: Sharing Earth Observation Resources* (15 July 2015), https://directory.eoportal.org/web/eoportal/satellite-missions/a/almaz; Cathleen Lewis, 'Zvezda Service Module Celebrates 15 Years in Orbit,' *Air and Space Blog* (19 August 2015), http://blog.nasm.si.edu/space/zvezda-celebrates-15-years.
68. The United States Department of Defense has only recently declassified documents on the Manned Orbiting Program, but the crossover between the Air Force and civilian programs of personnel, hardware and contractors has provided sufficient information to describe the program in broad strokes. The most famous, pre-declassification, recounting of the program was assembled in 'Astrospies.'

CHAPTER 13

Satellite Navigation and the Military-Civilian Dilemma: The Geopolitics of GPS and Its Rivals

Paul E. Ceruzzi

For the United States, a key step in the exploitation of outer space for military purposes was the preparation by the Douglas Aircraft Company's Project RAND report, issued in 1946, of a 'Preliminary Design of an Experimental World-Circling Spaceship.'[1] Published shortly after the United States and its allies learned details of the German V-2 ballistic missile and based on a thorough mathematical analysis of the challenges of orbiting a spacecraft, the report was prescient in foreseeing the many ways that scientists and the military would use such satellites once they became practical: weather forecasting, over-the-horizon communications, targeting and the scientific study of the solar system. Preceding the RAND report was the now-famous 1945 essay by Arthur C. Clarke (1917–2008) on the uses of satellites for communications, especially when placed in an orbit whose period matched the earth's rotation.[2]

Navigation was not among the uses listed in these papers. Nor was navigation prominent among the suggested uses of outer space by other popular science writers or science-fiction authors, who otherwise created a golden age of space science fiction and factual speculation in the decade and a half following the end of the Second World War.[3] Astroculture from that age concentrated on ventures to the moon, planets or beyond the solar system; space

P. E. Ceruzzi (✉)
Smithsonian National Air and Space Museum, Washington, DC, USA
e-mail: CeruzziP@si.edu

Alexander C. T. Geppert et al. (eds), *Militarizing Outer Space*
European Astroculture, vol. 3
https://doi.org/10.1057/978-1-349-95851-1_13

applications that looked back towards earth, including navigation, weather and communications, were hardly covered. Recent celebrations of the fiftieth anniversary of the *Star Trek* television series have noted that the crew of the starship Enterprise had no trouble locating the position of people or aliens on distant worlds, but this was treated as a mundane, almost trivial capability. We shall see that the development of a real capability, through GPS, was anything but mundane, although to the casual user of an automobile navigation system it feels that way. More recent science-fiction films and novels have addressed the dystopian visions of a world in which individuals are tracked and controlled by sinister governments or other entities. If these systems of surveillance and control use satellites, that is not the main focus of the narratives. Later in this chapter we shall address the issue of the extent to which satellite positioning systems like GPS are central to a global surveillance state, or not.

A half-century after that golden age, satellite positioning and navigation has become a central pillar of modern military activities, what in the United States has been called 'network-centric warfare,' or what I refer to as 'War 2.0.' The US Global Positioning System (GPS), a constellation of at least 24 satellites in medium-earth orbit, is a key part of a military network of reconnaissance, communications and weather satellites, tied in to similar ground- and aircraft-based systems. Together these tie the soldier to a world-encircling nexus in ways even the farsighted RAND report could not have imagined. The social use of GPS, as it enables applications found embedded into tablets, smartphones, automobiles and hobbyist 'drones,' is just as remarkable, although the underlying technology that enables these 'apps' is unknown to most users. Military users, including those who operate weaponized drones, are likewise often unaware of the combination of satellite, inertial and other positioning technologies that make such weapons practical. GPS has become an invisible piece of infrastructure, like clean water or electric power – taken for granted unless something disrupts it. That is a true measure of its effectiveness and its crucial place in society. It also indicates a failure of visionaries and science-fiction authors to anticipate what has become one of the primary uses of outer space today.[4]

Historians of technology have long known that developers of new technologies seldom foresee how their inventions eventually find a place in the world. Thomas Edison did not imagine the use of his phonograph for entertainment, nor Alexander Graham Bell the social uses of the telephone. Physicist Howard H. Aiken (1900–73), who designed one of the first digital computers in the 1940s, thought that only a few large computers would satisfy the computing needs of the entire United States. Even into the 1980s computer analysts did not think that inexpensive personal computers offered by Apple and other fledgling companies would be any threat to the IBM Corporation's large 'mainframe' product line.[5]

So too was the trajectory of the Global Positioning System. Its creators believed that the system would find wide use and garner a lot of support for many segments of society. However, the ways it is used by the American public and people all over the world was unforeseen, and those who operate and maintain GPS have had to adapt to that reality. This chapter places the evolution of GPS, including its penetration into markets little envisioned by its creators, in the context of post-Second World War geopolitics. It unpacks some of the details of US decisions on how to manage the provision of positioning information to civilian users without jeopardizing national security. The chapter shows that the rapid pace of electronics technology from the 1970s to the present took on a deterministic role in driving GPS – as it did computing and the Internet – into consumers' hands. GPS became integral to the daily social life of ordinary citizens in the manner similar to the way the telephone, radio and phonograph did in previous decades.

Allowing commercial ships and aircraft to use navigation aids developed by the military is hardly new. For centuries, lighthouses gave a beacon to commercial and navy ships. Seafaring nations published nautical charts and mathematical tables for all to use. In the twentieth century radio beacons and timing signals were available to civilian as well as military aviators. The designers of GPS had this model in mind, but the system's accuracy and global coverage, combined with advances in computing and microelectronics, changed the nature of that relationship. Those changes, plus shifting geopolitics after the end of the Cold War, also explain the proliferation of satellite positioning systems elsewhere, in spite of their high cost and in spite of US assurances that it will offer selected, high-quality GPS signals to the rest of the world for free. As of this writing, other nations and organization are fielding their own satellite-based systems: 'Galileo' by the European Union, 'IRNSS' by India, 'BeiDou' by China and 'QZSS' by Japan. At the height of the Cold War, the Soviet Union deployed a close copy of GPS, called 'GLONASS'; this system fell into disuse but has since been resurrected by Russia and it also gives worldwide coverage. These systems all adopt variations of GPS technology of employing precise atomic clocks onboard satellites to fix a position. What they also have in common is high cost. Their existence reveals that satellite positioning systems are critical to a nation's or the European Union's (EU) place in world affairs and that the US assurances of GPS availability to the world seem to be insufficient.[6]

I The military origins of GPS

The technology of satellite positioning has deep historical roots that go back to the seafarers of the eighteenth century and to the rapid technological advances made by the Allies and the Axis forces during and after the Second World War. An understanding of how GPS came to its current place as a critical global piece of infrastructure therefore requires a look at the historical antecedents on which it is based. Since at least the eighteenth century, nations have invested resources in the art and science of navigation. The well-known story of the British government's search for a method of determining longitude at sea was only one of many such endeavors. The Longitude Act of 1714 authorized prizes of up to £20,000 for solutions of the longitude problem – an indication of how important navigation was to the wealth and power of the nation.[7] The origins of the digital computer can be traced back to the nineteenth-century efforts of the English mathematician Charles Babbage (1791–1871), who sought to mechanize the production of navigational tables, for both commercial and navy use, after finding existing tables full of transcription errors.[8] The techniques developed in those years were based on celestial observations of the stars and other heavenly bodies, combined with accurate mechanical clocks synchronized to a time standard. Latitude was defined in relation to the earth's equator and longitude in reference to a Prime Meridian. Unlike the equator, the Prime Meridian's location is arbitrary. Several nations, including the United States, established their own Prime Meridians, but after 1884 Greenwich, England, was accepted as a global standard. Sailors used the mathematical tables mentioned above to translate those observations into coordinates of latitude and longitude. These efforts were seen as central to the United Kingdom's place in the world, as it used its mastery of the oceans to sustain a global empire.

In the twentieth century these celestial techniques were supplemented by radio beacons, which guided aircraft over land routes, and radio stations that broadcast precise time and weather information. Beginning in the 1930s, mechanical chronometers were replaced by time and frequency standards based on the vibrations of a quartz crystal, when excited by an electric current. Quartz oscillators were more accurate and could be integrated into aircraft and naval electronic systems, not only for navigation and positioning but also for radar, aircraft radios and other devices. Government-sponsored radio stations, WWV in the United States and CHU in Canada, used quartz oscillators to keep precise time, which they broadcast via powerful short wave transmitters. The use of the short wave bands meant that these broadcasts could be heard and used outside the boundaries of those countries, free of charge. Several European countries established similar broadcasts using other frequency bands.

After 1939 the demands for waging war across the globe transformed technologies and practices of navigation and positioning. The systems developed by the Allies during the war years and the V-2 ballistic missile developed by the German army enabled the technologies of GPS and other satellite systems that emerged 40 years later. The V-2's influence is well known; it was explicitly referenced by the RAND report. For RAND, Arthur C. Clarke and others, the V-2 pointed toward the possibility of placing instruments into earth orbit. The Soviet Union drew on V-2 technology to design the rocket that orbited Sputnik in 1957, and a descendant of the V-2 was used by the United States to place Explorer 1 into orbit the following year. Methods of navigation that went beyond the classical celestial techniques were also developed during the Second World War. Two in particular stand out: LORAN ('long-range navigation'), a ground-based radio system, contained elements that were later incorporated into GPS. Another, inertial navigation, was developed for the V-2 and refined to a high degree of sophistication in the postwar years by the United States and the Soviet Union. Although inertial navigation neither transmitted nor received radio signals of any kind, its central role in Cold War military strategy and policy ties it closely to the development of satellite navigation.

II LORAN, inertial guidance and the beginning of satellite navigation

LORAN was similar to several British radio-navigation systems, although it had a greater scope. The American adaptation of the British systems was supported by a wealthy American banker, Alfred Lee Loomis (1887–1975), who built a private research laboratory on his estate north of New York City. Research was later transferred to the Massachusetts Institute of Technology (MIT).[9] It operated on the principle of having pairs of radio transmitters located along coastlines, each broadcasting signals synchronized by quartz oscillators. A ship or aircraft would receive these signals and the navigator would note the time difference between the reception of each. That difference placed the craft along a hyperbola – one of the Conic Sections – defined as a line of constant difference between two points. By repeating this process using another, different pair of transmitters, the ship or aircraft's location could be determined by consulting charts overlaid with hyperbolas related to the stations. LORAN required training and exotic equipment onboard, plus the constant staffing of transmitting stations in remote locations. But it worked in any weather and did not require any transmission from the user that could give away position.[10] Satellite positioning systems like GPS operate on a different principle, but they do retain LORAN's fundamental concepts of transmitting synchronized time signals from widely separated transmitters and of not requiring any transmissions from the user equipment.

By 1945 LORAN stations provided coverage over the north Atlantic and north Pacific great circle routes. The United States continued its operation and made it available to commercial shipping and aircraft as well. LORAN was continually improved and the cumbersome charts, with their mazes of hyperbolic lines, were replaced by solid-state electronic receivers that gave the navigator his or her latitude and longitude. The modest New England fishing boats that were chronicled in the best-selling 1997 book and movie *The Perfect Storm*, for example, were equipped with compact, capable LORAN receivers.[11]

During the early development of the V-2 ballistic missile, German engineers working at Peenemünde incorporated gyroscopes and accelerometers onboard to stabilize the rocket as it ascended from the launch pad. A combination of internal and external radio-guided systems was developed to guide the rocket to its target. Onboard gyroscopes and accelerometers controlled attitude: pitch, roll and yaw.[12] Some of the V-2s used a radio beam to control drift and cut-off of thrust; other V-2s used internal gyros. Because internal gyroscopes were impervious to jamming or outside interference, that method was preferred, although the V-2 never achieved the kind of accuracy necessary to make it an effective weapon. After the war the United States and Soviet Union both mounted an intensive effort to improve the accuracy of these all-internal systems, which they called 'inertial' as they relied on an application of Newton's laws relating acceleration, velocity and position.[13] MIT researcher Charles Stark Draper (1901–87) was especially influential in driving the accuracy of these systems to a degree thought impractical in the immediate postwar years. Draper and his students not only provided inertial guidance systems for ballistic missiles, his instrumentation laboratory also designed the system used by the Apollo astronauts to visit the moon in 1969–72.

Inertial navigation's ability to determine velocity without interaction with the outside world made it desirable for ballistic missiles, whose trajectory could not be jammed by an enemy and for nuclear-powered submarines, which went to great lengths to hide their location under the sea. Practical inertial navigation for these applications was expensive, but its inherent qualities made the cost worthwhile. After 1945 suppliers developed inertial systems at much lower cost, with a slight relaxation of accuracy. Commercial airlines used them on many routes, especially over the Pacific where there were gaps in radio navigation.

Achieving high accuracy was expensive, as inertial gyroscopes and accelerometers tend to drift over time. Ballistic missiles were guided for only a short time, before drift became a significant factor, and commercial aircraft could supplement the information from the onboard inertial systems with radio-navigation aids as the craft approached a land mass. But submarines operated for long periods underwater. To correct for drift, they had to come close to the surface, get an accurate "fix" from another source and update their onboard inertial systems. The US Navy's response was to orbit a set of satellites, called TRANSIT, to work with the submarines. The first TRANSIT satellite was orbited in 1959, and the system remained in constant use until replaced by GPS in 1996. Like LORAN, it also used quartz oscillators onboard, to transmit timing and positioning information. A submarine would extend an antenna to receive the signals and locate its position by observing the change in frequency from the satellite due to the Doppler effect: the most-rapid change in frequency occurred as the satellite was closest to the submarine, similar to how one observes an ambulance or police siren changing pitch as it passes by.[14] As with all of the systems discussed so far, TRANSIT was also made available to civilians, including geologists who were working in deserts or remote areas where maps were unreliable. During the 1982 Falkland Islands War between the United Kingdom and Argentina, both navies fought one another with the help of TRANSIT.[15]

III Competing proposals to follow TRANSIT, LORAN and other positioning systems

The Global Positioning System was conceived, designed and implemented by the US Department of Defense (DoD). It was intended to be used by all three branches of service: the Army, Navy and Air Force, and to replace a myriad of incompatible systems that each service had deployed independently of each other. Civilian uses were also envisioned as part of the initial design. The success of TRANSIT, which found uses far beyond its initial purpose of guiding submarines, showed that space-based navigation could be as significant as space-based communications, weather or reconnaissance techniques. The concept of using precise time and frequency standards onboard satellites was proposed in several places, as were other techniques including an enhancement of TRANSIT's use of Doppler effects. One early impetus came from Ivan Getting (1912–2003) of the Raytheon Corporation, who proposed a three-dimensional extension of LORAN, using much more accurate atomic clocks, to guide intercontinental ballistic missiles (ICBM) to their targets. Getting's proposal was intended for a short-lived concept of basing Minuteman missiles on railroad cars, to prevent a Soviet first strike on fixed silos rendering the United States incapable of a response.[16] For ballistic missiles in fixed silos, inertial techniques were known to provide sufficient accuracy – even greater accuracy than desired by Congress.[17] But the very advantage of moving the missiles around on railroad cars meant that inertial techniques, which require a precise knowledge of the missile's launch point, were insufficient. By 1960 the concept of a 'Mobile Minuteman' had been cancelled, and Getting had moved from Raytheon to become the president of the Aerospace Corporation, a federally funded research arm of the US Air Force. The Corporation refined the earlier concept and in 1963 the Air Force designated the study as Project 621B.

The US Navy had already demonstrated the effectiveness of TRANSIT, which was able to fix a submarine's position to within 200 meters. By the late 1960s engineers at the Johns Hopkins University Applied Physics Laboratory in Maryland, where it was developed, were looking at ways of improving the technology to provide greater accuracy and coverage. At the same time, the Naval Research Laboratory (NRL), located in Washington, DC, was developing a system that placed atomic clocks on satellites, primarily to provide time signals but also suitable for positioning. Whereas the Air Force Project 621B envisioned a suite of satellites in geosynchronous orbit (at an altitude of around 36,700 kilometers), the NRL proposed a constellation of satellites in a lower orbit and inclined to the equator to give nearly worldwide coverage. The NRL successfully orbited two experimental 'TIMATION' satellites in 1967 and 1969, each carrying quartz oscillators, to test the concept. A third satellite in the series, called Navigation Technology Satellite 1 (NTS-1) and launched in 1974, was the first to carry an atomic clock, after the NRL purchased and modified a commercial Rubidium frequency standard from the

West German firm Efratom.[18] The rubidium clock turned out to be excessively sensitive to temperature, but a cesium clock installed on the follow-on satellite, NTS-2, worked well. At the same time the US Army was proposing a system, called 'SECOR,' which was intended to more accurately fix positions on earth to the worldwide latitude and longitude grid. Along with these military plans was an effort by the civilian Federal Aviation Administration (FAA) to enhance control of commercial air traffic, a service that the FAA felt was not well served by the military proposals.[19]

IV Convergence: a unified, space-based positioning system emerges, 1973–83

Students of Science, Technology and Society Studies (STS) are familiar with this complex mix of competing systems, each with its own approach to a general problem.[20] In analyses of the social shaping of technology, the genuine technical constraints of physics and engineering do not automatically determine how the system, when completed, will look. There is a lot of 'interpretive flexibility' before a combination of technical, social, economic and political factors leads to a convergence.[21] At that point the decisions are sealed in a 'black box,' hiding those factors from an outside observer. Such was the case with GPS, although we shall see that as other nations sought to create their own satellite positioning systems, they had cause to open the black box, if only slightly.

In 1972 the US Department of Defense decided to cut through this tangle and establish a unified system. Air Force Colonel Bradford Parkinson (1935–) was put in charge of the effort (Figure 13.1). With his Air Force background, Parkinson favored the 621B design, but he also saw advantages of the other competing concepts. At a marathon meeting in the Pentagon over the Labor Day weekend in 1973, Parkinson and his team established the basic architecture of what they initially called 'NAVSTAR,' now known as GPS. The basic architecture has persisted to the present day and it has also served as a basis for the European Galileo and Soviet/Russian GLONASS, although we shall see that there are differences.[22]

NAVSTAR was to consist of three basic components: a constellation of satellites in medium-earth (not geostationary) orbit; a ground control facility to manage the satellites; and user equipment. The constellation, of at least 24 satellites, would orbit at an altitude of about 17,500 kilometers, in circular orbits and in six orbital planes inclined 55 degrees relative to the equator.[23] The orbit placed the satellites far enough away so that irregularities in the earth's gravitational field would not significantly perturb the orbits. The combination of the six planes and inclination provided worldwide coverage and ensured also that each satellite would at some time pass over ground stations located on US territory. The satellites would carry atomic clocks onboard, all synchronized to the same time reference. So precise were these clocks that

Figure 13.1 Frank Butterfield (left) of the Aerospace Corporation, Air Force Colonel Bradford Parkinson (1935–) and Navy Commander Bill Huston (1929–2011) discuss the plans for GPS. This photo, probably taken at the Pentagon in the mid-1970s, was intended to show the combined Civilian, Air Force and Navy stakeholders in the new system. Prior to joining the Aerospace Corporation, Butterfield had worked at Magnavox, a principal supplier of commercial receiving equipment. A scale model of one of the satellites is between Parkinson and Huston, with a crude diagram of the concept on the blackboard in the back.
Source: Courtesy of Aerospace Corporation.

they had to be adjusted to account for both Einstein's Theory of Special Relativity, because they move rapidly with respect to the earth below and General Relativity, because at that altitude the force of gravity is weaker than it is on the earth's surface.[24] The ground stations would monitor the health of these clocks and establish the position of each satellite in space. The satellites would broadcast their position and the time to users, using a technique called 'pseudo-random' coding. GPS receivers would be entirely passive, with no need to transmit their position or any other information – a feature that had obvious military implications. As the question of surveillance and tracking will come up later in this chapter, it is important to note that for a user who maintains control of a receiver, it is impossible to be tracked using GPS alone.[25]

V 'Much to everyone's surprise': the question of dual use

Many of the design parameters discussed above were present in the Naval Research Laboratory's TIMATION experiments. The major exception, which came from the Air Force, was the choice of coding for the signals. Parkinson knew that the individual services would resist an attempt to unify the disparate navigation systems – each service had its own, tailored to its specific needs. LORAN and inertial navigation techniques were well established and working well. The Department of Defense recognized that it could gain support by bringing in civilian users from the beginning, not afterwards as was the case with LORAN or inertial navigation. The DoD had to balance that decision with a need to prevent civilian users from taking advantage of some of GPS's strongest assets. The architects of GPS addressed this issue by its choice of coding for the signals.

'Direct Sequence Spread Spectrum' was selected as the coding scheme, to make the system available to civilian users without jeopardizing its military effectiveness. The origins of this technique are obscure.[26] Since its adoption by GPS, the technique has become widespread, for example, used by American cell phones, by Wi-Fi and Bluetooth among other applications. The popular press recently discovered that Hedy Lamarr (1914–2000), a Viennese actress who moved to Hollywood in 1937 to pursue a film career, invented and patented a method of guiding a torpedo using a radio signal whose frequency "hopped" in a pseudo-random fashion, that is, the receiver on the torpedo and the transmitter on the ship or submarine had identical copies of the sequence of frequencies, but to an outsider the sequence *appeared* random.[27] Direct Sequence Spread Spectrum does not hop from one frequency to another, but it does use a wider bandwidth than traditional narrow-band radio. It superimposes a pseudo-random sequence of bits over the signal before transmission and receivers that also have this sequence are able to extract the original signal. To an outside listener who does not have the code, the signal is indistinguishable from background noise.

The initial GPS design employed two sets of codes. A short 'Coarse/Acquisition' (C/A) code was used to rapidly acquire the signal and give a moderately accurate position. A second 'precise' (P) code, much longer, gave more information. Initially only the C/A code was to be made available to civilian users.[28] It was projected that the code would allow one to fix a position to not less than 100 meters. For US military users, the P code was further encrypted and subsequent satellites transmitted other codes whose details remain classified. The designers hoped that by having these separate codes, GPS would be enthusiastically adopted by airlines, railroads, ships at sea and other users.

In spite of these efforts, Parkinson's group had trouble convincing users of the advantages of GPS and obtaining ongoing funds from the Department of Defense. Budget constraints led Parkinson's group to consider deploying a system with as few as nine satellites, using only the Coarse/Acquisition code.

The intent was to place units in the field and let users become familiar with the technology, even if the predicted accuracy was poor. When Parkinson's team tested this reduced configuration, however, they discovered a remarkable fact. In Parkinson's words, 'Much to everyone's surprise, [...] the unit performed almost as good as its more sophisticated counterparts, demonstrating accuracies in the 20–30 meter range.'[29]

This 'surprise' turned out to have enormous implications for satellite navigation in general, for the US military's sponsorship of GPS and for the social transformation effected by these systems in the twenty-first century. It also helps explain in part why the European Union and other nations have chosen to develop their own systems, even given the free worldwide availability of GPS. Again, in Parkinson's words, 'the Department of Defense was faced with a dilemma since the C/A code on which this equipment operated was to be generally available to anyone in the world who had access to the technology required to build a suitable receiver.'[30] That led to a study by the DoD along with the US National Security Council to consider the implications of foreign nations having such access. The result was to intentionally degrade the accuracy of the publicly available C/A code and to further restrict those who were to have access to the more accurate P code. An additional overlay of the Coarse/Acquisition code, called Selective Availability, was installed on the satellites. For the designers of the system the 'surprise' that the C/A code was so accurate was something they were proud of; with Selective Availability, they also felt that the restrictions they placed on the codes would still allow a robust civilian market, for ships, trucking and commercial aircraft, to develop.[31]

Although opposition to GPS remained, Parkinson was able to obtain funding to deploy the full 24-satellite constellation. More support came from another government office, which saw the worldwide coverage of the constellation as a way to monitor possible nuclear tests by the Soviet Union and other nations. In 1979 a 132 kilogram sensor was installed on the sixth satellite in order to detect such explosions. The need for such detection, coupled with funds from an office unconnected with the navigation team, further assured the future of GPS.[32]

Satellite launches continued into the 1980s and with each launch the system (still called NAVSTAR at the time) became more capable. In July 1983 the Collins Radio division of Rockwell International, a major avionics supplier, staged a dramatic demonstration of the system's capabilities, when a Rockwell Sabreliner business jet flew from Cedar Rapids, Iowa (the home of Collins Radio) to Le Bourget airport in Paris, where Charles Lindbergh had landed in 1927, navigating solely by the five satellites then in full operation. The aircraft made multiple stops along the way, not only to refuel but to wait until a sufficient number of satellites were in view to obtain a fix. The Sabreliner taxied to within 8 meters of a pre-surveyed stopping point at Le Bourget. The flight was well-publicized, making it clear that satellite navigation was real and that commercial aircraft could use it.[33]

VI Opening the black box, 1983–91

On 1 September 1983 a Korean Airlines Boeing 747, en route from Anchorage, Alaska, to Seoul, strayed over restricted Soviet territory and was shot down. All 269 passengers and crew onboard were killed. The tragedy inflamed the already hostile relations between the United States and the Soviet Union. The aircraft was equipped with several navigation aids, including a magnetic compass, radio-navigation equipment and an inertial system.[34] The most plausible explanation for the tragedy is that the crew failed to switch over to the aircraft's inertial navigation system as they approached the Asian continent, but because of either human error or poorly designed instrumentation, they thought they had selected it.[35] The crew continued on a course taking the aircraft over restricted territory. The cause has been attributed to a faulty 'user interface': a mislabeled selector switch that did not warn the crew that the aircraft's navigation had not in fact switched over to the inertial system. Others have placed the blame on the crew for not double-checking their position manually based on other information provided to them in the cockpit. The Soviets may have been confused by its awareness of a US military aircraft flying in the region, a four-engine RC-135 that resembled the Korean 747. In one sense the tragedy was inevitable, given the frequent and often-deliberate close encounters between Soviet and American ships and aircraft near each other's borders.[36] Among the responses from the Reagan administration was that the Global Positioning System, then under construction, would be made freely available for civilian use, to prevent such tragedies from occurring in the future. This pronouncement was part of a coordinated diplomatic and military US response to the event, aimed at emphasizing the openness of American society in stark contrast to the Soviet response, which alternated between denial, misinformation and contradictory statements.

Although US President Reagan did not know all of the details surrounding the tragedy, his statement implied that GPS was superior to the inertial navigation system the aircraft was carrying. That impression influenced the public perception of GPS as the 'next, new thing,' although in fact inertial navigation was, and remains, a very capable technology and is still in use. If the tragedy was caused by a poor design of the controls or a lack of attention by the crew, that could equally apply to any other system. The flight of the Sabreliner was publicized to the aviation community, but it was this tragic event that brought the existence of GPS into public awareness, generating favorable publicity for its commercial capabilities.

That tragedy was one of several developments that upset the balance between civilian and military use, which the Department of Defense had carefully crafted, by the late 1980s. Not for the first time would the 'black box' of GPS architecture be pried open. Also during that decade, engineers affiliated with several civilian government agencies were developing ways to circumvent Selective Availability. A primary technique was 'Differential GPS' (DGPS), based on installing a receiver at a carefully surveyed ground station, which computed the local errors of the system by comparing its surveyed location with the location given by the satellites. That information would generate a correction factor, which was sent to nearby receivers. Surveyors, who were among the first civilian users of GPS, adopted this technique. Because they were able to take readings at widely dispersed intervals of time, they were not as dependent on whatever satellites happened to be in view at a particular moment. They could take readings in the field and make the corrections later on, where they had access to powerful computers and other specialized equipment. Another variant was to send correction factors to a geostationary satellite, which could broadcast this information to receivers over a wide area (variants of this technique are the basis for several Asian satellite navigation systems, discussed later). In short, Selective Availability was being bypassed even as the full constellation of satellites was being launched. Recognizing that Selective Availability was not going to serve the military as intended, the Air Force developed alternate tactics to maintain control, mainly by developing ways to jam the civilian signal over selected areas at specified times, while retaining the full capability of the P-Code.[37]

In 1991, during the Persian Gulf War, the US military found the as-yet-unfinished GPS system to be very useful, given the treeless, featureless deserts of Kuwait and Iraq, and the lack of reliable maps of the area. The military had a shortage of GPS receivers and it responded by purchasing civilian units domestically and shipping them to the theater of war. The Air Force turned off Selective Availability and suddenly everyone's GPS receivers, not just the military's, gave better readings. Once it was turned off, the public reaction was swift and favorable. Under a directive from President Bill Clinton in May 2000, Selective Availability was permanently set to zero, with a guarantee that it would remain so.[38] That led to a dramatic increase in the popularity and novelty of civilian uses such as the popular hobby geocaching, a modern version of the established hobby of orienteering.[39] It also convinced the remaining skeptics among the military of the advantages of GPS over existing systems.

These incidents suggest that the civilian use of GPS escaped the controls that the US military establishment hoped to impose on it. The Department of Defense lost control, although the expectation remained that the DoD would continue paying for it. Like the Sorcerer's Apprentice, the DoD turned on a spigot, which became gusher that it could not turn off. This story has relevance to the debates among historians about whether technology is autonomous, socially constructed or somewhere in-between.[40] In this instance, there was a clear progression from an early set of as many as a dozen alternative architectures, with differing orbits, coding schemes and other parameters. By the mid-1970s these alternatives converged, resulting in the closure of the 'black box.' There was also the active direction of this process by the US Air Force, led by Parkinson, who was mindful of the need to convince his superiors at the Pentagon that GPS was first and foremost a military system that could serve the Pentagon's desire for better command-and-control of its operations.[41] Yet at the same time the trajectory of GPS veered off in unanticipated directions, including two direct interventions by US presidents, Ronald Reagan and Bill Clinton, neither of whom were knowledgeable of the technical issues related to navigation or satellites.

Historians of technology are familiar with this sequence of events. Donald MacKenzie's analysis of ballistic missile guidance argues that increasing accuracy, driven by Charles Stark Draper at MIT, was not always in alignment with US policy toward nuclear deterrence. Likewise, Janet Abbate's study of the origins of the Internet shows how designers of the Internet's predecessor, the Defense Department's ARPANET, were taken by surprise at the emergence and rapid adoption of e-mail, something they had not anticipated. Like the unanticipated uses of GPS, e-mail became a 'smashing success,' which ensured the widespread adoption of the ARPANET and eliminated any residual skepticism over its utility.[42]

One can find other interesting parallels between the trajectory of GPS and of the Internet. Both owe their origins to military needs, and both took form during the 1970s.[43] Both have become indispensable components to the infrastructure of a modern technoscientific society. Both rely on complex, cutting edge technology and both are expensive to construct, operate and maintain. However, unlike the Internet, GPS is largely invisible and its importance and place in society are poorly understood by the public. After 1987 control of the civilian portions of the Internet was transferred to the US National Science Foundation and after 1992 to commercial entities, with light direction from a quasi-government committee called 'ICANN.'[44] The United States is deeply involved with governance of the Internet and maintains its assertion of ownership. Debates over the future of the Internet are carried out in public, with other nations insisting on having their say.[45] By contrast, GPS remains under tight US government control. The Air Force manages the control and operation of the satellites from a secure control center at Schriever AFB, Colorado, while overall management is handled by the 'National Space-Based Positioning, Navigation and Timing Executive Committee,' which includes civilian as well as military members.[46] It would be difficult to imagine how any control of GPS could be transferred away from this committee or from the Air Force.

VII Alternatives to GPS: GLONASS, Galileo, Beidou

During the 1970s the Soviet Union began development of a similar system, called GLONASS.[47] Design began not long after the United States began its work. The first satellite launches took place in October 1982, and by 1991 there were 12 functioning satellites in orbit, enough to begin provisional operation. Its architecture closely mirrors GPS, with some exceptions. The orbital altitude is nearly the same, but the inclination is greater. The satellites carry atomic clocks, but the coding system is different.[48] Civilian as well as military uses were planned from the start. After the fall of the Soviet Union, the system was poorly maintained, but in recent years Russia has restored the full constellation of 24 satellites. Today, GLONASS is close to GPS in its coverage, accuracy and availability. Sporting goods stores now sell high-performance watches or personal devices made by Garmin and other American manufacturers, with GPS and GLONASS both available. Access to the European Galileo system, under construction as of this writing, will soon follow. These watches, used by long distance runners, mountain climbers and other outdoor enthusiasts, automatically select whichever system provides the best positioning information.[49]

As of 2020, satellite-based positioning systems are also under development in India, Japan and China. These have focused on regional rather than worldwide coverage. They deploy satellites in geosynchronous orbits, localized over their respective countries. A satellite placed in a 36,000 kilometer orbit has a period of 24 hours, matching the rotation of the earth. If it has an inclination of zero, it is 'geostationary': it appears in the same place in the sky at all times. That has advantages for communications, but their fixed position in the sky make geostationary orbit less suitable for navigation. If the orbit is inclined to the equator, the satellite is 'geosynchronous': it appears overhead but traces a lazy figure-8 pattern in the sky, making them more useful for fixing a position on the earth. The Chinese 'BeiDou' system is currently employing geosynchronous satellites, but it is transitioning from a localized to a global system similar to GPS and GLONASS.[50]

In October 2011 the European Space Agency (ESA) began launching satellites for its 'Galileo' system, under the direction of the European Commission (Figures 13.2 and 13.3). Galileo will be considered complete when it orbits a constellation of 24 satellites in three orbital planes, into medium-earth orbits slightly higher than the orbits of GPS. Launches have been proceeding steadily.[51] An initial operating capability was achieved in 2017, with 18 satellites in orbit. Galileo follows the basic GPS architecture, with several significant differences. These are: First, the system is under the control of civilian entities. It is funded by the European Union and managed by the European Commission. Military customers will be allowed to use the system, but it is fundamentally a civilian service. Second, Galileo will provide a search-and-rescue function, whereby a user in distress can send a signal to the satellites, which can in turn direct rescuers to a precise location. GPS is strictly a one-way system, although future-generation GPS satellites may have this capability. And like GPS, Galileo has, third, several codes that offer different degrees of precision, but unlike GPS, Galileo is planning to implement a method of licensing, enabling the EU to charge money for premium services.[52] Galileo is expensive and, like GPS, questions of its funding have at times threatened its future.[53] At this time it appears that these issues have

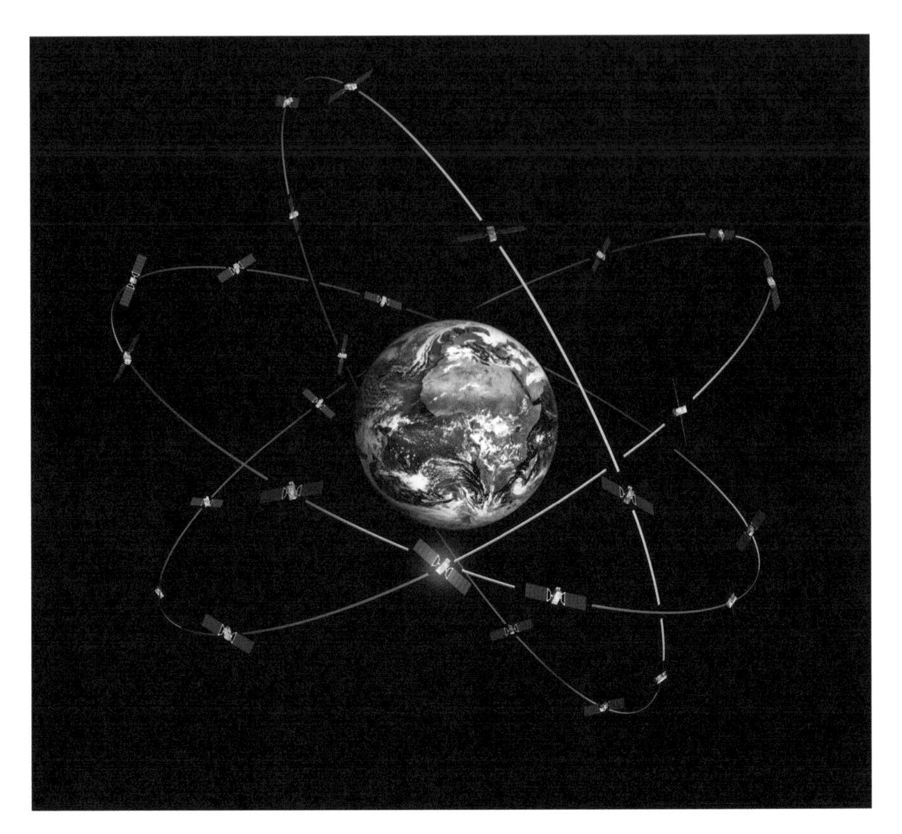

Figure 13.2 Galileo's constellation of 24 satellites along three orbital planes. In addition to the eight satellites in each orbital plane, the Galileo system will also orbit two spare satellites in each plane, for a planned total of 30 satellites to make up the full constellation. Artist's representation, not drawn to scale, 2014.
Source: Courtesy of European Space Agency/Pierre Carril.

been resolved. Initial tests of the technology indicate that it is working well. Future consumer and commercial products will incorporate Galileo receivers and select the best among Galileo, GLONASS and GPS according to local conditions.[54]

Construction of Galileo was not only threatened by its costs; it was also subjected to criticism by the United States, whose Defense Secretary Donald Rumsfeld was expressing hostility in 2002–03 toward the French for their refusal to support the American invasion of Iraq.[55] These remarks, including a comment by Rumsfeld that France and Germany were part of 'old Europe,' suggested that the proclamations by former Presidents Reagan and Clinton of the free use of GPS were in jeopardy. Whether intentional or not, Rumsfeld's comments probably helped Galileo overcome European skepticism. During the initial planning for Galileo, the precision of its signals and the promise of

Figure 13.3 This artwork, *Orbital Ascent* by German space artist Michael Najjar, depicts the November 2016 launch of an Ariane 5 rocket carrying four Galileo navigation satellites into orbit. The successful launch was a major step in establishing Galileo as a European counterpart to the American GPS system. The image evokes the surrealistic contrast between the advanced technology of the rocket and its payload, with the primeval Amazonian jungle of Kourou, its launch site. 202 × 132 cm, pigment print, 2016.
Source: Courtesy of Michael Najjar.

worldwide availability led to a fear that other nations could use Galileo's signals to target American assets. If the frequencies of those signals were close to those used by GPS, an attempt by the United States to jam Galileo in times of crisis might render its own military channels inoperable.[56] These disputes have been smoothed over, as frequencies were adjusted and as it became clear that global satellite positioning and navigation systems are infrastructures that most advanced technological nations wish to have. Nevertheless, the initial friction is further indication of how the US Air Force's response to the unexpected, albeit fortuitous, increase of accuracy of the civilian GPS signal did not solve the dual-use dilemma.[57]

VIII The microelectronics revolution: civilian and military

In 2007 the Apple Computer company introduced the iPhone. The iPhone was not the first portable device to integrate telephone, wireless computing and positioning information. Nor was the iPhone the first smartphone to incorporate a GPS receiver – Apple considered adding that capability but was constrained by licensing agreements with cell phone carriers to leave it out. However, Apple's attention to design and the tight integration of the other functions made the product a runaway success and it transformed the way consumers interact with computing and communication devices.[58] The name 'iPhone,' or its generic counterpart 'smartphone,' is a misnomer. These devices are in fact general purpose, high-performance computers, which are programmed to perform a number of tasks. Making telephone calls is only one of those uses; for many users it is not as frequently used as, say, sending short texts, taking photographs, or – critically for this narrative – locating one's position.

Behind the smartphone phenomenon is the often-studied, steady advance of the underlying microelectronics: increased performance, smaller size and increased memory capacity. The popular term for this trend is 'Moore's Law': named after Gordon Moore, a chemist at the Fairchild Camera and Instrument Corporation in Mountain View, California, who wrote a provocative editorial in 1965 that predicted the steady doubling of the number of active elements that could be placed on a silicon chip.[59]

Throughout the 1970s, as the Global Positioning System was taking form, its architects were aware of this law. Silicon integrated circuits were used on the Apollo missions to the moon from 1968 to 1972. Military-grade GPS receivers followed this law, from an 8 kilograms, $45,000 'Manpack' used by soldiers in the 1980s, to the brick-sized 'Precision Lightweight GPS Receiver' ('Plugger') used in the 1991 Persian Gulf War, to the handheld 'Defense Advanced GPS Receiver' ('Dagger') similar to those used by hikers and outdoor enthusiasts, only capable of receiving encrypted military signals.[60]

The significance of the iPhone and other handheld devices was not only that they took advantage of advances in microelectronics, but they also managed to integrate a variety of radio signals and computing power with one another. The social phenomenon unleashed by the smartphone is the result of its seamless integration of GPS signals with location data from Wi-Fi where available, cell phone signals, even onboard accelerometers and gyroscopes. And it is not just the integration of these location services, it is also the integration with what used to be contained on traditional paper maps and with what formerly was known as the 'Yellow Pages': information about businesses and their locations. GPS signals generally are not strong enough to penetrate walls of most buildings, but cell phone signals can and do. Cell phones do not cover the world, but most urban areas of the developed world have cellular coverage. One cannot be tracked solely by a one-way GPS receiver, but a cell phone provider has to know where a phone is to route a call. Mobile phone carriers in the United States not only can locate users, they must: The Federal Communications Commission requires that for 911 emergency calls, which today are seldom made from fixed wired phones. Modern smartphones are able to determine a position, by first using cell phone triangulation to locate the phone, then using that data to know which GPS satellites are overhead and then using the satellite signals to refine the phone's position. That is a much faster process than using the GPS Coarse/Acquisition signal alone. Wi-Fi, the use of which is restricted to smaller regions, also tracks receivers by triangulation.

The end result is that consumers who drive modern automobiles or who carry around smartphones can be tracked unless they are in remote areas where there is no cell or Wi-Fi coverage. They can be tracked by intelligence agencies, military services, law enforcement agencies, corporations developing consumer information, private investigators or worried parents keeping watch over their children. The list has almost no limit and the technology does not discriminate among military, other government, or corporate use. Scholars who follow the sociology of the militarized/surveillance state have noted this situation with alarm.[61] Their concerns are valid, but a person who is tracked in this manner does so because he or she voluntarily uses these devices. And although GPS is a critical part of the establishment of this condition, it is misleading to attribute the surveillance state solely to GPS alone.[62]

Where does this leave the US Air Force and the National Space-Based Positioning, Navigation and Timing Executive Committee, which oversees the continued maintenance and operation of GPS? Military applications of GPS are as critical as ever to the projection of American power around the world. Just as the satellites have come to symbolize the more complex relationship of domestic surveillance, so too does GPS stand in for what has become a major shift in military tactics, namely the rise of drone warfare. The Predator is the most famous of these drones, whose formal name is 'Unpiloted Aerial Vehicles' or UAVs. Unpiloted aircraft have been under development and in limited use for over half a century, but the Predator was the first to solve the many difficulties that prevented drone warfare from becoming part of the mainstream military strategy. Like the advent of the iPhone, the Predator's success has been due not to one single innovation but to its skillful *combination* of electro-optical systems, composite materials, lightweight engines – and GPS.[63] There are numerous other uses of GPS that produce "smart" weapons or other systems: glide bombs, air-dropped supplies, miniature helicopters that carry high-resolution cameras into hostile areas, etc.[64] For these military applications, the receivers are able to decode a variety of signals, the details of which remain classified, that are not available to civilians. These codes provide greater accuracy; they also are better able to operate in a hostile environment of intentional jamming and even inadvertent jamming caused by radar and radio sets used in the theater of combat.

The advent of this technology, called an 'invisible infrastructure' by Brad Parkinson, has served the US military well and that is no doubt a reason why copies of it are being developed elsewhere.[65] It explains why Russia has devoted resources to refurbish the former Soviet Union's GLONASS. It has also become a vital component to the modern global civilian economy, but as the British understood in the eighteenth century, it is not always possible or even necessary to distinguish between. Parkinson's statement reflects his pride in helping bring GPS to the world, but it also is an expression of his frustration that GPS is not recognized as the vital piece of infrastructure that it is. The US government has thus far not shown any reluctance to continue supporting it, but it remains an expensive system.[66]

Finally, Parkinson's description of satellite positioning as an invisible infrastructure helps explain why this ensemble of satellites, microelectronics, computer networks and related technologies was not foreseen by nearly all major writers of science fiction. One can find examples of specific technologies, for instance, the 'Communicator' used by Captain Kirk in the original *Star Trek* television series. But one looks in vain for a description of this combination in space-oriented science fiction.[67] Numerous works of astroculture depict a dystopia on earth, where inhabitants are tracked, monitored, controlled and otherwise manipulated by sinister governments or other entities. These go back at least as far as the 'telescreen' of George Orwell's novel *1984*. Writers of space science fiction have depicted dystopias of various forms, but usually these are located on other planets, not on earth. Satellite-based positioning, tracking and internetworked communications have outpaced the imaginations of all but a few science-fiction writers. The combination of GPS with other technologies, for both the US military and for civilian consumers, has had an unforeseen but enormous impact on the modern world. The failure to see it was not so much an indictment of the value or vision of astroculture as it is a testimony to the ability of technology to veer off in unforeseen directions, almost autonomously, as distasteful as that may seem.

Notes

1. Douglas Aircraft Corporation, *Preliminary Design of an Experimental World-Circling Spaceship*, Santa Monica: Douglas Aircraft Corporation, 1946; RAND Corporation Report SM-11827; available at http://www.rand.org/pubs/special_memoranda/SM11827.html. All Internet sources were last accessed on 15 July 2020.
2. Arthur C. Clarke, 'Extra-Terrestrial Relays: Can Rocket Stations Give World-Wide Radio Coverage?,' *Wireless World* 51.10 (October 1945), 305–8.
3. In his privately-circulated May 1945 essay 'The Space Station: Its Radio Applications,' Clarke mentioned 'navigational aids' as among the possible uses for a station, but left that use out of the published version. See Arthur C. Clarke, 'The Space Station: Its Radio Applications,' in John Logsdon, ed., *Exploring the Unknown: Selected Documents in the History of the US Civil Space Program*, vol. 1: *Using Space*, Washington, DC: NASA, 1997, 12–15, here 13. The web site 'technovelgy' is devoted to documenting how science fiction and science fact have historically intersected; navigation is not among the categories tracked by the site's authors. See 'Technovelgy: Where Science Meets Fiction,' http://www.technovelgy.com/ct/Science_List.asp. There is almost no mention of positioning, location or navigation in the literature surveyed. See also the contributions to David L. Ferro and Eric G. Swedin, eds, *Science Fiction and Computing: Essays on Interlinked Domains*, Jefferson: McFarland, 2011.
4. GPS, Galileo and the other satellite systems provide *positioning*, that is, the location of a ship, person or aircraft with reference to the earth's coordinate system of latitude, longitude and, for aircraft, altitude. *Navigation*, defined as the art of getting from one place in this coordinate system to another, uses this positioning data.
5. Richard Thomas DeLamarter, *Big Blue: IBM's Use and Abuse of Power*, New York: Dodd, Mead, 1986, xiii–xviii; see also Paul E. Ceruzzi, 'An Unforeseen Revolution: Computers and Expectations, 1935–1985,' in Joseph Corn, ed., *Imagining Tomorrow: History, Technology, and the American Future*, Cambridge, MA: MIT Press, 1986, 188–201.
6. Whereas most of these systems are still under development, any published descriptions of them will soon be out of date. See the following web sites: Russian Federation, Information and Analysis Center for Position, Navigation, and Timing, 'GLONASS News,' https://www.glonass-iac.ru/en; European Space Agency, 'The Future Galileo,' http://www.esa.int/Our_Activities/Navigation/The_future_Galileo; India, 'Indian Regional Navigation Satellite System,' http://irnss.isro.gov.in; Japan, Cabinet Office, 'Quasi-Zenith Satellite System,' http://qzss.go.jp/en.
7. Dava Sobel, *Longitude: The True Story of a Lone Genius Who Solved the Greatest Scientific Problem of His Time*, New York: Walker, 1995.
8. Philip and Emily Morrison, eds, *Charles Babbage and His Calculating Engines*, New York: Dover, 1961, xiv.
9. Andrew K. Johnston, Roger D. Connor, Carlene E. Stephens and Paul E. Ceruzzi, *Time and Navigation: The Untold Story of Getting from Here to There*, Washington, DC: Smithsonian Books, 2014, 89.

10. Ibid., 88–91; see also William J. Rankin, 'The Geography of Radionavigation and the Politics of Intangible Artifacts,' *Technology and Culture* 55.4 (July 2014), 622–74.
11. Sebastian Junger, *The Perfect Storm*, New York: Norton, 1997; *The Perfect Storm*, directed by Wolfgang Petersen, USA 2000 (Baltimore Pictures).
12. Walter Häussermann, 'On the Evolution of Rocket Navigation, Guidance, and Control (NGC) Systems,' in Ernst A. Steinhoff, ed., *The Eagle Has Returned*, San Diego: American Astronautical Society, 1976, 258–69.
13. Donald MacKenzie, *Inventing Accuracy: A Historical Sociology of Nuclear Missile Guidance*, Cambridge, MA: MIT Press, 1990.
14. Johnston et al., *Time and Navigation*, 142–5.
15. Bradford Parkinson, Thomas Stansell, Ronald Beard and Konstantine Gromov, 'A History of Satellite Navigation,' *Navigation: Journal of the Institute of Navigation* 42.1 (Spring 1995), 109–64, here 118.
16. Ivan A. Getting, *All in a Lifetime: Science in the Defense of Democracy*, New York: Vantage Press, 1989, 492–4, 578–9.
17. Ibid., 479–89.
18. Johnston et al., *Time and Navigation*, 155.
19. William J. Rankin, 'The Technological Culture of GPS: Is There Such a Thing as American Technoglobalism?,' paper presented at a workshop 'Technology, Strategy, and Power,' Yale University, New Haven, 23 March 2012; and idem, 'The Global Positioning System,' in Mark Monmonier, ed., *The History of Cartography*, vol. 6, part 2: *Cartography in the Twentieth Century*, Chicago: University of Chicago Press, 2015, 551–8.
20. MacKenzie, *Inventing Accuracy*; Wiebe E. Bijker, Thomas P. Hughes and Trevor Pinch, *The Social Construction of Technological Systems*, Cambridge, MA: MIT Press, 1989.
21. Ibid.; see also Ronald Kline and Trevor Pinch, 'The Social Construction of Technology,' in Donald MacKenzie and Judy Wajcman, eds, *The Social Shaping of Technology*, 2nd edn, Buckingham: Open University, 1999, 113–15.
22. NAVSTAR was not an acronym; see Parkinson, 'History of Satellite Navigation.'
23. Tom Logsdon, *The NAVSTAR Global Positioning System*, New York: Van Nostrand, 1992.
24. The two forces work in opposite directions but do not cancel each other out. The net offset is that the clocks on the satellites run faster than clocks on earth by a factor of about 4.5×10^{-10}.
25. Johnston et al., *Time and Navigation*, 156.
26. Robert A. Scholtz, 'The Origins of Spread-Spectrum Communications,' *IEEE Transactions on Communications*, COM-30.5 (May 1982), 822–54.
27. Richard Rhodes, *Hedy's Folly: The Life and Breakthrough Inventions of Hedy Lamarr, the Most Beautiful Woman in the World*, New York: Random House, 2011.
28. Bradford W. Parkinson and Stephen W. Gilbert, 'NAVSTAR: Global Positioning System – Ten Years Later,' *Proceedings of the IEEE* 71.10 (October 1983), 1177–86.
29. Ibid., 1184.
30. Ibid., 1186.

31. 'GPS History, Chronology, and Budgets: Appendix B,' in Scott Pace et al., *The Global Positioning System: Assessing National Policies*, Santa Monica: RAND Corporation, 1995, here 248.
32. 'GPS to Test Nuclear Detonation Sensor,' *Aviation Week and Space Technology* 111.9 (27 August 1979), 51.
33. Johnston et al., *Time and Navigation*, 166–7; Logsdon, *NAVSTAR Global Positioning System*, 166.
34. Cornelius T. Leondes, 'Inertial Navigation for Aircraft,' *Scientific American* 222.3 (March 1970), 80–6.
35. Asaf Degani, *Taming HAL: Designing Interfaces beyond 2001*, Basingstoke: Palgrave Macmillan, 2004, 49–65.
36. David F. Winkler, *Cold War at Sea: High Seas Confrontation between the United States and the Soviet Union*, Annapolis: Naval Institute Press, 2000, 176–210.
37. Parkinson, *History of Satellite Navigation*, 153.
38. Information taken from the official US government website 'Official U.S. Government Information about the Global Positioning System (GPS) and Related Topics'; see http://www.gps.gov/systems/gps/modernization/sa/.
39. The official web site is 'Geocaching'; see www.geocaching.com. This hobby would not have been feasible had Selective Availability remained in use.
40. Langdon Winner, *Autonomous Technology: Technics-out-of-Control as a Theme in Political Thought*, Cambridge, MA: MIT Press, 1977; Bijker et al., *Social Construction of Technological Systems.*
41. Paul N. Edwards, *The Closed World: Computers and the Politics of Discourse in Cold War America*, Cambridge, MA: MIT Press, 1997; MacKenzie, *Inventing Accuracy.*
42. Ibid.; Janet Abbate, *Inventing the Internet*, Cambridge, MA: MIT Press, 1999, 106–7.
43. The second volume in this trilogy on European astroculture is entirely devoted to the 1970s as a transition period; see Alexander C. T. Geppert, ed., *Limiting Outer Space: Astroculture after Apollo*, London: Palgrave Macmillan, 2018 (= *European Astroculture*, vol. 2).
44. Paul E. Ceruzzi, 'The Internet before Commercialization,' in idem and William Aspray, eds, *The Internet and American Business*, Cambridge, MA: MIT Press, 2008, 9–43.
45. Jack Goldsmith and Tim Wu, *Who Controls the Internet? Illusions of a Borderless World*, New York: Oxford University Press, 2006.
46. 'U.S. Space-Based Position, Navigation, and Timing Policy,' Fact Sheet, 15 December 2004; see http://www.gps.gov/policy/docs/. This document further states that 'The Executive Committee will be co-chaired by the Deputy Secretaries of the Department of Defense and the Department of Transportation or by their designated representatives. Its members will include representatives at the equivalent level from the Departments of State, Commerce and Homeland Security, the Joint Chiefs of Staff, the National Aeronautics and Space Administration, and from other Departments and Agencies as required. Components of the Executive Office of the President, including the Office of Management and Budget, the National Security Council staff, the Homeland

Security Council staff, the Office of Science and Technology Policy, and the National Economic Council staff, shall participate as observers to the Executive Committee.'

47. Russian Federation, Information and Analysis Center for Position, Navigation, and Timing, 'GLONASS News', https://www.glonass-iac.ru/en/guide.
48. Parkinson, *History of Satellite Navigation*, 156–7. GLONASS uses a scheme called 'Frequency Division Multiple Access,' to transmit time and position from the constellation of satellites. Whereas all GPS satellites transmit on the same two frequencies, GLONASS uses different frequencies for each satellite.
49. See, for example, the Garmin Corporation, 'Epix: Full-color Mapping on your Wrist,' https://buy.garmin.com/en-US/US/prod146065.html.
50. The relevant web sites are: European Space Agency, 'The Future Galileo,' http://www.esa.int/Our_Activities/Navigation/The_future_Galileo; India, 'Indian Regional Navigation Satellite System,' http://irnss.isro.gov.in; Japan, Cabinet Office, 'Quasi-Zenith Satellite System,' http://qzss.go.jp/en.
51. For a general history of Galileo, see 'European Commission, Growth, Sectors, Space, Galileo, the History of Galileo,' http://ec.europa.eu/growth/sectors/space/galileo/history/index_en.htm.
52. Ibid.
53. BBC News, 'Galileo "Compromise" is Emerging,' 23 November 2007, http://news.bbc.co.uk/2/hi/science/nature/7109971.stm.
54. Bradford Parkinson, private communication to the author.
55. Anne Applebaum, 'Here Comes the New Europe,' *Washington Post* (29 January 2003), A-21.
56. Iain Ross Ballantyne Bolton, 'Neo-realism and the Galileo and GPS Negotiations,' in Natalie Bormann and Michael Sheehan, eds, *Securing Outer Space*, London: Routledge, 2009, 186–204; see also Embassy of the United States, Dublin, Ireland, 'U.S., EU to Sign Landmark GPS-Galileo Agreement,' http://dublin.usembassy.gov/ireland/gps_galileo.html.
57. On dual use, see in particular the respective contributions by Alexander Geppert and Tilmann Siebeneichner, Michael Neufeld, Daniel Brandau and Regina Peldszus, Chapters 1, 2, 7 and 11 in this volume.
58. Paul E. Ceruzzi, *Computing: A Concise History*, Cambridge, MA: MIT Press, 2012, 177–206; see also Hiawatha Bray, *You Are Here: From the Compass to GPS*, New York: Basic Books, 2014.
59. Arnold Thackray, David C. Brock and Rachel Jones, *Moore's Law: The Life of Gordon Moore, Silicon Valley's Quiet Revolutionary*, New York: Basic Books, 2015.
60. Johnston et al., *Time and Navigation*, 163.
61. Caren Kaplan, Erik Loyer and Ezra Claytan Daniels, 'Precision Targets: GPS and the Militarization of Everyday Life,' *Canadian Journal of Communication* 38.3 (September 2013), 397–420.
62. Ibid., 406–8.
63. Richard Whittle, *Predator: The Secret Origins of the Drone Revolution*, New York: Henry Holt, 2014.
64. Johnston et al., *Time and Navigation*, 163.
65. Parkinson, private communication.

66. See, for example, 'GPS Modernization Stalls,' *Inside GNSS* (March/April 2014), http://www.insidegnss.com/node/3964; 'GAO blames Air Force, Raytheon for 1.1 B Cost Overrun,' *Real Clear Defense*, http://www.realclear-defense.com/2015/09/10/gao_blames_air_force_raytheon_for_11b_cost_over-run_273962.html.
67. Joshua Cuneo, '"Hello Computer": The Interplay of Star Trek and Modern Computing,' in Ferro and Swedin, *Science Fiction and Computing*, 131–48.

Epilogue

CHAPTER 14

What Is, and to What End Do We Study, European Astroculture?

Alexander C. T. Geppert

> While [the study of world history] accustoms man to think of himself as one with the entire past, and to press on with his conclusions into the distant future, at the same time it conceals the limitations of birth and death, by which man's life is so narrowly and oppressively circumscribed, and by an optical illusion extends his short existence into endlessness and imperceptibly makes the individual one with his species.
>
> Friedrich Schiller, 1798[1]

Militarizing Outer Space is the third and final volume in the *European Astroculture* trilogy. As it brings a long-term research program to a preliminary end, this finale seems a good moment to pause, step back and take an intellectual inventory of what has been achieved over the past decade or so – and what is still missing from the agenda. As historian Bernd Weisbrod (echoing Max Weber, of course) has reminded his students time and again, while the specific findings of any completed study are obviously key, it is part of the academic process that they always remain preliminary and incomplete. Hence, what matters even more than an original and substantiated argument is that the questions asked at the end be more insightful, more trenchant and simply "better" than those posed at the outset. Envisioned as a never-ending conversation among the living and with the dead, history never ends. It is those constantly changing questions and new takes which keep it running, as if it were its own perpetual motion machine. In that vein, what has been

Alexander C. T. Geppert (✉)
New York University, New York, USA
NYU Shanghai, Shanghai, China
e-mail: alexander.geppert@nyu.edu

Alexander C. T. Geppert et al. (eds), *Militarizing Outer Space*
European Astroculture, vol. 3
https://doi.org/10.1057/978-1-349-95851-1_14

successfully established by the endeavor, what has proven problematic – and what new questions have arisen?

A short, somewhat cursory definition, loosely thrown into the introduction to that first volume, has found a surprising degree of resonance. 'Astroculture,' it reads, 'comprises a heterogeneous array of images and artifacts, media and practices that all aim to ascribe meaning to outer space while stirring both the individual and the collective imagination.'[2] Spread over three volumes – *Imagining Outer Space* (published in hardcover in 2012, with a slightly revised paperback edition in 2018); *Limiting Outer Space* (hardcover 2018, paperback 2020), its sequel volume on the crisis-ridden 1970s after the American moon landings; and *Militarizing Outer Space*, the book at hand – altogether 44 contributions have employed this concept, some more rigidly than others. They explore the ways in which individuals, groups and societies over the course of the twentieth century have tried to make sense of the infinite stretches that surround earth and imbue the skies with a sense of spiritual meaning seemingly lost after Copernicus. More often than not, transforming life on earth figured even higher up on the agenda than humanizing the heavens, and the perspective from without – whether provided by envoys of humankind, satellites or an unprecedented plethora of aliens – helped to turn the world into a planet. Anything but esoteric, far-fetched or obscure, it is for this reason that the historical study of astroculture and space thought goes directly to the core of modern world-making or, as some would prefer to call it, modernity. Twentieth-century imagination, spurred by technological and political revolutions alike, no longer limited the *conditio humana* to terrestrial territory, as Hannah Arendt famously observed after Sputnik. Ironically, the quest to conquer infinity would eventually alienate 'man' from the world and diminish his stature to the extreme, as a consequence of a futile search for a stable and absolute Archimedean point outside earth, and despite the discovery of the globe as a whole along the way.[3]

The introduction to the first volume declared five distinct objectives. It linked the study of outer space to what is sometimes called the 'popular' imagination by proposing the above-quoted definition of astroculture; introduced Europe as a geographical focus to overcome the still dominant American/Soviet bipolarity of much of the more conventional space history; brought the relationship between 'science' and 'fiction' up for discussion; and related the history of twentieth-century astroculture to two other powerful beyonds – one temporal, the other transcendental – that is, futurity and spirituality. Sharing those conceptual aims, *Limiting Outer Space*, the second volume in the trilogy, focused on the reconfiguration of space utopias after the American moon landings, when an unbounded Space Age space craze gave way to post-Apollo fatigue and cosmic ennui. Turning the view to fights of fancy, *Militarizing Outer Space* examines the "dark" side of those long-standing space utopias and accompanying socio-technical imaginaries. It asks how outer space, often hailed as a site of heavenly renewal and otherworldly salvation, transformed from a promised sanctuary to a present threat, where the battles of the future were to be waged and the fate of humankind decided.

Thus, even though these works introduced additional vocabulary of analytical middle range including 'space fatigue,' the 'Post-Apollo paradox,' 'orbital space' and 'militant astroculture,' to name but a few, both the second and this third volume employ the conceptual agenda and adamantly follow the historiographical program laid out in the introduction to the first one. Taken as a whole, the trilogy pursues one single goal: to underscore the historical significance of outer space beyond a narrow Cold War context and to de-exoticize its overdue historicization using every trick in the proverbial (and present) book. Whether or not this objective has been fully met in every single chapter, it seems fair to state that the trilogy, together with about 50 other scholarly publications in English, French and German, has succeeded in making 'astroculture' a new field of modern (European) historiography, just as promised at the end of that first introduction and evidenced by around two dozen book reviews so far.[4]

What then have we learned, not only from the fourteen contributions in the volume at hand but also from the larger enterprise as a whole? Three remarks will suffice. First, one of the most salient and oft-discussed features of astroculture is the constant oscillation between spacebound imaginaries and spaceflight realities, between fiction and technology or – particularly in the case of this volume – between heavenly utopias and apocalyptic battlefields. To come to terms with this to and fro, back and forth between different ontological claims, operating with 'science fiction' as an analytical category has proven more detrimental than productive. Evoking 'science' and 'fiction' in order to identify different components of astroculture almost inevitably leads to either imprecise or muddled references to the narrative genre by the same name, that is science fiction. Further, much space-related science fiction would more adequately be classified as 'technology fiction.' Attentive readers will have noticed that there is, therefore, rather limited use of 'science fiction' as a category throughout the trilogy.

Second, contributions to the present volume remind us in particular that all astroculture is deeply political, more so than originally somewhat naïvely assumed. Occasionally astroculture can be downright controversial and vehemently contested, but it is hardly ever the enthrallment of outer space and the promise of its future conquest as such which heated minds. Nonetheless, astroculture's public repercussions and political ramifications go far beyond glitzy, if superficial 'let's colonize the moon and inhabit the solar system' propaganda. Examples such as East and West German systems competition transposed onto the postwar heavens, the SDI skirmishes fought throughout the 1980s, or the still popular myth of the European space effort being exclusively restricted to a civilian, non-military cause demonstrate the explanatory power of an approach that highlights something ostensibly as soft as the imaginary as a driving force behind twentieth-century spaceflight. The complementary notion of 'space thought' can help to see, unpack and eventually situate the philosophical foundations, political promises and intellectual investments in space in a longer tradition within the modern history of ideas.

And third, the mind-boggling number and ubiquity of astrocultural images and futuristic imaginaries, the widespread enthusiasm of which they are both cause and expression as well as the deep-rooted emotions which they evoke to the present day corroborate the overall diagnosis that the production and appropriation of outer space is indeed a key technopolitical project of twentieth-century modernity. As such, it is by no means limited to the spacefaring champions of the world, the former two superpowers, but is indeed of worldwide appeal, as historians have begun to realize in recent years. Astroculture is not limited to human representations in or of outer space. Neither is it synonymous with what is sometimes dubbed 'space culture.' Rather, at its core are complex negotiation processes to fill the infinite void that surrounds earth with meaning and make sense of its extraterrestrial beyond. The product of such processes is outer space as we think we know it, even if only a tiny few have ever been out there to see it with their own eyes.

What, on the other hand, has been achieved less well? Almost ironically, the seemingly most straightforward of the five objectives – 'introducing Europe' – has proven the greatest challenge in technical and practical terms. Critics have rightly pointed out that 'Europe' here is often synonymous with 'Western Europe,' at times even conflated with 'the West.' As a consequence, all three volumes and the European Astroculture project at large have been in danger of replicating Cold War Europe's bifurcated geography without properly integrating its Eastern parts into the account, clearly a blind spot. Why is that so? One problem, familiar to all historians of Europe, is the insurmountable vagueness of the term itself, another its scope. Is Europe a continent, a concept or both? Even though early on the choice was made to concentrate on Western Europe, there is, admittedly, far too little, if anything at all, on Italy, Spain, Greece, Portugal, Austria, Benelux, Scandinavia and many other European states and regions. If the existence of a distinct Space Age Europe should hopefully have been established as an undebatable factum by now, the envisaged Europeanization of its historical genesis remains uneven and far from complete.

What's more, identifying, reconstructing and historicizing hitherto little-known building blocks of astroculture has been a rewarding and equally fascinating task. Fortunately this young, yet flourishing field of historical research now features long-forgotten space *personae* and amateur lobby groups, sounds of space and space toys, obscure TV shows and B movies, reports of UFO sightings and alien encounter narratives all previously considered fringe, if not entirely irrelevant to serious historiographical attention, let alone collectively understood as forming part of a larger historical undertaking. Building up such a particular archaeology of astroculture and space thought is necessary as an indispensable first step towards proper historicization, yet it is neither sufficient nor conclusive. Maybe it is the inherent fascination of space lurking in the background that obstructs a sober analytical view, yet every single manifestation of astroculture is obviously the product of an act of local creativity embedded within an ever-fluctuating web of knowledge. As such, it is subject not only to processes of translation and adaptation, but also to deliberate disengagement, even demarcation from one setting and context to another.

Two concrete conclusions follow. First, how do both individual 'acts' and larger bodies of astroculture relate and compare to each other, both in their making and what they "do"? Is it, for instance, possible to identify the characteristics of a particularly European space aesthetics fully cognizant of its Soviet and American counterparts while trying to emancipate itself and establish a distinct visual autonomy? And, second, when engaging in that type of archaeological genealogy and recovery, it is all too easily forgotten that the study of astroculture is not an end in itself, but first and foremost of interest to historians of modernity because it drives the fabrication and accouterment of space. It is the latter that motivates the former, not vice versa. Astroculture is the *explanans*, outer space the *explanandum*.

Last but not least, what are those pointed, more perspicacious and hopefully productive questions that this moment allows one to pose, at the end of an epilogue that concludes the trilogy's final volume? Again, three brief remarks must do, not least as these preliminary conclusions are to pave the way for future research on the planetization of earth by means of space. The first question regards what some observers have hastily been calling the Second Space Age, that is the one said to be currently unfolding before our very eyes. History may rhyme at times but it does not repeat, Tilmann Siebeneichner and I noted in the introduction to this volume, struck by parallels between Ronald Reagan's 'Star Wars' fantasies and the 45th US president's derivative and equally ridiculed effort to create a new US Space Force. A similarly propagandist revival can be observed elsewhere. For years, the three super rich 'astropreneurs' of the present – Jeff Bezos (1964–; Amazon/Blue Origin), Richard Branson (1950–; Virgin Galactic) and Elon Musk (1971–; Tesla/Space X) – have unremittingly attempted to sell their private expansion dreams as technoscientific rescue missions for a world threatened by the climate catastrophe. Humankind must venture into the abyss or will become extinct, their mantra goes. Space historians cannot help but recognize long-shelved astrofuturist tropes behind such rehashed escape-and-conquer schemes. While any sensationalist talk of a 'new Space Race' or even a Second Space Age can easily be refuted as historically unsubstantiated, historians must nonetheless ask themselves why the promise of space is undergoing such a resurgence. What does it say about the present that stale expansion phantasms have again become ubiquitous? And how are historians to historicize something prior to its becoming historical?

Second, if the envisaged Europeanization of space history appears far from complete, then this sense of tentativeness applies all the more to astroculture and space thought around the world. There is no doubt that astroculture has become a global, if not globally unified, phenomenon in the early twenty-first century. Yet what precisely that entails for heavily invested spacefaring nations such as present-day China, Japan and India and their respective cultural histories is entirely unclear. If there is any justification for any new Space Age speak, then it is in the Global South. The analytical, empirical and intellectual gains to be made from globalizing the history of outer space, spaceflight and space exploration are gigantic. Distinguishing between different overlapping, yet related, national trajectories and identifying a multiplicity of Space Ages, from *Frau im Mond* by way of Lutz Kayser's Orbital Transport- und Raketen Aktiengesellschaft (OTRAG) in Zaire to the Chinese Chang'e-4, is an enormous task that will complicate existing periodizations and eventually help complete the historical account.

Finally, while much Second Space Age propaganda amounts to no more than cheap talk, intuitively its advocates do have a point. In his 1789 inaugural lecture on the nature and value of universal history, Friedrich Schiller pleaded for a maximum comprehensive approach which, in its planetary inclusion, is not unlike Hannah Arendt's. 'From the total sum of [...] events, the universal historian picks out those which have had on the contemporary state of the world and on the condition of the generation now alive an influence which is essential, undeniable, and easy to discern,' Schiller suggested, and then works backward, from the 'most recent state of the world up to the origin of things.'[5] Both modes, to face the biggest picture possible and to study the history of the present top-down, can coalesce in a planetary perspective on the making of our planetized present. The role of space in today's 'most recent state of the world' commonly referred to as globalization is twofold. Communication, navigation and reconnaissance depend on an omnipresent, albeit largely invisible infrastructure positioned in low-earth orbit. At the same time, such technological prowess is complemented by what – for lack of a better term – can be termed a planetary consciousness, prefigured by the normalized perspective of an extraterrestrial observer. Without this imagined view from above and from without, astroculture loses its cultural significance, devolving toward internal conversations and fantasies, mere amusement and entertainment. It is in stepping back that the corpus of astroculture takes on a weight appropriate to its cultural presence. The proliferation of 'images and artifacts, media and practices' that all 'aim to ascribe meaning to outer space while stirring both the individual and the collective imagination' allows uncommon insight into human society at a particular historical moment. There is no doubt as to the pivotal significance of astroculture for how we became who we are.

Notes

1. Friedrich Schiller, 'Was heißt und zu welchem Ende studiert man Universalgeschichte? Eine akademische Antrittsrede' [1789], in Otto Dann et al., eds, *Friedrich Schiller: Werke und Briefe in zwölf Bänden*, vol. 6: *Historische Schriften und Erzählungen I*, Frankfurt am Main: Deutscher Klassiker Verlag, 2000, 411–31, here 429: 'Indem [die Beschäftigung mit der Weltgeschichte] den Menschen gewöhnt, sich mit der ganzen Vergangenheit zusammen zu fassen, und mit seinen Schlüssenin die ferne Zukunft voraus zu eilen: so verbirgt sie die Grenzen von Geburt und Tod, die das Leben des Menschen so eng und so drückend umschließen, so breitet sie optisch täuschend sein kurzes Dasein in einen unendlichen Raum aus, und führt das Individuum unvermerkt in die Gattung hinuiber.' For the English translation, see idem, 'The Nature and Value of Universal History: An Inaugural Lecture [1789],' *History and Theory* 11.3 (1972), 321–34, here 333.
2. Alexander C. T. Geppert, 'European Astrofuturism, Cosmic Provincialism: Historicizing the Space Age,' in idem, ed., *Imagining Outer Space: European Astroculture in the Twentieth Century*, Basingstoke: Palgrave Macmillan, 2012, 3–23, here 8 (= *European Astroculture*, vol. 1).
3. Hannah Arendt, 'Man's Conquest of Space,' *American Scholar* 32.4 (Fall 1963), 527–40, here 538, 540; see also idem, *The Human Condition*, Chicago: University of Chicago Press, 1958, 1–2, 261–4.
4. See this volume's bibliography for a complete list of project publications, in addition to my *The Future in the Stars: Europe, Astroculture and the Age of Space* and other forthcoming monographs. Selected reviews of *Imagining Outer Space* and *Limiting Outer Space* include, in chronological order, Helmut Mayer, 'Raumkultur,' *Frankfurter Allgemeine Zeitung* 155 (6 July 2012), 28; Janet Vertesi, *Quest: The History of Spaceflight Quarterly* 19.4 (2012), 58–9; Anke Ortlepp, *H-Soz-u-Kult* (3 August 2012); Jon Agar, *British Journal for the History of Science* 46.2 (June 2013), 352–4; Roger D. Launius, *Technology and Culture* 54.3 (July 2013), 689–90; Pamela Gossin, *Isis* 104.3 (September 2013), 641–3; De Witt Douglas Kilgore, 'Exploring Astroculture,' *Science Fiction Studies* 41.2 (July 2014), 447–50; Jason Beery, *European Review of History/Revue européenne d'histoire* 21.6 (April 2014), 919–21; A. Ebert, *popcultureshelf.com* (2 December 2018); Barry Kent, *The Observatory* 139.1270 (June 2019), 121–3; David Baneke, *Isis* 110.3 (September 2019), 656–7; and De Witt Douglas Kilgore, 'Scaling Back Astroculture,' *Science Fiction Studies* 46.3 (November 2019), 620–3. See also Catherine Radtka and Alexander C. T. Geppert, 'L'Astroculture européenne, terrain de recherche,' *Humanités Spatiales* (May 2016), available at https://humanites-spatiales.fr/lastroculture-europeenne-terrain-de-recherche (last accessed 15 July 2020).
5. Schiller, 'Was heißt und zu welchem Ende studiert man Universalgeschichte?,' 425–6; idem, 'Nature and Value of Universal History,' 331.

Bibliography

Publications of the Emmy Noether Research Group 'The Future in the Stars: European Astroculture and Extraterrestrial Life in the Twentieth Century'

Brandau, Daniel, 'Cultivating the Cosmos: Spaceflight Thought in Imperial Germany,' *History and Technology* 28.3 (September 2012), 225–54.

———, 'Die Plausibilität des Fortschritts: Deutsche Raumfahrtvorstellungen im Jahre 1928,' in Uwe Fraunholz and Anke Woschech, eds, *Technology Fiction: Technische Visionen und Utopien in der Hochmoderne*, Bielefeld: transcript, 2012, 65–91.

———, 'Demarcations in the Void: Early Satellites and the Making of Outer Space,' *Historical Social Research* 40.1 (March 2015), 239–64.

———, *Raketenträume: Raumfahrt- und Technikenthusiasmus in Deutschland 1923–1963*, Paderborn: Schöningh, 2019.

Bruggmann, Jana, 'Der Weltraum im Zeitalter seiner technischen Reproduzierbarkeit: Das wissenschaftliche Theater der Berliner Urania, 1889–1905,' *Technikgeschichte* 84.4 (2017), 305–28.

———, 'Die Erd' ist aufgegangen,' *Die Zeit* 23 (29 May 2019), 19.

———, '"Earthrise" *avant la lettre*: Camille Flammarion, Bruno H. Bürgel und die Genese einer extraterrestrischen Perspektive,' in Philipp Auchter, Boris Buzek, Mateusz Cwik and Philipp Theisohn, eds, *Des Sirius goldne Küsten: Astronomie und Weltraumfiktion*, Paderborn: Wilhelm Fink, 2019, 221–50.

Geppert, Alexander C. T., 'Flights of Fancy: Outer Space and the European Imagination, 1923–1969,' in Steven J. Dick and Roger D. Launius, eds, *Societal Impact of Spaceflight*, Washington, DC: NASA, 2007, 585–600.

———, 'Space *Personae*: Cosmopolitan Networks of Peripheral Knowledge, 1927–1957,' *Journal of Modern European History* 6.2 (2008), 262–86 [German: 'Männer des Universums: Europäische Weltraumvisionäre und die Genese einer

Alexander C. T. Geppert et al. (eds), *Militarizing Outer Space*
European Astroculture, vol. 3
https://doi.org/10.1057/978-1-349-95851-1

"kosmopolitischen Internationale", 1927–1957,' in Sascha Mamczak and Wolfgang Jeschke, eds, *Das Science Fiction Jahr 2010*, Munich: Heyne, 2010, 789–824].

———, 'Anfang – oder Ende des planetarischen Zeitalters? Der Sputnikschock als Realitätseffekt, 1945–1957,' in Igor J. Polianski and Matthias Schwartz, eds, *Die Spur des Sputnik: Kulturhistorische Expeditionen ins kosmische Zeitalter*, Frankfurt am Main: Campus, 2009, 74–94.

———, 'European Astrofuturism, Cosmic Provincialism: Historicizing the Space Age,' in idem, ed., *Imagining Outer Space: European Astroculture in the Twentieth Century*, Basingstoke: Palgrave Macmillan, 2012, 3–24.

———, 'Rethinking the Space Age: Astroculture and Technoscience,' *History and Technology* 28.3 (September 2012), 219–23.

———, 'Extraterrestrial Encounters: UFOs, Science and the Quest for Transcendence, 1947–1972,' *History and Technology* 28.3 (September 2012), 335–62.

———, 'Storming the Heavens: Soviet Astroculture, 1957–1969,' *H-Soz-u-Kult* (17 July 2013).

———, 'Infrastrukturen der Weltraumimagination: Außenstationen im 20. Jahrhundert,' in Bundeskunsthalle, ed., *Outer Space: Der Weltraum zwischen Kunst und Wissenschaft*, Bonn: Nicolai, 2014, 124–7.

———, 'Die Zeit des Weltraumzeitalters, 1942–1972,' in idem and Till Kössler, eds, *Obsession der Gegenwart: Zeit im 20. Jahrhundert*, Göttingen: Vandenhoeck & Ruprecht, 2015, 218–50 (= *Geschichte und Gesellschaft*. Sonderheft 25).

———, 'L'Astroculture européenne, terrain de recherche,' *Humanités Spatiales* (May 2016) (with Catherine Radtka).

———, 'Imaginary Infrastructures and the Making of Outer Space,' in Lukas Feireiss and Michael Najjar, eds, *Planetary Echoes: Exploring the Implications of Human Settlement in Outer Space*, Leipzig: Spector, 2018, 123–34.

———, 'The Post-Apollo Paradox: Envisioning Limits During the Planetized 1970s,' in idem, ed., *Limiting Outer Space: Astroculture after Apollo*, London: Palgrave Macmillan, 2018, 3–26.

———, 'Phantasie, Projekt, Produkt: Der Weltraum des 20. Jahrhunderts,' *Aus Politik und Zeitgeschichte* 69.29–30 (July 2019), 19–25.

———, *The Future in the Stars: Europe, Astroculture and the Age of Space*, forthcoming.

Geppert, Alexander C. T., ed. *Imagining Outer Space: European Astroculture in the Twentieth Century*, Basingstoke: Palgrave Macmillan, 2012 (2nd edn, London: Palgrave Macmillan, 2018) (= *European Astroculture*, vol. 1).

———, ed., *Astroculture and Technoscience*, London: Routledge, 2012 (= *History and Technology* 28.3).

———, ed., *Limiting Outer Space: Astroculture after Apollo*, London: Palgrave Macmillan, 2018 (= *European Astroculture*, vol. 2).

Geppert, Alexander C. T. and Tilmann Siebeneichner, eds, *Berliner Welträume im frühen 20. Jahrhundert*, Baden-Baden: Nomos, 2017 (= *Technikgeschichte* 84.4).

———, '*Lieux de l'Avenir*: Zur Lokalgeschichte des Weltraumdenkens,' *Technikgeschichte* 84.4 (2017), 285–304.

Macauley, William R., 'Inscribing Scientific Knowledge: Interstellar Communication, NASA's Pioneer Plaque and Contact with Cultures of the Imagination, 1971–72,' in Alexander C. T. Geppert, ed., *Imagining Outer Space: European Astroculture in the Twentieth Century*, Basingstoke: Palgrave Macmillan, 2012, 285–303.

———, 'Crafting the Future: Envisioning Space Exploration in Post-war Britain,' *History and Technology* 28.3 (September 2012), 281–309.
Macauley, William R., ed., *Sounds of Space*, Leiden: Leiden University Press, 2014 (= *Journal of Sonic Studies* 8).
Siebeneichner, Tilmann, 'Exploring the Heavens: Space Technology and Religious Implications,' *Technology and Culture* 55.4 (2014), 988–94.
———, 'Europas Griff nach den Sternen: Das Weltraumlabor Spacelab, 1973–1998,' *Themenportal Europäische Geschichte* (2015).
———, '*Interstellar:* Wiedergeburt im Weltraum. Zukunftsvorstellungen seit dem späten 20. Jahrhundert,' *Zeitgeschichte-online* (1 March 2015).
———, 'Die "Narren von Tegel": Technische Innovation und ihre Inszenierung auf dem Berliner Raketenflugplatz, 1930–1934,' *Technikgeschichte* 84.4 (2017), 353–80.
———, 'Spacelab: Peace, Progress and European Politics in Outer Space, 1973–85,' in Alexander C. T. Geppert, ed., *Limiting Outer Space: Astroculture after Apollo*, London: Palgrave Macmillan, 2018, 259–82.
———, 'Auf einer Briefmarke in den Weltraum: Spacelab und das Versprechen einer Zukunft in den Sternen,' *Comparativ* 28.3 (March 2018), 90–103.

Conference Reports

Aselmeyer, Norman, 'Stellare Kriege,' *Technikgeschichte* 81.4 (2014), 371–8.
Blaß, Björn, 'UFO History,' *H-Soz-u-Kult* (4 March 2012).
Boyce-Jacino, Katherine, 'Embattled Heavens: The Militarization of Space in Science, Fiction, and Politics,' *Foundation: The International Review of Science Fiction* 43.118 (Fall 2014), 96–100.
Geppert, Alexander C. T., 'Space in Europe, Europe in Space: Symposium on 20th-Century Astroculture,' *NASA History News & Notes* 25.2 (May 2008), 1–12.
———, 'Imagining Outer Space: European Astroculture in the Twentieth Century,' *ZiF: Mitteilungen* 3 (2008), 14–18.
———, 'Imagining Outer Space, 1900–2000,' *H-Soz-u-Kult* (16 April 2008).
Haake, Ruth, 'Berliner Welträume im 20. Jahrhundert,' *H-Soz-u-Kult* (14 October 2015).
Mehl, Friederike, 'Berlin Symposium on Outer Space and the End of Utopia in the 1970s,' *NASA History News & Notes* 29.2–3 (2012), 1–15.
———, 'Envisioning Limits: Outer Space and the End of Utopia,' *H-Soz-u-Kult* (9 July 2012).
Rauchhaupt, Ulf von, 'Als der größte Großraum zum Schlachtfeld wurde. Die Raumfahrt zwischen Politik, Technik und Science-Fiction: Eine Tagung widmet sich der dunklen Seite der Astrokultur in der Zeit des Kalten Krieges,' *Frankfurter Allgemeine Zeitung* (16 April 2014), N3.
Reichard, Tom, 'Battlefield Cosmos: The Militarization of Space, 1942–1990,' *NASA History News & Notes* 31.3 (2014), 20–1.
———, 'Embattled Heavens: The Militarization of Space in Science, Fiction, and Politics,' *H-Soz-u-Kult* (8 August 2014).
Rippert, Katja, 'The Sonic Dimension of Outer Space, 1940–1980,' *NASA History News & Notes* 30.2 (2013), 17–20.

———, 'Sounds of Space,' *H-Soz-u-Kult* (26 April 2013).
Seifert-Hartz, Constanze, 'Futuring the Stars: Europe in the Age of Space,' *H-Soz-u-Kult* (21 July 2016).
———, 'The Final Countdown: Europe in the Age of Space,' *NASA History News & Notes* 33.4 (2016), 25–7.
Wünsche, Nadja, 'Für Männer interessant, für Frauen nur verrückt: Jeder Erdteil hat seine Tassen, auch die lockeren,' *Frankfurter Allgemeine Zeitung* (5 October 2011), N3.

Bibliographies

Arnold, David C., 'Space and War,' in Dennis Showalter et al., eds, *Military History: Oxford Bibliographies Online*, Oxford: Oxford University Press, 2017, http://www.oxfordbibliographies.com/view/document/obo-9780199791279/obo-9780199791279-0168.xml (last modified 28 February 2017).
Brauch, Hans Günter and Rainer Fischbach, eds, *Military Use of Outer Space: A Research Bibliography*, Stuttgart: AG Friedensforschung und Europäische Sicherheitspolitik (AFES), Institut für Politikwissenschaft, Universität Stuttgart, 1986.
———, *Militärische Nutzung des Weltraums: Eine Bibliographie*, Berlin: Arno Spitz, 1988.
Harnly, Caroline D. and David A. Tyckoson, eds, *Space Weapons*, Phoenix: Oryx Press, 1985.
Hübner, Thomas, ed., *Raumfahrt-Bibliographie: Ein Verzeichnis nichttechnischer deutschsprachiger Literatur von 1923 bis 1997*, Hörstel: Raumfahrt-Info-Dienst, 1998.
Lawrence, Robert M., *Strategic Defense Initiative: Bibliography and Research Guide*, Boulder: Westview Press, 1987.
Pisano, Dominick A. and Cathleen S. Lewis, eds, *Air and Space History: An Annotated Bibliography*, New York: Garland, 1988.
Schiller, Volker, *Strategic Defense Initiative (SDI): Auswahlbibliographie (1979–1987)*, Bonn: Deutscher Bundestag, 1988.
Stewart, Alva W., *The Strategic Defense Initiative: A Brief Bibliography*, Monticello: Vance Bibliographies, 1987.
Weber, Gerhard, *Militarisierung des Weltraums: Krieg der Sterne, SDI. Auswahlbibliographie*, 2 vols, Jena: Universitätsbibliothek, 1986/1988.
Wong, Wilson W. S. and James Gordon Fergusson, *Military Space Power: A Guide to the Issues*, Santa Barbara: Praeger, 2010.

Dictionaries, Encyclopedias and Handbooks

The Arms Control Association, ed., *Star Wars Quotes*, Washington, DC: The Arms Control Association, 1986.
Boffey, Philip M., William J. Broad, Leslie H. Gelb, Charles Mohr and Holcomb B. Noble, *Claiming the Heavens: The New York Times Complete Guide to the Star Wars Debate*, New York: Times Books, 1988.

Chapman, Bert, *Space Warfare and Defense: A Historical Encyclopedia and Research Guide*, Santa Barbara: ABC-CLIO, 2008.
Dasch, E. Julius, ed., *A Dictionary of Space Exploration*, Oxford: Oxford University Press, 2005.
David, James E., *Conducting Post-World War II National Security Research in Executive Branch Records: A Comprehensive Guide*, Westport: Greenwood Press, 2001.
Dickens, Peter and James S. Ormrod, eds, *The Palgrave Handbook of Society, Culture and Outer Space*, Basingstoke: Palgrave Macmillan, 2016.
Dickson, Paul, *A Dictionary of the Space Age*, Baltimore: Johns Hopkins University Press, 2009.
Johnson, Stephen B., ed., *Space Exploration and Humanity: A Historical Encyclopedia*, 2 vols, Santa Barbara: ABC-CLIO, 2010.
King-Hele, Desmond, Doreen M. C. Walker, J. Alan Pilkington, Alan N. Winterbottom, Harry Hiller and Geoffrey E. Gerry, eds, *The RAE Table of Earth Satellites 1957–1986: Compiled at the Royal Aircraft Establishment, Farnborough, Hants, England*, 3rd edn, New York: Stockton Press, 1987.
Loosbrock, John F., ed., *Space Weapons: A Handbook of Military Astronautics*, New York: Praeger, 1959.
Robertson, Ann E., *Militarization of Space*, New York: Facts on File, 2011.
Schrogl, Kai-Uwe, Peter L. Hays, Jana Robinson, Denis Moura and Christina Giannopapa, eds, *Handbook of Space Security*, 2 vols, New York: Springer, 2015.
Tobias, Russell R. and David G. Fisher, eds, *USA in Space*, 3rd edn, Pasadena: Salem Press, 2006.
Verger, Fernand, Isabelle Sourbès-Verger and Raymond Ghirardi, *The Cambridge Encyclopedia of Space: Missions, Applications and Exploration*, Cambridge: Cambridge University Press, 2003.
Waldman, Harry, *The Dictionary of SDI*, Wilmington: Scholarly Resources, 1988.
Walsh, Patrick J., *Spaceflight: A Historical Encyclopedia*, 3 vols, Santa Barbara: Greenwood Press, 2010.

Filmography[1]

La Lune à un mètre, directed by Georges Méliès, FR 1898 (Star Film Company).
Voyage dans la lune, directed by Georges Méliès, FR 1902 (Star Film Company).
Himmelskibet (*A Trip to Mars*), directed by Holger-Madsen, DK 1918 (Nordisk Film).
The First Men in the Moon, directed by Bruce Gordon and J. L. V. Leigh, UK 1919 (Gaumont British).
Aelita, directed by Yakov Protazanov, USSR 1924 (Mezhrabpom-Rus).
Mezhplanetnaya revolyutsiya (*Interplanetary Revolution*), directed by Zenon Komissarenko, Yury Merkulov and Nikolay Khodataev, USSR 1924 (Mezhrabpom-Rus).
Frau im Mond, directed by Fritz Lang, DE 1929 (Ufa).

[1]Films are in chronological order.

Things to Come, directed by William Cameron Menzies, UK 1936 (London Film Productions).
Destination Moon, directed by Irving Pichel, USA 1950 (George Pal Productions).
The Day the Earth Stood Still, directed by Robert Wise, USA 1951 (Twentieth Century Fox).
When Worlds Collide, directed by Rudolph Maté, USA 1951 (Paramount Pictures).
Invaders from Mars, directed by William Cameron Menzies, USA 1953 (Edward L. Alperson Productions).
War of the Worlds, directed by Byron Haskin, USA 1953 (Paramount Pictures).
Devil Girl from Mars, directed by David MacDonald, UK 1954 (Danziger Productions Ltd.).
Earth vs. the Flying Saucers, directed by Fred F. Sears, USA 1956 (Columbia Pictures).
Fire Maidens from Outer Space, directed by Cy Roth, UK 1956 (Criterion Films).
Forbidden Planet, directed by Fred M. Wilcox, USA 1956 (Metro-Goldwyn-Mayer).
Doroga k zvezdam (*Road to the Stars*), directed by Pavel Klushantsev, USSR 1957 (Leningrad Popular Science Film Studio).
War of the Satellites, directed by Roger Corman, USA 1958 (Allied Artists).
Nebo zovyot (*Battle beyond the Sun*), directed by Mikhail Karyukov and Aleksandr Kozyr, USSR 1959 (A. P. Dovzenko Filmstudio/Mosfilm).
Battle in Outer Space, directed by Ishirō Honda, JP 1959 (Toho).
Plan 9 from Outer Space, directed by Ed Wood, USA 1959 (Reynolds Pictures).
Der schweigende Stern (*First Spaceship on Venus*), directed by Kurt Maetzig, DDR/PL 1960 (VEB DEFA-Studio für Spielfilme/DEFA Gruppe Roter Kreis/Film Polski/Iluzjon).
Space Men, directed by Antonio Margheriti, IT 1960 (Titanus/Ultra Film).
Dr. No (*James Bond*), directed by Terence Young, UK 1962 (United Artists).
La Jetée, directed by Chris Marker, FR 1962 (Argos Films/Radio-Télévision Français).
Dr. Strangelove or: How I Learned to Stop Worrying and Love the Bomb, directed by Stanley Kubrick, UK/USA 1964 (Columbia Pictures).
Fail-Safe, directed by Sidney Lumet, USA 1964 (Columbia Pictures).
Terrore nello spazio (*Planet of the Vampires*), directed by Mario Bava, IT/ESP/USA 1965 (Italian International Film/Castilla Cooperative Cinematográfica/American International Pictures).
Raumpatrouille – Die phantastischen Abenteuer des Raumschiffes Orion (*Space Patrol Orion*), television series created by Theo Mezger and Michael Braun, 7 episodes, BRD 1966 (Bavaria Atelier).
You Only Live Twice (*James Bond*), directed by Lewis Gilbert, UK 1967 (United Artists).
2001: A Space Odyssey, directed by Stanley Kubrick, USA 1968 (Metro-Goldwyn-Mayer).
Barbarella, directed by Roger Vadim, FR/IT 1968 (Marianne Productions).
U.F.O. – S.H.A.D.O., television series created by Gerry Anderson and Sylvia Anderson, 26 episodes, UK 1970–73 (ITC Films).
Die Delegation (*The Delegation*), directed by Rainer Erler, BRD 1970 (Bavaria Film).
Diamonds Are Forever (*James Bond*), directed by Guy Hamilton, UK 1971 (United Artists).
Solaris, directed by Andrey Tarkovsky, USSR 1972 (Creative Unit of Writers & Cinema Workers/Mosfilm/Unit Four).

Dark Star, directed by John Carpenter, USA 1974 (Jack H. Harris Enterprises).
Moskva – Kassiopeya (*Moscow – Cassiopeia*), directed by Richard Viktorov, USSR 1974 (Gorky Studio).
Bol'shoye kosmicheskoye puteshestviye (*The Big Space Travel*), directed by Valentin Selivanov, USSR 1975 (Gorky Studio).
Otroki vo vselennoy (*Teens in the Universe*), directed by Richard Viktorov, USSR 1975 (Gorky Studio).
The Man Who Fell to Earth, directed by Nicolas Roeg, UK 1976 (British Lion Films).
Close Encounters of the Third Kind, directed by Steven Spielberg, USA 1977 (Columbia Pictures).
Space Battleship Yamato, directed by Leiji Matsumoto, JP 1977 (Office Academy).
Star Wars, directed by George Lucas, USA. Episode IV: *A New Hope* (1977); Episode V: *The Empire Strikes Back* (1980); Episode VI: *Return of the Jedi* (1983) (Lucasfilm).
Uchû kara no messêji (*Message from Space*), directed by Kinji Fukasaku, JP 1978 (Toei Company).
Moonraker (*James Bond*), directed by Lewis Gilbert, USA/UK/FR 1979 (United Artists).
Alien, directed by Ridley Scott, UK/USA 1979 (Brandywine/Twentieth Century Fox).
Test Pilota Pirxa (*Pilot Pirx's Inquest*), directed by Marek Piestrak, PL/USSR 1979 (Przedsiebiorstwo Realizacji Filmów 'Zespoły Filmowe'/Filmistuudio 'Tallinnfilm'/Dovzhenko Film Studios).
Il était une fois... l'espace (*Once upon a Time... Space*), television series created by Albert Barillé and Eiken, 26 episodes, FR/JP/CA 1982 (Procidis).
The Day After, directed by Nicholas Meyer, USA 1983 (ABC Circle Films).
Le Dernier combat, directed by Luc Besson, FR 1983 (Les Films du Loup).
Aliens, directed by James Cameron, USA 1986 (Brandywine).
Kin-Dza-Dza!, directed by Georgiy Daneliya, USSR 1986 (Mosfilm).
Starcom: The U.S. Space Force, television series created by Brynne Stephens, 13 episodes, USA 1987 (DIC Animation City).
Na srebrnym globie (*On the Silver Globe*), directed by Andrzej Żuławski, PL 1988 (Zespół Filmowy 'Kadr').
Es ist nicht leicht, ein Gott zu sein (*Hard to Be a God*), directed by Peter Fleischmann, BRD/FR/USSR/CH 1989 (B. A. Produktion/Garance).
Starship Troopers, directed by Paul Verhoeven, USA 1997 (Touchstone Pictures).

Literature

Abrahamson, James A. and Carl Sagan, 'Weapons in Space: A "Star Wars" Debate,' *The Futurist* 19.5 (October 1985), 15–17.
Adams, Carsbie C., *Space Flight: Satellites, Spaceships, Space Stations, and Space Travel Explained*, New York: McGraw-Hill, 1958.
Aftergood, Steven, 'Nuclear Space Mishaps and Star Wars,' *Bulletin of the Atomic Scientists* 42.8 (October 1986), 40–3.
Agnew, John A., *Geopolitics: Re-visioning World Politics*, London: Routledge, 1998 (2nd edn 2003).

Alic, John A., 'The Dual Use of Technology: Concepts and Policies,' *Technology in Society* 16.2 (April–June 1994), 155–72.

Althainz, Peter, Mario Birkholz, Horst Gölzenleuchter, Felix Liebke, Rainer Rudert and Jürgen Scheffran, *Militarisierung des Weltraums*, Münster: Bund demokratischer Wissenschaftler, 1984.

Amis, Martin, *Invasion of the Space Invaders*, London: Hutchinson, 1982.

Anders, Günther, *Der Blick vom Mond: Reflexionen über Weltraumflüge*, Munich: C. H. Beck, 1970.

Andersen, Christian Ulrik, '*Monopoly* and the Logic of Sensation in *Spacewar!*,' in Olga Goriunova, ed., *Fun and Software: Exploring Pleasure, Paradox, and Pain in Computing*, London: Bloomsbury, 2014, 197–212.

Arbatov, Alexei and Vladimir Dvorkin, eds, *Outer Space: Weapons, Diplomacy, and Security*, Washington, DC: Carnegie Endowment for International Peace, 2010.

Arbess, Daniel, 'Star Wars and Outer Space Law,' *Bulletin of the Atomic Scientists* 41.9 (October 1985), 19–22.

Arendt, Hannah, *The Human Condition*, Chicago: University of Chicago Press, 1958.

———, 'The Conquest of Space and the Stature of Man' [1963], in idem, *Between Past and Future: Eight Exercises in Political Thought*, New York: Penguin, 2006, 260–74.

Baker, David, *The Rocket: The History and Development of Rocket and Missile Technology*, New York: Crown, 1978.

———, *The Shape of Wars to Come*, Cambridge: Patrick Stephens, 1981.

Barbier, Colette, 'The French Decision to Develop a Military Nuclear Programme in the 1950s,' *Diplomacy & Statecraft* 4.1 (March 1993), 103–13.

Baucom, Donald R., *The Origins of SDI, 1944–1983*, Lawrence: University Press of Kansas, 1992.

Bauman, Zygmunt, 'Reconnaissance Wars of the Planetary Frontierland,' *Theory, Culture & Society* 19.4 (August 2002), 81–90.

Baumgarten-Crusius, Artur, *Die Rakete als Weltfriedenstaube*, Leipzig: Verband der Raketen-Forscher und -Förderer/Roßberg'sche Buchdruckerei, 1931.

Beard, Edmund, *Developing the ICBM: A Study in Bureaucratic Politics*, New York: Columbia University Press, 1976.

Beck, Peter J., *The War of the Worlds from H. G. Wells to Orson Welles, Jeff Wayne, Steven Spielberg and beyond*, London: Bloomsbury, 2016.

Becker-Schaum, Christoph, Philipp Gassert, Martin A. Klimke, Wilfried Mausbach and Marianne Zepp, eds, *The Nuclear Crisis: The Arms Race, Cold War Anxiety, and the German Peace Movement of the 1980s*, Oxford: Berghahn, 2016.

Becklake, John, 'The British Black Knight Rocket,' *Journal of the British Interplanetary Society* 43.7 (July 1990), 283–90.

Benford, Gregory, ed., *Popular Mechanics: The Amazing Weapons That Never Were: Robots, Flying Tanks and Other Machines of War*, New York: Hearst Books, 2012.

Benjamin, Marina, *Rocket Dreams: How the Space Age Shaped Our Vision of a World beyond*, New York: Free Press, 2003.

Bernhard, Patrick and Holger Nehring, eds, *Den Kalten Krieg denken: Beiträge zur sozialen Ideengeschichte*, Essen: Klartext, 2014.

Bertram, Christoph, 'Schlachtfeld Weltall?,' *Deutsches Allgemeines Sonntagsblatt* (27 August 1978).

———, 'Strategic Defense and the Western Alliance,' *Daedalus* 114.3 (Summer 1985), 279–96.

Bille, Matt and Erika Lishock, *The First Space Race: Launching the World's First Satellites*, College Station: Texas A&M University Press, 2004.

Bjork, Rebecca S., *The Strategic Defense Initiative: Symbolic Containment of the Nuclear Threat*, Albany: SUNY Press, 1992.

Black, Jeremy, *The Cold War: A Military History*, London: Bloomsbury, 2015.

Bloomfield, Lincoln P., 'The Politics of Outer Space,' *Bulletin of the Atomic Scientists* 19.5 (May 1963), 12–14.

Bollenhöfener, Klaus, Klaus Farin and Dierk Spreen, eds, *Spurensuche im All: Perry Rhodan Studies*, Berlin: Thomas Tilsner, 2003.

Bormann, Natalie and Michael Sheehan, eds, *Securing Outer Space*, London: Routledge, 2009.

Borrowman, Gerald L., 'Soviet Military Activities in Space,' *Journal of the British Interplanetary Society* 35.2 (February 1982), 86–92.

Boyer, Paul S., *By the Bomb's Early Light: American Thought and Culture at the Dawn of the Atomic Age*, New York: Pantheon, 1985 (2nd edn, Chapel Hill: University of North Carolina Press, 1994).

———, *Fallout: A Historian Reflects on America's Half-Century Encounter with Nuclear Weapons*, Columbus: Ohio State University Press, 1998.

Boyer, Yves, 'Le Projet américain de défense dans l'espace ou comment l'idéologie transcende la technologie,' *Politique étrangère* 49.2 (1984), 365–75.

———, 'Raketenabwehr im Weltraum: Antwort auf eine moralische Frage oder Reform der Strategie?,' *Europa-Archiv* 40.15 (10 August 1985), 467–74.

Boym, Svetlana, 'Kosmos: Remembrances of the Future,' in Adam Bartos and Svetlana Boym, *Kosmos: A Portrait of the Russian Space Age*, New York: Princeton Architectural Press, 2001, 82–99.

Brachet, Gérard and Deloffre, Bernard, 'Space for Defence: A European Vision,' *Space Policy* 22.2 (May 2006), 92–9.

Brand, Stewart, 'Spacewar: Fanatic Life and Symbolic Death Among the Computer Bums,' *Rolling Stone* 123 (7 December 1972), 50–6.

Brandstetter, Thomas, 'Der Staub und das Leben: Szenarien des nuklearen Winters,' *Archiv für Mediengeschichte* 5 (2005), 149–56.

Brauch, Hans Günter, *Angriff aus dem All: Der Rüstungswettlauf im Weltraum*, Bonn: Dietz, 1984.

Brauch, Hans Günter, ed., *Star Wars and European Defence: Implications for Europe. Perceptions and Assessments*, Basingstoke: Macmillan, 1987.

Braun, Wernher von, 'Crossing the Last Frontier,' *Collier's* 27.12 (22 March 1952), 24–9, 72–3.

———, 'Space Superiority as a Means for Achieving World Peace,' *Ordnance* 37.197 (March/April 1953), 770–5.

———, 'Reminiscences of German Rocketry,' *Journal of the British Interplanetary Society* 15.3 (May/June 1956), 125–45.

Brennan, D. G., 'Why Outer Space Control?,' *Bulletin of the Atomic Scientists* 15.5 (May 1959), 198–202.

Broad, William J., *Star Warriors: A Penetrating Look into the Lives of the Young Scientists Behind Our Space Age Weaponry*, New York: Simon & Schuster, 1985.
———, 'Science Fiction Authors Choose Sides in "Star Wars",' *New York Times* (26 February 1985), C1.
———, 'Reagan's "Star Wars" Bid: Many Ideas Converging,' *New York Times* (4 March 1985), 1, 8.
Brünner, Christian and Alexander Soucek, eds, *Outer Space in Society, Politics and Law*, New York: Springer, 2011.
Bulkeley, Rip, *The Sputniks Crisis and Early United States Space Policy*, Bloomington: University of Indiana Press, 1991.
Burrows, William E., *Deep Black: Space Espionage and National Security*, New York: Random House, 1986.
———, *This New Ocean: The Story of the First Space Age*, New York: Random House, 1998.
Bury, Helen, *Eisenhower and the Cold War Arms Race: 'Open Skies' and the Military-Industrial Complex*, London: I. B. Tauris, 2014.
Bush, Vannevar, *Modern Arms and Free Men*, New York: Simon & Schuster, 1949.
Calder, Nigel, *Nuclear Nightmares: An Investigation into Possible Wars*, New York: Viking, 1980.
Caldicott, Helen and Craig Eisendrath, *War in Heaven: The Arms Race in Outer Space*, New York: New Press, 2007.
Canan, James, *War in Space*, New York: Harper & Row, 1982.
Cantril, Hadley, *The Invasion from Mars: A Study in the Psychology of Panic. With the Complete Script of the Famous Orson Welles Broadcast*, Princeton: Princeton University Press, 1940.
Carter, Dale, *The Final Frontier: The Rise and Fall of the American Rocket State*, London: Verso, 1988.
Ceruzzi, Paul E., *Computing: A Concise History*, Cambridge, MA: MIT Press, 2012.
———, *GPS*, Cambridge, MA: MIT Press, 2018.
Chapman, John L., *Atlas: The Story of a Missile*, New York: Harper, 1960.
Chayes, Abram and Jerome Bert Wiesner, eds, *ABM: An Evaluation of the Decision to Deploy an Antiballistic Missile System*, New York: Harper & Row, 1969.
Chun, Clayton K. S., 'Expanding the High Frontier: Space Weapons in History,' *Astropolitics* 2.1 (Spring 2004), 63–78.
———, *Defending Space: US Anti-satellite Warfare and Space Weaponry*, New York: Osprey, 2006.
Clarke, Arthur C., 'Peacetime Uses for V2,' *Wireless World* (February 1945), 58.
———, 'Extra-Terrestrial Relays: Can Rocket Stations Give World-Wide Radio Coverage?,' *Wireless World* 51.10 (October 1945), 305–8.
———, 'The Rocket and the Future of Warfare,' *Royal Air Force Quarterly* 17.2 (March 1946), 61–9.
———, 'The Social Consequences of Communications Satellites,' in idem, *Voices from the Sky: Previews of the Coming Space Age*, New York: Pyramid Books, 1967, 113–22.
———, 'Star Wars and Star Peace,' *Interdisciplinary Science Review* 12.3 (September 1987), 272–7 (also published as idem, 'Star Wars, Star Peace,' *The Daily Telegraph* [3 January 1987]).

Clarke, Ignatius F., *Voices Prophesying War: Future Wars 1763–3749* [1966], 2nd edn, Oxford: Oxford University Press, 1992.

———, 'Future-War Fiction: The First Main Phase, 1871–1900,' *Science Fiction Studies* 24.3 (November 1997), 387–412.

Cleator, Philip E., *Rockets Through Space: or The Dawn of Interplanetary Travel*, London: George Allen & Unwin, 1936.

Cloud, John, 'Imaging the World in a Barrel: CORONA and the Clandestine Convergence of the Earth Sciences,' *Social Studies of Science* 31.2 (April 2001), 231–51.

Clynes, Manfred E. and Nathan S. Kline, 'Cyborgs and Space,' *Astronautics* 9 (September 1960), 26–7, 74–6.

———, 'Drugs, Space, and Cybernetics: Evolution to Cyborgs,' in Bernard E. Flaherty, ed., *Psychophysiological Aspects of Space Flight*, New York: Columbia University Press, 1961, 345–71.

Cockroft, John, *Atomic Challenge: A Symposium*, London: Winchester, 1947.

Collins, Martin, 'One World… One Telephone: Iridium, One Look at the Making of a Global Age,' *History and Technology* 21.3 (September 2005), 301–24.

———, *A Telephone for the World: Iridium, Motorola, and the Making of a Global Age*, Baltimore: Johns Hopkins University Press, 2018.

Conze, Eckart, 'Modernitätsskepsis und die Utopie der Sicherheit: NATO-Nachrüstung und Friedensbewegung in der Geschichte der Bundesrepublik,' *Zeithistorische Forschungen/Studies in Contemporary History* 7.2 (April 2010), 220–39.

Conze, Eckart, Martin Klimke and Jeremy Varon, eds, *Nuclear Threats, Nuclear Fear, and the Cold War of the 1980s*, Cambridge: Cambridge University Press, 2016.

Conze, Werner, Reinhard Stumpf and Michael Geyer, 'Militarismus,' in Otto Brunner, Werner Conze and Reinhart Koselleck, eds, *Geschichtliche Grundbegriffe: Historisches Lexikon zur politisch-sozialen Sprache in Deutschland*, vol. 4, Stuttgart: Klett-Cotta, 1978, 1–47.

Cosgrove, Denis E., *Apollo's Eye: A Cartographic Genealogy of the Earth in the Western Imagination*, Baltimore: Johns Hopkins University Press, 2001.

———, *Geography and Vision: Seeing, Imagining and Representing the World*, London: I. B. Tauris, 2008.

Couston, Mireille and Louis Pilandon, *L'Europe puissance spatiale*, Brussels: Bruylant, 1991.

Daalder, Ivo H., *The SDI Challenge to Europe*, Cambridge, MA: Ballinger, 1987.

Davenport, Richard P., *Strategies for Space: Past, Present and Future*, Newport: Naval War College, 1988.

David, James E., *Spies and Shuttles: NASA's Secret Relationships with the DoD and CIA*, Gainesville: University Press of Florida, 2015.

Day, Dwayne A., John M. Logsdon and Brian Latell, eds, *Eye in the Sky: The Story of the Corona Spy Satellites*, Washington, DC: Smithsonian Institution Press, 1998.

DeBlois, Bruce M., 'The Advent of Space Weapons,' *Astropolitics* 1.1 (Spring 2003), 29–53.

DeBlois, Bruce M., Richard L. Garwin, R. Scott Kemp and Jeremy C. Marwell, 'Space Weapons: Crossing the U.S. Rubicon,' *International Security* 29.2 (Fall 2004), 50–84.

Deudney, Daniel, *Space: The High Frontier in Perspective*, Washington, DC: Worldwatch Institute, 1982.
———, 'Krieg oder Frieden im Weltraum,' *Europa-Archiv: Zeitschrift für internationale Politik* 37.18 (September 1982), 553–62.
———, *Whole Earth Security: A Geopolitics of Peace*, Washington, DC: Worldwatch Institute, 1983.
———, 'Forging Missiles into Spaceships,' *World Policy Journal* 2.2 (Spring 1985), 271–303.
———, *Dark Skies: Space Expansionism, Planetary Geopolitics, and the Ends of Humanity*, Oxford: Oxford University Press, 2020.
'Deutsche Raketen für Nasser/Rüstung: 36, 135 und 333,' *Der Spiegel* 17.19 (8 May 1963), 56–71.
DeVorkin, David H., *Science with a Vengeance: How the Military Created the US Space Sciences after World War II*, New York: Springer, 1992.
Diamond, Edwin, *The Rise and Fall of the Space Age*, Garden City: Doubleday, 1964.
Dick, Steven J., ed., *Remembering the Space Age: Proceedings of the Fiftieth Anniversary Conference*, Washington, DC: NASA, 2008.
Dick, Steven J. and Roger D. Launius, eds, *Critical Issues in the History of Spaceflight*, Washington, DC: NASA, 2006.
———, *Societal Impact of Spaceflight*, Washington, DC: NASA, 2007.
Dickens, Peter and James S. Ormrod, *Cosmic Society: Towards a Sociology of the Universe*, London: Routledge, 2007.
———, eds, *The Palgrave Handbook of Society, Culture and Outer Space*, Basingstoke: Palgrave Macmillan, 2016.
Disch, Thomas M., 'The Road to Heaven: Science Fiction and the Militarization of Space,' *The Nation* 242 (10 May 1986), 650–6.
Divine, Robert A., *The Sputnik Challenge: Eisenhower's Response to the Soviet Satellite*, Oxford: Oxford University Press, 1993.
Dolman, Everett C., *Astropolitik: Classical Geopolitics in the Space Age*, London: Frank Cass, 2002.
Dornberger, Walter, *V2: Der Schuß ins All*, Bechtle: Esslingen, 1952 (*Peenemünde: Die Geschichte der V-Waffen*, 19th edn, Frankfurt am Main: Ullstein, 2013).
———, 'Military Utilization of Space,' *Aviation Week & Space Technology* 75.12 (18 September 1961), 57–61.
Douhet, Giulio, *Il dominio dell'aria: Saggio sull'arte della guerra aerea* [1921], 2nd edn, Rome: Istituto nazionale fascista di cultura, 1927.
Driscoll, Robert W., *Engineering Man for Space: The Cyborg Study*, Farmingdale: United Aircraft Corporate Systems Center, 1963.
Ducrocq, Albert, *Les Armes secrètes allemandes*, Paris: Berger-Levrault, 1947.
———, *Les Armes de demain*, Paris: Berger-Levrault, 1948.
Duelfer, Charles, 'War and Space: Space Is Already Militarized!,' *Analog Science Fiction/Science Fact* 99.5 (May 1979), 148–65.
Duric, Mira, *The Strategic Defence Initiative: US Policy and the Soviet Union*, Aldershot: Ashgate, 2003.
Dürr, Hans-Peter, 'Die forschungspolitischen Auswirkungen von SDI,' *Gewerkschaftliche Monatshefte* 12 (1985), 725–37.

Duvall, Raymond and Jonathan Havercroft, 'Taking Sovereignty out of This World: Space Weapons and Empire of the Future,' *Review of International Studies* 34.4 (October 2008), 755–75.

Echternkamp, Jörg and Hans-Hubertus Mack, eds, *Geschichte ohne Grenzen? Europäische Dimensionen der Militärgeschichte vom 19. Jahrhundert bis heute*, Berlin: de Gruyter, 2016.

Edelson, Edward, 'Space Weapons: The Science Behind the Big Debate,' *Popular Science* 225 (July 1984), 53–9, 100.

Edgerton, David, *England and the Aeroplane: An Essay on a Militant and Technological Nation*, Basingstoke: Macmillan, 1991.

———, *Warfare State: Britain, 1920–1970*, Cambridge: Cambridge University Press, 2005.

———, 'War, Reconstruction, and the Nationalization of Britain, 1939–1951,' *Past & Present* 210/s.6 (January 2011), 29–46.

———, *Britain's War Machine: Weapons, Resources, and Experts in the Second World War*, London: Allen Lane, 2011.

Edgington, Ryan, 'An "All-Seeing Flying Eye": V-2 Rockets and the Promises of Earth Photography,' *History and Technology* 28.3 (September 2012), 363–71.

Edwards, Paul N., *The Closed World: Computers and the Politics of Discourse in Cold War America*, Cambridge, MA: MIT Press, 1997.

Eisler, Steven, *Space Wars: Worlds and Weapons*, London: Octopus, 1979.

Emme, Eugene M., ed., *The Impact of Air Power: National Security and World Politics*, Princeton: Van Nostrand, 1959.

———, ed., *The History of Rocket Technology: Essays on Research, Development, and Utility*, Detroit: Wayne State University Press, 1964 (= *Technology and Culture* 4.4).

'"Enemey [*sic*] 'Moons' Approaching at Height of 20,000 Miles…" Our Military Expert Reviews New War Horror Research,' *Civil and Military Gazette* (20 January 1949).

Engel, Rolf, *Moskau militarisiert den Weltraum*, Landshut: Verlag politisches Archiv, 1979.

Erickson, Paul, Judy L. Klein, Lorraine Daston, Rebecca Lemov, Thomas Sturm and Michael D. Gordin, *How Reason Almost Lost Its Mind: The Strange Career of Cold War Rationality*, Chicago: University of Chicago Press, 2013.

Fischer, Peter, *The Origins of the Federal Republic of Germany's Space Policy 1959–1965: European and National Dimensions*, Noordwijk: ESA, 1994.

FitzGerald, Frances, *Way Out There in the Blue: Reagan, Star Wars and the End of the Cold War*, New York: Simon & Schuster, 2000.

Flynn, Nigel, *War in Space*, New York: Exeter, 1986.

Förster, Stig, ed., *An der Schwelle zum Totalen Krieg: Die militärische Debatte über den Krieg der Zukunft, 1919–1939*, Paderborn: Schöningh, 2002.

Franklin, H. Bruce, *War Stars: The Superweapon and the American Imagination*, Oxford: Oxford University Press, 1988 (2nd edn, Amherst: University of Massachusetts Press, 2008).

Freedman, Lawrence, *The Future of War: A History*, New York: PublicAffairs, 2017.

Friedman, George and Meredith, *The Future of War: Power, Technology, and American World Dominance in the Twenty-First Century*, New York: Crown, 1996.

Gaddis, John Lewis, 'The Emerging Post-revisionist Synthesis on the Origins of the Cold War,' *Diplomatic History* 7.3 (July 1983), 171–90.

———, 'The Long Peace: Elements of Stability in the Postwar International System,' *International Security* 10.4 (Spring 1986), 99–142.

———, *The Long Peace: Inquiries into the History of the Cold War*, New York: Oxford University Press, 1987.

———, 'The Cold War, the Long Peace, and the Future,' *Diplomatic History* 16.2 (April 1992), 234–46.

Gainor, Christopher, 'The Atlas and the Air Force: Reassessing the Beginnings of America's First Intercontinental Ballistic Missile,' *Technology and Culture* 54.2 (April 2013), 346–70.

———, *The Bomb and America's Missile Age*, Baltimore: Johns Hopkins University Press, 2018.

Galison, Peter and Barton Bernstein, 'In Any Light: Scientists and the Decision to Build the Superbomb, 1952–1954,' *Historical Studies in the Physical and Biological Sciences* 19.2 (1989), 267–347.

Garcin, Thierry, *Les Enjeux stratégiques de l'espace*, Brussels: Bruylant, 2001.

Gatland, Kenneth W., *Astronautics in the Sixties: A Survey of Current Technology and Future Development*, London: Iliffe Books, 1962.

Gavin, Francis J., *Nuclear Statecraft: History and Strategy in America's Atomic Age*, Ithaca: Cornell University Press, 2012.

Gavin, James M., *War and Peace in the Space Age*, New York: Harper, 1958.

Gerner, Josef, *Information aus dem Weltraum: Die neue Dimension des Gefechts*, Herford: Mittler, 1990.

Geyer, Michael, 'The Militarization of Europe, 1914–1945,' in John R. Gillis, ed., *Militarization of the Western World*, New Brunswick: Rutgers University Press, 1989, 65–102.

———, 'Cold War Angst: The Case of West-German Opposition to Rearmament and Nuclear Weapons,' in Hanna Schissler, ed., *The Miracle Years: A Cultural History of West Germany, 1949–1968*, Princeton: Princeton University Press, 2001, 376–408.

Gillespie, Paul G. and Grant T. Weller, eds, *Harnessing the Heavens: National Defense Through Space*, Chicago: Imprint Publications, 2008.

Gillis, John R., ed., *The Militarization of the Western World*, New Brunswick: Rutgers University Press, 1989.

Gire, Bruno and Jacques Schibler, 'The French National Space Programme, 1950–1975,' *Journal of the British Interplanetary Society* 40.8 (August 1987), 51–66.

Goldsen, Joseph M., ed., *Outer Space in World Politics*, New York: Frederick A. Praeger, 1963.

Golovine, Michael N., *Conflict in Space: A Pattern of War in a New Dimension*, London: Temple Press, 1962.

———, 'Space: The New Spearhead,' *Spaceflight* 7.6 (November 1965), 181–3.

Gomel, Elana, 'Gods Like Men: Soviet Science Fiction and the Utopian Self,' *Science Fiction Studies* 31.3 (November 2004), 358–77.

Gorin, Peter A., 'Zenit: The First Soviet Photo-Reconnaissance Satellite,' *Journal of the British Interplanetary Society* 50.11 (November 1997), 441–8.

Graetz, J. M., 'The Origin of *Spacewar!*,' *Creative Computing Video & Arcade Game* (Spring 1983), 78–84.

Graham, Daniel O., *High Frontier: A New National Strategy*, Washington, DC: The Heritage Foundation, 1982.

———, *High Frontier: A Strategy for National Survival*, New York: Tom Doherty, 1983.

Grant, Matthew and Benjamin Ziemann, *Understanding the Imaginary War: Culture, Thought and Nuclear Conflict, 1945–90*, Manchester: Manchester University Press, 2016.

Gray, Chris Hables, '"There Will Be War!": Future War Fantasies and Militaristic Science Fiction in the 1980s,' *Science Fiction Studies* 21.3 (November 1994), 315–36.

———, 'Drones, War, and Technological Seduction,' *Technology and Culture* 59.4 (October 2018), 954–62.

Gray, Chris Hables, ed., *The Cyborg Handbook*, New York: Routledge, 1995.

Greiner, Bernd, 'Strategien, Krisen und Kriegsgefahr: SDI im Spiegel der Kuba-Krise,' in Gunnar Lindström, ed., *Bewaffnung des Weltraums: Ursachen – Gefahren – Folgen*, Berlin: Dietrich Reimer, 1986, 38–64.

———, 'Zwischenbilanzen zum Kalten Krieg,' *Mittelweg 36* 16.3 (June/July 2007), 51–8.

Greiner, Bernd, Christian Th. Müller and Dierk Walter, eds, *Angst im Kalten Krieg*, Hamburg: Hamburger Edition, 2009.

Greiner, Bernd, Tim B. Müller and Klaas Voß, eds, *Erbe des Kalten Krieges*, Hamburg: Hamburger Edition, 2013.

Grewell, Greg, 'Colonizing the Universe: Science Fictions Then, Now, and in the (Imagined) Future,' *Rocky Mountain Review of Language and Literature* 55.2 (Fall 2001), 25–47.

Guerrier, Steven W. and Wayne C. Thompson, eds, *Perspectives on Strategic Defense*, Boulder: Westview Press, 1987.

Gupta, Rakesh, 'US Militarization in Outer Space,' *International Studies* 22.2 (April 1985), 153–68.

Hacker, Barton C., 'The Machines of War: Western Military Technology 1850–2000,' *History and Technology* 21.3 (September 2005), 255–300.

Hackett, John, *The Third World War: August 1985*, New York: Macmillan, 1978.

Haley, Andrew G., *Space Law and Government*, New York: Meredith, 1963.

Hall, John, 'Space Wars,' *Radio Times* (21–27 October 1978), 92–101.

Hall, R. Cargill, 'The Eisenhower Administration and the Cold War: Framing American Astronautics to Serve National Security,' *Prologue* 27.1 (Spring 1995), 58–72.

Handberg, Roger, *Seeking New World Vistas: The Militarization of Space*, Westport: Praeger, 2000.

———, 'Military Space Policy: Debating the Future,' *Astropolitics* 2.1 (Spring 2004), 79–89.

Hantke, Steffen, '*Raumpatrouille*: The Cold War, the "Citizen in Uniform", and West German Television,' *Science Fiction Studies* 31.1 (March 2004), 63–80.

———, 'Military Culture,' in Rob Latham, ed., *The Oxford Handbook of Science Fiction*, Oxford: Oxford University Press, 2014, 329–39.

———, *Monsters in the Machine: Science Fiction Film and the Militarization of America after World War II*, Jackson: University Press of Mississippi, 2016.

Haraway, Donna, 'A Manifesto for Cyborgs: Science, Technology and Socialist Feminism in the 1980s,' *Socialist Review* 80.2 (March–April 1985), 65–107.

Harvey, Brian, *Race into Space: The Soviet Space Programme*, Chichester: Horwood, 1988.

Hayward, Keith, *British Military Space Programmes*, London: Royal United Services Institution, 1996.

Hebert, Karl D., 'Regulation of Space Weapons: Ensuring Stability and Continued Use of Outer Space,' *Astropolitics* 12.1 (March 2014), 1–26.

Hecht, Gabrielle, ed., *Entangled Geographies: Empire and Technopolitics in the Global Cold War*, Cambridge, MA: MIT Press, 2011.

Heinlein, Robert A., *Have Space Suit – Will Travel*, New York: Ace, 1958.

———, *Starship Troopers*, New York: G. P. Putnam's Sons, 1959.

———, *The Puppet Masters*, London: Hodder & Stoughton, 1987.

Hesse, Markus, *Europäische Weltraumpolitik: Sicherheitspolitische Aspekte*, Berlin: Duncker & Humblot, 2012.

Hilger, Josef, *Raumpatrouille: Die phantastischen Abenteuer des Raumschiffes ORION*, Berlin: Schwarzkopf & Schwarzkopf, 2000 (2nd edn 2005).

Hill, Charles N., *A Vertical Empire: The History of the UK Rocket and Space Programme, 1950–1971*, London: Imperial College Press, 2001 (2nd edn 2012).

Hoeppener-Flatow, Wilhelm, 'Kriegswaffen der Zukunft,' *Neue Preußische Kreuz-Zeitung* 85.283 (9 October 1932), 7.

Högenauer, Ernst, 'Mit Hyperschall ins All: Europas Beitrag zur Eroberung des Weltraums,' *Das Parlament* 36.33/34 (16/23 August 1986), 12.

Holloway, David, 'The Strategic Defense Initiative and the Soviet Union,' *Daedalus* 114.3 (Summer 1985), 257–78.

Hoplites [pseud.], 'La France dans la "guerre des étoiles",' *Le Monde* (6 March 1985), 1, 4.

Horn, Eva, *Zukunft als Katastrophe*, Frankfurt am Main: Fischer, 2014.

Horrigan, Brian, 'Popular Culture and Visions of the Future in Space, 1901–2001,' in Bruce Sinclair, ed., *New Perspectives on Technology and American Culture*, Philadelphia: American Philosophical Society, 1986, 49–67.

Hudson, Heather E., *Communication Satellites: Their Development and Impact*, New York: Free Press, 1990.

Hughes, Robert C., *SDI: A View from Europe*, Washington, DC: National Defense University Press, 1990.

Hulett, Louisa S., *From Cold Wars to Star Wars: Debates over Defense and Detente*, Lanham: University Press of America, 1988.

Hunter, Kerry L., *The Reign of Fantasy: The Political Roots of Reagan's Star Wars Policy*, New York: Peter Lang, 1992.

Huwart, Olivier, *Du V2 à Véronique: La naissance des fusées françaises*, Paris: Marines éditions, 2004.

Hüppauf, Bernd, *Was ist Krieg? Zur Grundlegung einer Kulturgeschichte des Kriegs*, Bielefeld: transcript, 2013.

Jaramillo, Cesar, 'The Multifaceted Nature of Space Security Challenges,' *Space Policy* 33.2 (August 2015), 63–6.

Jasani, Bhupendra, *Outer Space: Battlefield of the Future?*, London: Taylor & Francis, 1978.

———, 'Space: Battlefield of the Future?,' *Futures* 14.5 (October 1982), 435–47.

Jasani, Bhupendra and Christopher Lee, *Countdown to Space War*, London: Taylor & Francis, 1984.
Johansen, Anatol, 'Droht ein Krieg im Weltraum? Verhandlungen über "Killer-satelliten" in Helsinki,' *Frankfurter Allgemeine Zeitung* (8 June 1978), 4.
Johnson, John, *Das deutsche Wunder 193?*, Leipzig: Weicher, 1930.
Johnson, Michael Peter, *Mission Control: Inventing the Groundwork of Spaceflight*, Gainesville: University Press of Florida, 2015.
Johnson, Oris B., 'Space: Today's First Line of Defense,' *Air University Review* 20.1 (November–December 1968), 95–102.
Johnson-Freese, Joan, *Space as a Strategic Asset*, New York: Columbia University Press, 2007.
———, *Heavenly Ambitions: America's Quest to Dominate Space*, Philadelphia: University of Pennsylvania Press, 2009.
———, *Space Warfare in the 21st Century: Arming the Heavens*, London: Routledge, 2017.
Jones, Greta, 'The Mushroom-Shaped Cloud: British Scientists' Opposition to Nuclear Weapons Policy, 1945–57,' *Annals of Science* 43.1 (January 1986), 1–26.
Jones, Robert A., 'They Came in Peace for All Mankind: Popular Culture as a Reflection of Public Attitudes to Space,' *Space Policy* 20.1 (February 2004), 45–8.
Jungk, Robert, 'Die dunkle Seite des Mondes: Setzen Amerikas Militärs die Raumfahrt unter Druck?,' *Die Zeit* 30 (25 July 1969), 39.
Kagan, Robert, *Of Paradise and Power: America and Europe in the New World Order*, New York: Alfred A. Knopf, 2003.
Kahn, Herman, *On Thermonuclear War*, Princeton: Princeton University Press, 1960.
Kalic, Sean N., 'Reagan's SDI Announcement and the European Reaction: Diplomacy in the Last Decade of the Cold War,' in Leopoldo Nuti, ed., *The Crisis of Détente in Europe: From Helsinki to Gorbachev, 1975–1985*, London: Routledge, 2009, 99–110.
———, *US Presidents and the Militarization of Space, 1946–1967*, College Station: Texas A&M University Press, 2012.
Kaplan, Caren, Erik Loyer and Ezra Claytan Daniels, 'Precision Targets: GPS and the Militarization of Everyday Life,' *Canadian Journal of Communication* 38.3 (November 2013), 397–420.
Karafantis, Layne, *Under Control: Constructing the Nerve Centers of the Cold War*, PhD thesis, Baltimore: Johns Hopkins University, 2016.
Karas, Thomas, *The New High Ground: Strategies and Weapons of Space Age War*, New York: Simon & Schuster, 1983.
Karl, Wilfried, ed., *Rüstungskooperation und Technologiepolitik als Problem der westeuropäischen Integration*, Opladen: Leske + Budrich, 1994.
Kay, William D., 'NASA and Space History,' *Technology and Culture* 40.1 (January 1999), 120–7.
Kilgore, De Witt Douglas, *Astrofuturism: Science, Race, and Visions of Utopia in Space*, Philadelphia: University of Pennsylvania Press, 2003.
'Die Killer werden immer besser: Der Wettlauf um die militärische Vorherrschaft im Weltraum hat begonnen,' *Welt am Sonntag* 32 (9 August 1981), 2.
Kirby, Stephen and Gordon Robson, eds, *The Militarisation of Space*, Brighton: Wheatsheaf Books, 1987.

Klein, John J., *Understanding Space Strategy: The Art of War in Space*, London: Routledge, 2019.

Kniesche, Thomas W., 'Germans to the Final Frontier: Science Fiction, Popular Culture, and the Military in 1960s Germany: The Case of *Raumpatrouille*,' *New German Critique* 101 (Summer 2007), 157–85.

Krepon, Michael and Julia Thompson, eds, *Anti-satellite Weapons, Deterrence and Sino-American Space Relations*, Washington, DC: Stimson, 2013.

Kretschmer, Thomas and Uwe Wiemken, eds, *Militärische Nutzung des Weltraums: Grundlagen und Optionen*, Frankfurt am Main: Report-Verlag, 2004.

Krieger, F. J., *Behind the Sputniks: A Survey of Soviet Space Science*, Washington, DC: Public Affairs Press, 1958.

Krige, John, 'What Is "Military" Technology? Two Cases of US-European Scientific and Technological Collaboration in the 1950s,' in Francis H. Heller and John R. Gillingham, eds, *The United States and the Integration of Europe: Legacies of the Postwar Era*, New York: St. Martin's Press, 1996, 307–38.

———, 'NATO and the Strengthening of Western Science in the Post-Sputnik Era,' *Minerva* 38.1 (January 2002), 81–108.

———, *American Hegemony and the Postwar Reconstruction of Science in Europe*, Cambridge, MA: MIT Press, 2006.

Krige, John and Arturo Russo, *Reflections on Europe in Space*, Noordwijk: ESA, 1994.

———, *Europe in Space 1960–1973*, Noordwijk: ESA, 1994.

Krige, John, Arturo Russo and Lorenza Sebesta, *A History of the European Space Agency*, vol. 1: *The Story of ESRO and ELDO, 1958–1987*, vol. 2: *The Story of ESA, 1973–1987*, Noordwijk: ESA, 2000.

Kubbig, Bernd W., ed., *Die militärische Eroberung des* Weltraums, 2 vols, Frankfurt am Main: Suhrkamp, 1990.

———, *Wissen als Machtfaktor im Kalten Krieg: Naturwissenschaftler und die Raketenabwehr der USA*, Frankfurt am Main: Campus, 2004.

Kühne, Thomas and Benjamin Ziemann, eds, *Was ist Militärgeschichte?*, Paderborn: Schöningh, 2000.

Kunze, Harald, ed., *Friedenskampf gegen USA-Weltraumrüstung: Protokoll des interdisziplinären wissenschaftlichen Kolloquiums vom 31.1.1986 in Jena zum Thema 'Die Strategie der USA zur Militarisierung des Weltraums und der Kampf der Friedenskräfte*,' Jena: Friedrich-Schiller-Universität, 1986.

Lakoff, Sanford A. and Herbert F. York, *A Shield in Space? Technology, Politics, and the Strategic Defense Initiative*, Berkeley: University of California Press, 1989.

Launius, Roger D. and Dennis R. Jenkins, eds, *To Reach the High Frontier: A History of U.S. Launch Vehicles*, Lexington: University of Kentucky Press, 2002.

Lee, Christopher, *War in Space*, London: Hamish Hamilton, 1986.

Lehne, Stefan, 'Die militärische Nutzung des Weltraums: Aspekte und Perspektiven,' *Österreichische Militärische Zeitschrift* 23.1 (January/February 1985), 42–7.

Lellouche, Pierre, *L'Avenir de la guerre*, Paris: Mazarine, 1985.

Leveridge, Rosalind, *Fantastic Voyages of the Cinematic Imagination: George Méliès's* Trip to the Moon, Albany: SUNY Press, 2012.

Ley, Willy, *Rockets, Missiles and Space Travel*, New York: Viking, 1944 (*Rockets, Missiles, and Men in Space*, 2nd edn, New York: Viking, 1968).

———, 'How We Could Wage a War from Man-Made Stars,' *Look* 15.16 (31 July 1951), 20–5.

Lindström, Gunnar, ed., *Bewaffnung des Weltraums: Ursachen – Gefahren – Folgen*, Berlin: Dietrich Reimer, 1986.

Linenthal, Edward Tabor, *Symbolic Defense: The Cultural Significance of the Strategic Defense Initiative*, Urbana: University of Illinois Press, 1989.

Liu, Hao and Fabio Tronchetti, 'United Nations Resolution 69/32 on the "No First Placement of Weapons in Space": A Step Forward in the Prevention of an Arms Race in Outer Space?,' *Space Policy* 38 (November 2016), 64–7.

Logsdon, John M., *John F. Kennedy and the Race to the Moon*, Basingstoke: Palgrave Macmillan, 2010.

———, *After Apollo? Richard Nixon and the American Space Program*, Basingstoke: Palgrave Macmillan, 2015.

———, *Ronald Reagan and the Space Frontier*, London: Palgrave Macmillan, 2019.

Logsdon, Tom, *The Navstar Global Positioning System*, New York: Van Nostrand, 1992.

Long, Franklin A., Donald Hafner and Jeffrey Boutwell, *Weapons in Space*, New York: Norton, 1986.

Longden, Norman and Duc Guyenne, eds, *Twenty Years of European Cooperation in Space: An ESA Report*, Noordwijk: ESA, 1984.

Lovell, Bernard, *The Origins and International Economics of Space Exploration*, Edinburgh: Edinburgh University Press, 1973.

Lucas, Michael, 'SDI and Europe,' *World Policy Journal* 3.2 (Spring 1986), 219–49.

Lutz, Dieter S., *Auf dem Weg zur Militarisierung des Weltraums? Analysen und Materialien*, Hamburg: Institut für Friedensforschung und Sicherheitspolitik, 1984.

Lützenkirchen, Willy and Egmont R. Koch, 'Der Himmel hängt voller Spione: Die Militärs erobern den Weltraum,' *X – Unsere Welt heute* 4.8 (August 1972), 56–61.

MacDonald, Fraser, 'Geopolitics and "The Vision Thing": Regarding Britain and America's First Nuclear Missile,' *Transactions of the Institute of British Geographers* 31.1 (March 2006), 53–71.

———, 'Anti-*Astropolitik*: Outer Space and the Orbit of Geography,' *Progress in Human Geography* 31.5 (October 2007), 592–615.

———, 'Space and the Atom: On the Popular Geopolitics of Cold War Rocketry,' *Geopolitics* 13.4 (Winter 2008), 611–34.

Macho, Thomas, 'Technische Utopien und Katastrophenängste,' *Gegenworte* 10 (Fall 2002), 12–14.

MacKenzie, Donald, *Inventing Accuracy: A Historical Sociology of Nuclear Missile Guidance*, Cambridge, MA: MIT Press, 1990.

Macko, Stanley J., *Satellite Tracking*, New York: John F. Rider, 1962.

Mackowski, Maura Phillips, *Testing the Limits: Aviation Medicine and the Origins of Manned Space Flight*, College Station: Texas A&M University Press, 2006.

Madders, Kevin, *A New Force at a New Frontier: Europe's Development in the Space Field in the Light of Its Main Actors, Policies, Law and Activities from Its Beginnings up to the Present*, Cambridge: Cambridge University Press, 1997.

Mahnken, Thomas G., Joseph Maiolo and David Stevenson, eds, *Arms Races in International Politics: From the Nineteenth to the Twenty-First Century*, Oxford: Oxford University Press, 2016.

Maier, Charles S., *Once within Borders: Territories of Power, Wealth, and Belonging since 1500*, Cambridge, MA: Harvard University Press, 2016.

Maier, Charles S., ed., *The Cold War in Europe: Era of a Divided Continent* [1991], 3rd edn, Princeton: Markus Wiener, 1996.

Majsova, Natalija, 'Outer Space and Cyberspace: An Outline of Where and How to Think of Outer Space in Video Games,' *Teorija in praksa* 51.1 (February 2014), 106–22.

Mandelbaum, W. Adam, *The Psychic Battlefield: A History of the Military-Occult Complex*, New York: St. Martin's Press, 2000.

Manno, Jack, *Arming the Heavens: The Hidden Military Agenda for Space, 1945–1995*, New York: Dodd, Mead, 1984.

Martin, James, *Telecommunications and the Computer*, Englewood Cliffs: Prentice-Hall, 1976.

Maurer, Eva, Julia Richers, Monica Rüthers and Carmen Scheide, eds, *Soviet Space Culture: Cosmic Enthusiasm in Socialist Societies*, Basingstoke: Palgrave Macmillan, 2011.

McAleer, Neil, *Odyssey: The Authorised Biography of Arthur C. Clarke*, London: Gollancz, 1992 (*Visionary: The Odyssey of Sir Arthur C. Clarke*, 2nd edn, Baltimore: The Clarke Project, 2010).

McCuen, Gary E., ed., *Militarizing Space*, Hudson: Gary E. McCuen Publications, 1989.

McCurdy, Howard E., *Space and the American Imagination*, Washington, DC: Smithsonian Institution Press, 1997 (2nd edn, Baltimore: Johns Hopkins University Press, 2011).

McDougall, Walter A., 'Technocracy and Statecraft in the Space Age: Toward the History of a Saltation,' *American Historical Review* 87.4 (October 1982), 1010–40.

———, 'Space-Age Europe: Gaullism, Euro-Gaullism, and the American Dilemma,' *Technology and Culture* 26.2 (April 1985), 179–203.

———, 'Sputnik, the Space Race, and the Cold War,' *Bulletin of the Atomic Scientists* 41.5 (May 1985), 20–5.

———, *...The Heavens and the Earth: A Political History of the Space Age*, New York: Basic Books, 1985.

McLean, Alasdair W. M., *Western European Military Space Policy*, Aldershot: Dartmouth, 1992.

McLean, Alasdair W. M. and Fraser Lovie, *Europe's Final Frontier: The Search for Security Through Space*, Commack: Nova Science, 1997.

McLeod, Ken, 'Space Oddities: Aliens, Futurism and Meaning in Popular Music,' *Popular Music* 22.3 (October 2003), 337–55.

McNeill, William Hardy, *The Pursuit of Power: Technology, Armed Force, and Society since A.D. 1000*, Chicago: University of Chicago Press, 1982.

Mellor, Felicity, 'Colliding Worlds: Asteroid Research and the Legitimization of War in Space,' *Social Studies of Science* 37.4 (August 2007), 499–531.

Menon, P. K., 'Demilitarization of Outer Space,' *International Journal on World Peace* 4.2 (April–June 1987), 127–50.

Meyer, David S., 'Star Wars, *Star Wars*, and American Political Culture,' *Journal of Popular Culture* 26.2 (Fall 1992), 99–115.

Michaud, Michael A. G., *Reaching for the High Frontier: The American Pro-Space Movement, 1972–84*, New York: Praeger, 1986.

Mieczkowski, Yanek, *Eisenhower's Sputnik Moment: The Race for Space and World Prestige*, Ithaca: Cornell University Press, 2013.

Millard, Doug, *Satellite: Innovation in Orbit*, London: Reaktion, 2017.
Miller, Ryder W., *From Narnia to a Space Odyssey: The War of Ideas between Arthur C. Clarke and C. S. Lewis*, New York: iBooks, 2003.
Milner, Greg, *Pinpoint: How GPS Is Changing Technology, Culture, and Our Minds*, New York: Norton, 2016.
Mindell, David A., *Digital Apollo: Human and Machine in Spaceflight*, Cambridge, MA: MIT Press, 2008 (2nd edn 2011).
Mineiro, Michael C., *Space Technology Export Controls and International Cooperation in Outer Space*, New York: Springer, 2012.
Mitter, Rana and Patrick Major, eds, *Across the Blocs: Cold War Cultural and Social History*, London: Frank Cass, 2004.
Moltz, James Clay, *The Politics of Space Security: Strategic Restraint and the Pursuit of National Interests*, Stanford: Stanford University Press, 2008.
———, *Crowded Orbits: Conflict and Cooperation in Space*, New York: Columbia University Press, 2014.
Monchaux, Nicholas de, *Spacesuit: Fashioning Apollo*, Cambridge, MA: MIT Press, 2011.
Monnens, Devin and Martin Goldberg, 'Space Odyssey: The Long Journey of *Spacewar!* from MIT to Computer Labs Around the World,' *Kinephanos* 6 (June 2015), 124–47.
Mowthorpe, Matthew, 'US Military Space Policy 1945–92,' *Space Policy* 18.1 (February 2002), 25–36.
———, *The Militarization and Weaponization of Space*, Lanham: Lexington Books, 2004.
Mueller, Karl P., 'Totem and Taboo: Depolarizing the Space Weaponization Debate,' *Astropolitics* 1.1 (Spring 2003), 4–28.
Mutschler, Max M., *Arms Control in Space: Exploring Conditions for Preventive Arms Control*, Basingstoke: Palgrave Macmillan, 2013.
Mutschler, Max M. and Christophe Venet, 'The European Union as an Emerging Actor in Space Security?,' *Space Policy* 28.2 (May 2012), 118–24.
Nast, Mirjam, *'Perry Rhodan' lesen: Zur Serialität der Lektürepraktiken einer Heftromanserie*, Bielefeld: transcript, 2017.
Neufeld, Jacob, *Ballistic Missiles in the United States Air Force 1945–1960*, Washington, DC: United States Air Force, 1990.
Neufeld, Michael J., *The Rocket and the Reich: Peenemünde and the Coming of the Ballistic Missile Era*, New York: Free Press, 1995.
———, 'The End of the Army Space Program: Interservice Rivalry and the Transfer of the von Braun Group to NASA, 1958–1959,' *Journal of Military History* 69.3 (July 2005), 737–58.
———, '"Space Superiority": Wernher von Braun's Campaign for a Nuclear-Armed Space Station, 1946–1956,' *Space Policy* 22.1 (February 2006), 57–62.
———, 'Wernher von Braun's Ultimate Weapon,' *Bulletin of the Atomic Scientists* 63.4 (July 2007), 50–78.
———, *Von Braun: Dreamer of Space, Engineer of War*, New York: Alfred A. Knopf, 2007.
———, 'The Three Heroes of Spaceflight: The Rise of the Tsiolkovskii – Goddard – Oberth Interpretation and Its Current Validity,' *Quest: The History of Spaceflight Quarterly* 19.4 (2012), 4–13.

———, *Spaceflight: A Concise History*, Cambridge, MA: MIT Press, 2018.

Newhouse, John, *The Nuclear Age: From Hiroshima to Star Wars*, London: Joseph, 1989.

———, *War and Peace in the Nuclear Age*, New York: Alfred A. Knopf, 1989.

Norris, Pat, *Spies in the Sky: Surveillance Satellites in War and Peace*, New York: Springer, 2007.

Oberth, Hermann, *Die Rakete zu den Planetenräumen*, Munich: Oldenbourg, 1923.

———, *Wege zur Raumschiffahrt*, Munich: Oldenbourg, 1929.

———, 'Die 3 Gesichter der Rakete,' *Die Gartenlaube* 43 (23 October 1930), 887–91.

Oder, Frederic C. E., James C. Fitzpatrick and Paul E. Worthman, *The CORONA Story*, Washington, DC: National Reconnaissance Office, 1987.

Odishaw, Hugh, ed., *The Challenges of Space*, Chicago: University of Chicago Press, 1962.

O'Hanlon, Michael E., *Neither Star Wars nor Sanctuary: Constraining the Military Uses of Space*, Washington, DC: Brookings Institution Press, 2004.

Ordway, Frederick I., Mitchell R. Sharpe and Ronald C. Wakeford, 'Project Horizon: An Early Study of a Lunar Outpost,' *Acta Astronautica* 17.10 (October 1988), 1105–21.

Oreskes, Naomi and John Krige, eds, *Science and Technology in the Global Cold War*, Cambridge, MA: MIT Press, 2014.

Orr, Jackie, 'The Militarization of Inner Space,' *Critical Sociology* 30.2 (March 2004), 451–81.

Østreng, Willy, *Politics in High Latitudes: The Svalbard Archipelago*, London: Hurst, 1978.

Ó Tuathail, Gearóid, *Critical Geopolitics: The Politics of Writing Global Space*, Minneapolis: University of Minnesota Press, 1996.

Pace, Scott, 'Security in Space,' *Space Policy* 33.2 (August 2015), 51–5.

Paracha, Sobia, 'Military Dimensions of the Indian Space Program,' *Astropolitics* 11.3 (September 2013), 156–86.

Paradiso, Nunzia, *The EU Dual Approach to Security and Space: Twenty Years of European Policy Making*, Vienna: European Space Policy Institute, 2013.

Parker, Geoffrey, ed., *The Cambridge History of Warfare*, Cambridge: Cambridge University Press, 2005.

Parkinson, Bob, ed., *Interplanetary: A History of the British Interplanetary Society*, London: British Interplanetary Society, 2008.

Parkinson, Bradford W., Thomas Stansell, Ronald Beard and Konstantine Gromov, 'A History of Satellite Navigation,' *Navigation* 42.1 (Spring 1995), 109–64.

Parks, Lisa, *Cultures in Orbit: Satellites and the Televisual*, Durham, NC: Duke University Press, 2005.

Paul, Günter, 'Die Militarisierung des Weltraums: Eine Analyse der sowjetischen Raumfahrt,' *Frankfurter Allgemeine Zeitung* (17 October 1979), 9.

———, *Aufmarsch im Weltall: Die Kriege der Zukunft werden im Weltraum entschieden*, Bonn: Keil, 1980.

———, 'Stabilisierend – destabilisierend: Die militärische Nutzung des Weltraums,' *Frankfurter Allgemeine Zeitung* (30 November 1984), 11.

Payne, Keith B., ed., *Laser Weapons in Space: Policy and Doctrine*, Boulder: Westview Press, 1983.

Peebles, Curtis, *Battle for Space*, Poole: Blandford, 1983.
———, *The Corona Project: America's First Spy Satellites*, Annapolis: Naval Institute Press, 1997.
———, *High Frontier: The U.S. Air Force and* the *Military Space Program*, Washington, DC: Air Force History and Museums Program, 1997.
Pendray, G. Edward, 'Next Stop the Moon,' *Collier's* 21.36 (7 September 1946), 11–13, 75.
———, 'Threshold of the Rocket Age: Today's Military Research Points the Way to Tomorrow's Peacetime Employment of High-Speed, Far-Ranging Craft,' *New York Times Magazine* (29 May 1949), 22–3.
Peoples, Columba, '*Sputnik* and "Skill Thinking" Revisited: Technological Determinism in American Responses to the Soviet Missile Threat,' *Cold War History* 8.1 (February 2008), 55–75.
———, 'The Growing "Securitization" of Outer Space,' *Space Policy* 26.4 (November 2010), 205–8.
———, *Justifying Ballistic Missile Defense: Technology, Security and Culture*, Cambridge: Cambridge University Press, 2010.
Perry, Robert L., *The Ballistic Missile Decisions*, Santa Monica: RAND, 1967.
Philpott, Trevor, 'Guided Robots Go to War,' *Picture Post* 57.8 (22 November 1952), 15–19.
Pickering, Andrew, 'Cyborg History and the World War II Regime,' *Perspectives on Science* 3.1 (Spring 1995), 1–48.
Polianski, Igor J. and Matthias Schwartz, eds, *Die Spur des Sputnik: Kulturhistorische Expeditionen ins kosmische Zeitalter*, Frankfurt am Main: Campus, 2009.
Pressler, Larry, *Star Wars: The Strategic Defense Initiative Debates in Congress*, New York: Praeger, 1986.
Preston, Bob, Dana J. Johnson, Sean J. A. Edwards, Michael Miller and Calvin Shipbaugh, *Space Weapons, Earth Wars*, Santa Monica: RAND, 2002.
Price, David H., 'Militarizing Space,' in Roberto J. González, Hugh Gusterson and Gustaaf Houtman, eds, *Militarization: A Reader*, Durham, NC: Duke University Press, 2019, 316–19.
Proske, Rüdiger, 'Schlachtfeld Weltraum: Die neuen Fronten werden am Himmel aufgebaut,' *Geo Special 'Weltraum'* 8.3 (1983), 42–51.
'Pushbutton Defense for Air War,' *Life Magazine* 42.6 (11 February 1957), 62–7.
Quistgaard, Erik, 'ESA and Europe's Future in Space,' *ESA Bulletin* 39 (August 1984), 20–2.
Reagan, Ronald, 'Address to the Nation on Defense and National Security,' 23 March 1983, Ronald Reagan Presidential Library, https://www.reaganlibrary.gov/research/speeches/32383d, accessed 15 July 2020.
'Reagan's Star Wars,' *New York Review of Books* 31.7 (26 April 1984), 47–52.
Redfield, Peter, 'Beneath a Modern Sky: Space Technology and Its Place on the Ground,' *Science, Technology and Human Values* 21.3 (Summer 1996), 251–74.
———, *Space in the Tropics: From Convicts to Rockets in French Guiana*, Berkeley: University of California Press, 2000.
———, 'The Half-Life of Empire in Outer Space,' *Social Studies of Science* 32.5/6 (October/December 2002), 791–825.
Rehm, Georg W., 'Politisch-militärische Aspekte der Weltraumfahrt,' *Weltraumfahrt* 13.4 (July/August 1962), 107–10.

Reinke, Niklas, *Geschichte der deutschen Raumfahrtpolitik: Konzepte, Einflußfaktoren und Interdependenzen 1923–2002*, Berlin: de Gruyter, 2004.
Reiss, Edward, *The Strategic Defense Initiative*, Cambridge: Cambridge University Press, 1992.
Rhea, John, *SDI: What Could Happen. Eight Possible Star Wars Scenarios*, Harrisburg: Stackpole Books, 1988.
Rhodes, Richard, *Dark Sun: The Making of the Hydrogen Bomb*, New York: Simon & Schuster, 1995.
Richardson, Robert S., 'Rocket Blitz from the Moon,' *Collier's* 122 (23 October 1948), 24–5, 44–5.
———, 'Militärbasis auf dem Mond?,' *Die Weltwoche* 17.812 (3 June 1949), 9.
Richelson, Jeffrey T., *America's Secret Eyes in Space: The U.S. Keyhole Spy Satellite Program*, New York: Harper & Row, 1990.
———, *America's Space Sentinels: DSP Satellites and National Security*, Lawrence: University Press of Kansas, 1999 (2nd edn 2012).
Ritchie, David, *Spacewar*, New York: Atheneum, 1982.
Robin, Ron, *The Making of the Cold War Enemy: Culture and Politics in the Military-Intellectual Complex*, Princeton: Princeton University Press, 2001.
Robinson, Jana, 'Transparency and Confidence-Building Measures for Space Security,' *Space Policy* 27.1 (February 2011), 27–37.
Rohrwild, Karlheinz, 'Von der Weltraumrakete zur Kriegswaffe, 1900–1936,' in Dieter B. Herrmann and Christian Gritzner, eds, *Beiträge zur Geschichte der Raumfahrt: Ausgewählte Vorträge der Raumfahrthistorischen Kolloquien 1986–2015*, Berlin: trafo Wissenschaftsverlag, 2017, 101–243.
Roland, Alex, 'Barnstorming in Space: The Rise and Fall of the Romantic Era of Spaceflight, 1957–1986,' in Radford Byerly Jr., ed., *Space Policy Reconsidered*, Boulder: Westview Press, 1989, 33–52.
———, *The Military-Industrial Complex*, Washington, DC: American Historical Association, 2001.
———, 'The Military-Industrial Complex: Lobby and Trope,' in Andrew J. Bacevich, ed., *The Long War: A New History of U.S. National Security Policy since World War II*, New York: Columbia University Press, 2007, 335–70.
Roma, Alfredo, 'Drones and Popularisation of Space,' *Space Policy* 41 (2017), 65–7.
Roman, Peter J., *Eisenhower and the Missile Gap*, Ithaca: Cornell University Press, 1995.
Rosenberg, David Alan, 'American Atomic Strategy and the Hydrogen Bomb Decision,' *Journal of American History* 66.1 (June 1979), 62–87.
———, 'The Origins of Overkill: Nuclear Weapons and American Strategy, 1945–1960,' *International Security* 7.4 (Spring 1983), 3–71.
Rosenfeld, Albert, 'Pitfalls and Perils out There,' *Life Magazine* 57.14 (2 October 1964), 112–24.
Ross, H. E., 'Lunar Spacesuit,' *Journal of the British Interplanetary Society* 9.1 (January 1950), 23–37.
Rougeron, Camille, 'L'Espace et son utilisation militaire,' *Science et Vie* 77 (1966), 121–7.
Ruffner, Kevin C., ed., *CORONA: America's First Satellite Program*, Washington, DC: CIA, 1995.

Russo, Arturo, 'Science in Space vs Space Science: The European Utilisation of Spacelab,' *History and Technology* 16.2 (October 1999), 137–78.

Saegesser, Lee D., *Space: The High Ground*, unpublished manuscript, Washington, DC: NASA History Office, ca. 1983.

Sagan, Carl, 'Why "Star Wars" Concept Is a Useless Crock,' *National Catholic Reporter* (10 May 1985), 14.

Salin, Patrick A., 'Privatization and Militarization in the Space Business Environment,' *Space Policy* 17.1 (February 2001), 19–26.

Salkeld, Robert, *War and Space*, Englewood Cliffs: Prentice-Hall, 1970.

Sanders, Steven, ed., *The Philosophy of Science Fiction Film*, Lexington: University Press of Kentucky, 2007.

Sandner, David, 'Shooting for the Moon: Méliès, Verne, Wells, and the Imperial Satire,' *Extrapolation* 39.1 (April 1998), 5–25.

Sänger, Eugen, *Raketenbomber* (*America Bomber*), Trauen: s.n., 1940.

———, 'Militärische Bedeutung der Raumfahrt,' *Außenpolitik* 9.7 (July 1958), 433–41.

———, *Raumfahrt – Technische Überwindung des Krieges: Aktuelle Aspekte der Überschall-Luftfahrt und Raumfahrt*, Hamburg: Rowohlt, 1958.

———, *Raumfahrt: heute – morgen – übermorgen*, Düsseldorf: Econ, 1963.

Sänger, Eugen and Irene Sänger-Bredt, *Über einen Raketenantrieb für Fernbomber*, Ainring: Deutsche Forschungsanstalt für Segelflug e.V. 'Ernst Udet,' 1944.

Schlosser, Eric, *Command and Control: Nuclear Weapons, the Damascus Accident, and the Illusion of Safety*, New York: Penguin, 2013.

Schreiber, Wolfgang, 'Die Bedeutung der Erforschung und Nutzung des Weltraums für die militärische Sicherheit,' *Europa-Archiv* 41.21 (1986), 629–38.

Sebesta, Lorenza, 'Choosing Its Own Way: European Cooperation in Space. Europe as a Third Way between Science's Universalism and US Hegemony,' *Journal of European Integration History* 12.2 (2006), 27–55.

Sebesta, Lorenza and Filippo Pigliacelli, *La terra vista dall'alto: Breve storia della militarizzazione dello spazio*, Rome: Carocci editore, 2008.

Seed, David, *American Science Fiction and the Cold War: Literature and Film*, Edinburgh: Edinburgh University Press, 1999.

Seed, David, ed., *Future Wars: The Anticipations and the Fears*, Liverpool: Liverpool University Press, 2012.

Seidel, Robert W., 'From Glow to Flow: A History of Military Laser Research and Development,' *Historical Studies in the Physical and Biological Sciences* 18.1 (1987), 111–47.

Self, Mary R., *History of the Development of Guided Missiles, 1946–1950*, Dayton: Air Material Command, 1951.

Sheehan, Neil, *A Fiery Peace in a Cold War: Bernard Schriever and the Ultimate Weapon*, New York: Random House, 2009.

Sherry, Michael S., *In the Shadow of War: The United States since the 1930s*, New Haven: Yale University Press, 1995.

Siddiqi, Asif A., *Challenge to Apollo: The Soviet Union and the Space Race, 1945–1974*, Washington, DC: NASA, 2000.

———, 'The Almaz Space Station Complex: A History, 1964–1992,' *Journal of the British Interplanetary Society* 54.11/12 (November/December 2001), 361–415.

———, *The Red Rockets' Glare: Spaceflight and the Soviet Imagination, 1857–1957*, Cambridge: Cambridge University Press, 2010.

———, 'Fighting Each Other: The N-1, Soviet Big Science, and the Cold War at Home,' in Naomi Oreskes and John Krige, eds, *Science and Technology in the Global Cold War*, Cambridge, MA: MIT Press, 2014, 189–225.

'"Der Sieg im Weltraum ist möglich",' *Der Spiegel* 38.46 (12 November 1984), 136–47.

Simes, Dimitri K., 'The Military and Militarism in Soviet Society,' *International Security* 6.3 (Winter 1981–82), 118–35.

Simonova, Alexandra V., 'Formirovanie kosmicheskoi mifologiii kak faktora razvitiya nauchnykh issledovaii kosmosa v SSSR i Rossii,' *Sociologiia vlasti* 12.4 (December 2014), 156–72.

Slack, Edward R., 'A Brief History of Satellite Communications,' *Pacific Telecommunications Review* 22.3 (Fall 2001), 7–20.

Slayton, Rebecca, 'From Death Rays to Light Sabers: Making Lasers Surgically Precise,' *Technology and Culture* 52.1 (January 2011), 45–74.

———, *Arguments That Count: Physics, Computing, and Missile Defense, 1949–2012*, Cambridge, MA: MIT Press, 2013.

Slotten, Hugh R., 'Satellite Communications, Globalization, and the Cold War,' *Technology and Culture* 43.2 (April 2002), 315–50.

Slotkin, Arthur L., *Doing the Impossible: George E. Mueller and the Management of NASA's Human Spaceflight Program*, New York: Springer, 2012.

Slusser, George E. and Eric S. Rabkin, eds, *Fights of Fancy: Armed Conflict in Science Fiction and Fantasy*, Athens: University of Georgia Press, 1993.

Smith, Michael G., *Rockets and Revolution: A Cultural History of Early Spaceflight*, Lincoln: University of Nebraska Press, 2014.

Sobchack, Vivian, *Screening Space: The American Science Fiction Film*, 2nd edn, New Brunswick: Rutgers University Press, 1997.

Solovey, Mark, 'Science and the State During the Cold War: Blurred Boundaries and a Contested Legacy,' *Social Studies of Science* 31.2 (April 2001), 165–70.

Solovey, Mark and Hamilton Cravens, eds, *Cold War Social Science: Knowledge Production, Liberal Democracy, and Human Nature*, Basingstoke: Palgrave Macmillan, 2012.

Sontag, Susan, 'The Imagination of Disaster,' *Commentary* 40.4 (October 1965), 42–8.

Soviet Military Space Doctrine, Washington, DC: Defense Intelligence Agency, 1984.

Spinrad, Norman, 'Quand "La Guerre des étoiles" devient réalité,' *Le Monde Diplomatique* 46.544 (July 1999), 28.

Spires, David N., *Beyond Horizons: A Half Century of Air Force Space Leadership*, Maxwell: Air University Press, 1998.

Springer, Anthony M., 'Securing the High Ground: The Army's Quest for the Moon,' *Quest: The History of Spaceflight Quarterly* 7.2 (Summer/Fall 1999), 342–8.

Stapledon, Olaf, *Last and First Men: A Story of the Near and Far Future*, London: Methuen Publishing, 1930.

———, 'Interplanetary Man?,' *Journal of the British Interplanetary Society* 7.6 (November 1948), 213–33.

Stares, Paul B., *The Militarization of Space: U.S. Policy, 1945–1984*, Ithaca: Cornell University Press, 1985.

Steinberg, Gerald M., 'The Militarization of Space: From Passive Support to Active Weapons Systems,' *Futures* 14.5 (October 1982), 374–92.

Stillson, Albert C., 'The Military Control of Outer Space,' *Journal of International Affairs* 13.1 (Spring 1959), 70–7.

Stine, G. Harry, *Confrontation in Space*, Englewood Cliffs: Prentice-Hall, 1981.

Strangl, Raimund, *Raketen in der Kriegstechnik*, Dorfen: Barbara-Verlag Hugo Meiler, 1931.

Sun-Tzu (Sunzi), *The Art of War*, New York: Penguin, 2003.

Suri, Jeremi, 'Conflict and Co-operation in the Cold War: New Directions in Contemporary Historical Research,' *Journal of Contemporary History* 46.1 (January 2011), 5–9.

Sweetman, John, *Cavalry of the Clouds: Air War over Europe 1914–1918*, Stroud: History Press, 2010.

Teilhard de Chardin, Pierre, *L'Avenir de l'homme*, Paris: Editions du Seuil, 1959 (Eng. *The Future of Man*, London: William Collins, 1964).

Teller, Edward and Carl Sagan, 'The Case for SDI/The Case Against SDI,' *Discover* 6 (September 1985), 66–74.

Thomas, Kenneth S. and Harold J. McMann, *U.S. Spacesuits*, New York: Springer, 2006 (2nd edn 2012).

Thompson, E. P., 'The Real Meaning of Star Wars,' *Nation* 240 (9 March 1985), 257, 273–5.

Thompson, E. P., ed., *Star Wars: Science-Fiction Fantasy or Serious Probability?*, Harmondsworth: Penguin, 1985.

Thompson, E. P. and Ben Thompson, *Star Wars: Self-Destruct Incorporated*, London: Merlin, 1985.

Thompson, Nicholas, 'Nuclear War and Nuclear Fear in the 1970s and 1980s,' *Journal of Contemporary History* 46.1 (January 2011), 136–49.

Tompkins, Andrew S., *Better Active than Radioactive! Anti-nuclear Protest in 1970s France and West Germany*, Oxford: Oxford University Press, 2016.

'Todlos glücklich,' *Der Spiegel* 20.53 (26 December 1966), 89–101.

Trischler, Helmuth, *Luft- und Raumfahrtforschung in Deutschland 1900–1970: Politische Geschichte einer Wissenschaft*, Frankfurt am Main: Campus, 1992.

———, *The 'Triple Helix' of Space: German Space Activities in a European Perspective*, Noordwijk: ESA, 2002.

———, 'Contesting Europe in Space,' in Martin Kohlrausch and Helmuth Trischler, *Building Europe on Expertise: Innovators, Organizers, Networkers*, Basingstoke: Palgrave Macmillan, 2014, 243–75.

Trischler, Helmuth and Hans Weinberger, 'Engineering Europe: Big Technologies and Military Systems in the Making of Twentieth Century Europe,' *History and Technology* 21.1 (March 2005), 49–83.

Trischler, Helmuth, Kai-Uwe Schrogl and Andrea Kuhn, eds, *Ein Jahrhundert im Flug: Luft- und Raumfahrtforschung in Deutschland 1907–2007*, Frankfurt am Main: Campus, 2007.

Tsipis, Kosta, 'Laser Weapons,' *Scientific American* 245.5 (December 1981), 51–7.

Turchetti, Simone and Peder Roberts, eds, *The Surveillance Imperative: The Rise of the Geosciences During the Cold War*, Basingstoke: Palgrave Macmillan, 2014.

Turner, Fred, *From Counterculture to Cyberculture: Stewart Brand, the Whole Earth Network, and the Rise of Digital Utopianism*, Chicago: University of Chicago Press, 2006.

Twigge, Stephen Robert, *The Early Development of Guided Weapons in the United Kingdom, 1940–1960*, Chur: Harwood, 1993.

US Army Weapons Command, Directorate of R & D, Future Weapons Office, *Meanderings of a Weapon Oriented Mind When Applied in a Vacuum Such as on the Moon*, Rock Island: US Army Weapons Command, 1965.

Van Dyke, Vernon, *Pride and Power: The Rationale of the Space Program*, Urbana: University of Illinois Press, 1964.

Van Riper, A. Bowdoin, *Rockets and Missiles: The Life Story of a Technology*, Westport: Greenwood Press, 2004.

Veldman, Meredith, *Fantasy, the Bomb and the Greening of Britain: Romantic Protest, 1945–1980*, Cambridge: Cambridge University Press, 1994.

Velupillai, David, 'Europe's Equatorial Launch Site,' *Flight International* (17 February 1979), 467–71.

Villain, Jacques, *Satellites espions: histoire de l'espace militaire mondial*, Paris: Vuibert, 2009.

Vowinckel, Annette, Marcus M. Payk and Thomas Lindenberger, eds, *Cold War Cultures: Perspectives on Eastern and Western European Societies*, Oxford: Berghahn, 2012.

Voute, Cesar, 'A European Military Space Community: Reality or Dream,' *Space Policy* 2 (August 1986), 206–22.

Waggoner, A. G., 'Department of Defense Space Program,' in Hugh Odishaw, ed., *The Challenges of Space*, Chicago: University of Chicago Press, 1962, 195–203.

Wald, Priscilla, 'The "Hidden Tyrant": Propaganda, Brainwashing, and Psycho-Politics in the Cold War Period,' in Jonathan Auerbach and Russ Castronovo, eds, *The Oxford Handbook of Propaganda Studies*, Oxford: Oxford University Press, 2013, 109–30.

Walker, Chuck and Joel Powell, *ATLAS: The Ultimate Weapon*, Burlington: Apogee, 2005.

Weapons in Space, vol. 1: *Concepts and Technologies*, vol. 2: *Implications for Security*, Cambridge, MA: American Academy of Arts and Sciences, 1985 (= *Daedalus* 114.2/3).

Weart, Spencer R., *Nuclear Fear: A History of Images*, Cambridge, MA: Harvard University Press, 1988 (*The Rise of Nuclear Fear*, 2nd edn 2012).

Weisbrode, Kenneth and Heather H. Yeung, 'How We Lost the Sky,' *New York Times* (23 July 2018).

Weldes, Jutta, ed., *To Seek out New Worlds: Science Fiction and World Politics*, Basingstoke: Palgrave Macmillan, 2003.

Wells, H. G., 'The War of the Worlds,' *Pearson's Magazine* 3/4.16–24 (April–December 1897), 363–73, 487–96, 587–96, 108–19, 221–32, 329–39, 447–56, 558–68, 736–45.

———, *The War of the Worlds*, London: William Heinemann, 1898.

'Der Weltraum wird zum Schlachtfeld,' *Der Spiegel* 35.14 (30 March 1981), 156–77.

Werber, Niels, 'Selbstbeschreibungen des Politischen – in Serie: Perry Rhodan 1961–2018,' *Kulturwissenschaftliche Zeitschrift* 3.1 (2018), 75–98.

Werth, Karsten, *Ersatzkrieg im Weltraum: Das US-Raumfahrtprogramm in der Öffentlichkeit der 1960er Jahre*, Frankfurt am Main: Campus, 2006.

Westad, Odd Arne, *The Global Cold War: Third World Interventions and the Making of Our Times*, Cambridge: Cambridge University Press, 2005.

Westfahl, Gary, *The Spacesuit Film: A History, 1918–1969*, Jefferson: McFarland, 2012.

Westwick, Peter J., '"Space-Strike Weapons" and the Soviet Response to SDI,' *Diplomatic History* 32.5 (November 2008), 955–79.

Weyer, Johannes, 'European Star Wars: The Emergence of Space Technology through the Interaction of Military and Civilian Interest-Groups,' in Everett Mendelsohn, Merritt Roe Smith and Peter Weingart, eds, *Science, Technology and the Military*, Dordrecht: Kluwer, 1988, 243–88.

———, *Akteurstrategien und strukturelle Eigendynamiken: Raumfahrt in Westdeutschland 1945–1965*, Göttingen: Schwartz, 1993.

Whittle, Richard, *Predator: The Secret Origins of the Drone Revolution*, New York: Henry Holt, 2014.

Whyte, Neil and Philip Gummett, 'The Military and Early United Kingdom Space Policy,' *Contemporary Record* 8.2 (Fall 1994), 343–69.

———, 'Far beyond the Bounds of Science: The Making of the United Kingdom's First Space Policy,' *Minerva* 35.2 (Summer 1997), 139–69.

Wicken, Olav, 'Cold War in Space Research: Ionospheric Research and Military Communication in Norwegian Politics,' in John Peter Collett, ed., *Making Sense of Space: History of Norwegian Space Activities*, Oslo: Scandinavian University Press, 1995, 41–73.

Wilkinson, John, T. B. Millar and Marie-France Garaud, 'Foreign Perspectives on the SDI,' *Daedalus* 114.3 (Summer 1985), 297–313.

Williams, Geoffrey Lee and Alan Lee Williams, *The European Defence Initiative: Europe's Bid for Equality*, New York: St. Martin's Press, 1986.

Winter, Frank H., *Prelude to the Space Age: The Rocket Societies, 1924–1940*, Washington, DC: Smithsonian Institution Press, 1983.

Wittner, Lawrence S., *Confronting the Bomb: A Short History of the World Nuclear Disarmament Movement*, Stanford: Stanford University Press, 2009.

Wolf, Dieter O. A., Hubertus M. Hoose and Manfred A. Dauses, *Die Militarisierung des Weltraums: Rüstungswettlauf in der vierten Dimension*, Koblenz: Bernard und Graefe, 1983.

Wolfe, Audra J., *Competing with the Soviets: Science, Technology, and the State in Cold War America*, Baltimore: Johns Hopkins University Press, 2013.

Wolfe, Tom, *The Right Stuff*, New York: Farrar, Straus & Giroux, 1979.

'World Will Be Ruled from Skies Above,' *Life Magazine* 54.20 (17 May 1963), 4.

Yergin, Daniel, *Shattered Peace: The Origins of the Cold War and the National Security State*, Boston: Houghton Mifflin, 1977.

Yost, David S., 'Western Europe and the U.S. Strategic Defense Initiative,' *Journal of International Affairs* 41.2 (Summer 1988), 269–323.

Zaloga, Steven J., *Target America: The Soviet Union and the Strategic Arms Race, 1945–1964*, Novato: Presidio Press, 1993.

———, *The Kremlin's Nuclear Sword: The Rise and Fall of Russia's Strategic Nuclear Forces, 1945–2000*, Washington, DC: Smithsonian Institution Press, 2002.

Zedalis, Rex and Catherine Wade, 'Anti-satellite Weapons and the Outer Space Treaty of 1967,' *California Western International Law Journal* 8.3 (Summer 1978), 454–82.

Zettler, Hans, 'Militärs beherrschen den Weltraum,' *Frankfurter Allgemeine Zeitung* (15 February 1978), 27–8.

Index

Page numbers appearing in italics refer to illustrations. A page reference in the form 'n' indicates a note number; for example, 204n59 is note 59 on page 204. Index entries that begin with a number are indexed as if the number were spelled out; for example, *2001: A Space Odyssey* is alphabetized as 'Two thousand and one.''

Alexander C. T. Geppert et al. (eds), *Militarizing Outer Space*
European Astroculture, vol. 3
https://doi.org/10.1057/978-1-349-95851-1

M

U

Printed in the United States
by Baker & Taylor Publisher Services